Quantification in LC and GC

Edited by
Hans-Joachim Kuss and
Stavros Kromidas

Related Titles

H.-J. Hübschmann

Handbook of GC/MS

Fundamentals and Applications

2008
ISBN: 978-3-527-31427-0

M. McMaster

HPLC

A Practical User's Guide

2007
ISBN: 978-0-471-75401-5

E. Prichard, V. Barwick

Quality Assurance in Analytical Chemistry

2007
ISBN: 978-0-470-01204-8

E. Hahn-Deinstrop

Applied Thin-Layer Chromatography

Best Practice and Avoidance of Mistakes

2007
ISBN: 978-3-527-31553-6

Y. V. Kazakevich, R. LoBrutto

HPLC for Pharmaceutical Scientists

2007
ISBN: 978-0-471-68162-5

Kromidas, S. (ed.)

HPLC Made to Measure

A Practical Handbook for Optimization

2006
ISBN: 978-3-527-31377-8

Quantification in LC and GC

A Practical Guide to Good Chromatographic Data

Edited by
Hans-Joachim Kuss and Stavros Kromidas

WILEY-VCH

WILEY-VCH Verlag GmbH & Co. KGaA

The Editors

Dr. Hans-Joachim Kuss
Psychatrische Klinik der
Ludwig-Maximilians-Universität
Nussbaumstrasse 7
80336 München
Germany

Dr. Stavros Kromidas
Rosenstrasse 16
66125 Saarbrücken
Germany

Library of Congress Card No.: applied for

British Library Cataloguing-in-Publication Data
A catalogue record for this book is available from
the British Library.

**Bibliographic information published by
the Deutsche Nationalbibliothek**
The Deutsche Nationalbibliothek lists this
publication in the Deutsche Nationalbibliografie;
detailed bibliographic data are available on the
Internet at http://dnb.d-nb.de

© 2009 WILEY-VCH Verlag GmbH & Co. KGaA,
Weinheim

Printed in the Federal Republic of Germany
Printed on acid-free paper

Cover Design   Schulz Grafik-Design, Fußgönheim
Typesetting   Manuela Treindl, Laaber
Printing   betz-druck GmbH, Darmstadt
Bookbinding   Litges & Dopf Buchbinderei GmbH,
Heppenheim

ISBN: 978-3-527-32301-2

Contents

Quantification in LC and GC: A Practical Guide to Good Chromatographic Data
Edited by Hans-Joachim Kuss and Stavros Kromidas
Copyright © 2009 WILEY-VCH Verlag GmbH & Co. KGaA, Weinheim
ISBN: 978-3-527-32301-2

Preface

With chromatography as many substances as possible should be separated from each other as well as possible. A poor separation is not "repairable" later. However, we have the impression that the following evaluation and assessment of the results are not always treated with due caution under everyday stress. One likes to rely on the computer programs – finally the integration software is validated and this is "covered" by the manufacturer's certificate. It is hardly registered as a problem that the integration algorithms have changed only insignificantly for 30 years and that the area of peaks merged together is always determined simply by perpendicular drop. The assessment of the results is often based on consideration of the questions: "Does the c.v. match?" and "Is my value in spec?" Sometimes this may suffice, sometimes not.

We would like to deal with the following questions in this book: How accurate are the integrated areas or heights of commercial systems? Can one check this at all? How important are the integration parameters, what effect do they have on results? However, we also wanted to have a look at the questions: "How can I really judge a result?" How correct can a result be in reality?" and finally: "What do the authorities have to say on this subject, what must I do, what am I free to do?" To this end we have invited experienced colleagues who have made their knowledge available in various areas. Our many thanks are due to these colleagues.

On the one hand, it is the aim of the book to show that one perhaps should not always blindly trust integration systems and that it is worthwhile developing a deeper appreciation of integrating. On the other hand, we want to make a contribution to the critical analysis of chromatographic results, primarily for quantitative statements used as a basis for decisions. We hope to give the reader many suggestions and ideas for finding his own "right" evaluation and assessment practice. We thank the WILEY-VCH publishing house, and particularly Lesley Belfit and Frank Weinreich, for very close cooperation.

Saarbrücken and München,
August 2009

Hans-Joachim Kuss
Stavros Kromidas

Quantification in LC and GC: A Practical Guide to Good Chromatographic Data
Edited by Hans-Joachim Kuss and Stavros Kromidas
Copyright © 2009 WILEY-VCH Verlag GmbH & Co. KGaA, Weinheim
ISBN: 978-3-527-32301-2

List of Contributors

Werner Engewald
University of Leipzig
Institute of Analytical Chemistry
Linnéstrasse 3
04103 Leipzig
Germany

Joachim Ermer
Sanofi-Aventis Deutschland GmbH
Frankfurt Chemistry Industriepark
Hoechst Building D710
Room 202
65926 Frankfurt am Main
Germany

Daniela Held
PSS Polymer Standards Service
GmbH
In der Dalheimer Wiese 5
55120 Mainz
Germany

Heiko Herrmann
Dionex GmbH
Am Wörtzgarten 10
65510 Idstein
Germany

Mike Hillebrand
Sanofi-Aventis Deutschland GmbH
Industriepark Höchst
65926 Frankfurt am Main
Germany

Peter Kilz
PSS Polymer Standards Service
GmbH
In der Dalheimer Wiese 5
55120 Mainz
Germany

Detlef Jensen
Dionex GmbH
Am Wörtzgarten 10
65510 Idstein
Germany

Hartmut Kirchherr
Medizinisches Labor Bremen
Haferwende 12
28357 Bremen
Germany

Stavros Kromidas
Rosenstrasse 16
66125 Saarbrücken
Germany

Quantification in LC and GC: A Practical Guide to Good Chromatographic Data
Edited by Hans-Joachim Kuss and Stavros Kromidas
Copyright © 2009 WILEY-VCH Verlag GmbH & Co. KGaA, Weinheim
ISBN: 978-3-527-32301-2

Hans-Joachim Kuss
Psychiatrische Klinik der
Ludwig-Maximilians-Universität
München
Nussbaumstrasse 7
80336 München
Germany

Uwe D. Neue
Waters Corporation
34 Maple Street
Milford, MA 01757
USA

Linda Ng
Center for Drug Evaluation and
Research
US Food and Drug Administration
Fishers Lane 5600
Rockville, MD 20857
USA

Ulrich Panne
Humboldt University &
BAM Federal Institute for
Materials Research and Testing &
Instrumental Analytical Chemistry
Richard-Willstaetter-Strasse 11
12489 Berlin
Germany

Ulrich Rose
15, Rue Himmerich
67000 Strasbourg
France

Daniel Stauffer
F. Hoffmann-La Roche AG
PSHW
Bau 93/1.18
4070 Basel
Switzerland

Structure of the Book

Part 1: Evaluation and Estimation of Chromatographic Data

In Part 1 it is shown, how chromatographic data are obtained and how the results are evaluated.

In Chapter 1 (Hans-Joachim Kuss and Daniel Stauffer) it is described how a chromatogram arises and which factors influence the shape, the height and the area of the peaks. The most important equations for Gaussian or exponential modified Gaussian (EMG) peaks are given.

In Chapter 2 (Daniel Stauffer and Hans-Joachim Kuss) the integration parameters are explained: What do they mean and what are they called in different integration systems.

Chapter 3 (Hans-Joachim Kuss) deals mainly with the integration and integration errors. The question considered is the accuracy of commercial integration systems. Furthermore, the possibility of checking this using a chromatogram simulation is described with Gaussian peaks to verify the results.

Chapter 4 (Uwe Neue) treats the simulation of EMG-peaks and gives the theoretical background for peaks in gradient hplc.

Chapter 5 (Hans-Joachim Kuss) describes the special difficulties of the integration of asymmetric (EMG)-peaks.

Chapter 6 (Mike Hillebrand) treats the possibility of "recalculating" merged peaks with the deconvolution method.

Chapter 7 (Hans-Joachim Kuss) treats the calculations for evaluation and calibration, including weighted calibration.

The general criteria for the judgement of analytical data are introduced systematically and analyzed critically in Chapter 8 (Joachim Ermer).

Most of the chromatograms and/or tables in Chapters 3 to 8 can be found on the supplementary CD. The examples are ordered following the book's table of contents. Each time a remark such as "Chromatogram.cdf" or "Table.xls" is found in the text, the corresponding file can be found on the CD.

Part 1 concludes with Chapter 9 (Ulrich Panne) which considers the metrological aspects of analytical data together with a detailed discussion of the measuring uncertainty.

Quantification in LC and GC: A Practical Guide to Good Chromatographic Data
Edited by Hans-Joachim Kuss and Stavros Kromidas
Copyright © 2009 WILEY-VCH Verlag GmbH & Co. KGaA, Weinheim
ISBN: 978-3-527-32301-2

Part 2: Characterization of the Evaluation of Different Chromatographic Modes

The evaluation of chromatographic results depends on the chromatographic techniques used. An overview is given of some special conditions for: Gas chromatography in Chapter 10 (Werner Engewald), LCMS coupling in Chapter 11 (Hartmut Kirchherr), ion chromatography in Chapter 12 (Heiko Herrmann and Detlef Jensen) and gel permeation chromatography in Chapter 13 (Daniela Held and Peter Kilz).

Part 3: Requirements for Chromatographic Data Analysis from the Viewpoint of Organisations and Public Authorities

The area regulated most strongly is certainly the pharmaceutical one. The requirements are explained from three different viewpoints: the US guidelines in Chapter 14 (Linda Ng), the EU guidelines in Chapter 15 (Ulrich Rose) and the view of the pharmaceutical industry in Chapter 16 (Joachim Ermer). Instead of an epilogue, the everyday dealing with analysis results is discussed in the final Chapter 17 (Stavros Kromidas).

The individual chapters were written to represent closed modules, but "jumping" between them is possible at any time. It is the aim of the editors to introduce a book of ideas and to give a lot of suggestions so that one can possibly find one's own solutions. Therefore, the divergent opinions of the authors on topics were accepted. Some repetitions were also accepted so as not to impair the harmony in the textual context. The reader may profit from the different representation of the topics and the individual weighting of the authors.

Part 1
Evaluation and Estimation of Chromatographic Data

Quantification in LC and GC: A Practical Guide to Good Chromatographic Data
Edited by Hans-Joachim Kuss and Stavros Kromidas
Copyright © 2009 WILEY-VCH Verlag GmbH & Co. KGaA, Weinheim
ISBN: 978-3-527-32301-2

1
Evaluating Chromatograms

Hans-Joachim Kuss and Daniel Stauffer

Chromatography is, in principle, a dilution process. In HPLC analysis, on dissolving the substances to be analyzed in an eluent and then injecting 20 μl, the peak volume exiting the column is greater than 20 μl. This is a consequence of the chromatographic conditions. Depending on the column dimensions, there is a critical injection volume, which, if exceeded, leads to additional peak broadening. If one dissolves the substances to be analyzed in a weaker eluent, as is generally done when the gradient elution technique is used, then the injection volume that can be tolerated is increased significantly. The efficiency (Fig. 1.1) of the column is defined by the plate number N and the selectivity of the separation of two components is given by the separation factor α.

The **time taken for an eluent or carrier gas molecule to run through the column without retention is the mobile phase hold-up time** t_M. This often corresponds approximately to the time of the appearance of the first peak in the chromatogram. If one uses too strong an eluent, all components to be separated are eluted at the hold-up time because they are not held back on the column. Differences in the interactions of the substances with the stationary phase lead to different retention times, t_S, in the stationary phase. To achieve a separation, t_S must be greater than t_M.

$$t_R = t_M + t_S \tag{1.1}$$

$$k = \frac{t_S}{t_M} \tag{1.2}$$

$$u = \frac{L}{t_M} \tag{1.3}$$

The sum of t_M and t_S yields the measured retention time t_R (Fig. 1.2) and the quotient of t_S and t_M is the retention factor k, which is not dependent on the column dimensions or the flow, as is t_R. With increasing pressure the flow and

Quantification in LC and GC: A Practical Guide to Good Chromatographic Data
Edited by Hans-Joachim Kuss and Stavros Kromidas
Copyright © 2009 WILEY-VCH Verlag GmbH & Co. KGaA, Weinheim
ISBN: 978-3-527-32301-2

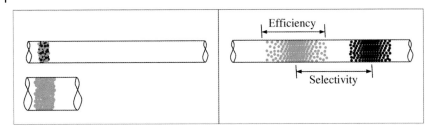

Figure 1.1 Efficiency is dependent on the peak width, selectivity on the retention factor quotient.

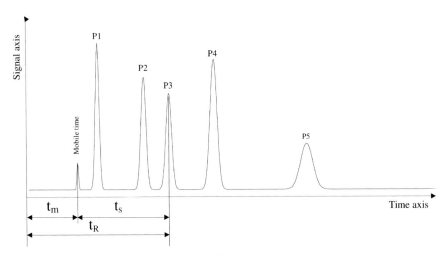

Figure 1.2 Retention times are measured at the apex of each peak. Unretained components elute at the mobile phase hold-up time.

the linear velocity increase in parallel. The linear velocity u is calculated by dividing the column length, L, by t_M.

1.1
Efficiency

The ideal peak shape is described by a symmetric Gaussian peak (Fig. 1.3), characterized by the retention time t_R and the standard deviation σ of the retention time, which is a measure of the peak width.

$$N = \frac{t_R^2}{\sigma^2} \tag{1.4}$$

$$w_{50\%} = 2.35\,\sigma \tag{1.5}$$

$$w_{13\%} = 4\,\sigma \tag{1.6}$$

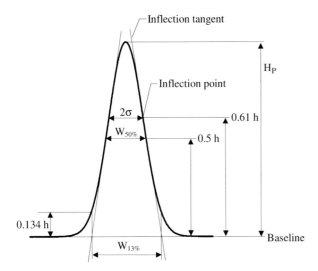

Figure 1.3 Characteristic parameters of a Gaussian peak.

Besides the well known graphic methods, N can be calculated using the area A and height H:

$$N = 2\pi \left(\frac{t_R H_P}{A} \right)^2 \tag{1.7}$$

Combining Eqs. (1.4) and (1.7) it can be seen that σ can be determined from the retention time, area and height.

$$\frac{t_R^2}{\sigma^2} = 2\pi \frac{t_R^2 H_P^2}{A^2} \tag{1.8}$$

$$\sigma = \frac{A}{H_P \sqrt{2\pi}} \tag{1.9}$$

These equations only apply to isocratic separations. For low k values, one usually finds smaller plate numbers than for higher k values. The extra column volume V_{ec} leads to an additional constant peak broadening. The smaller earlier peaks are influenced more strongly by this than the later peaks.

1.2
EMG Model

In practice, peaks often have a tailing, which can be described as a Gaussian peak with an overlaid exponential function (exponential modified Gauss function

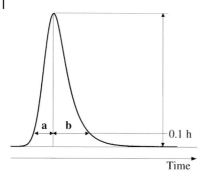

Figure 1.4 The asymmetry factor is measured at 10% of the height.

(EMG) [1]). The increase in the Gaussian peak is influenced less by the exponential tail with the time constant τ than by the decreasing portion of the peak. The retention time is shifted to a slightly higher value, as is the peak width.

$$N = \frac{t_R^2}{\sigma^2 + \tau^2} \tag{1.10}$$

The values of σ and τ are not determined by the integration method used. The extent of the tailing, which is measured at 10% of the peak height, is represented by the asymmetry factor A_f (Fig. 1.4).

$$A_f = \frac{a}{b} \tag{1.11}$$

According to the EMG model, the plate number is calculated as:

$$N = 41.7 \frac{\left[\dfrac{t_R}{a + b}\right]^2}{\dfrac{b}{a} + 1.25} \tag{1.12}$$

In the range $\tau/\sigma = 0.5$–3 Tamisier-Karolak [2] found an empirical equation:

$$\frac{\tau}{\sigma} = -1.03 \left(\frac{b}{a}\right) - 0.97 \tag{1.13}$$

This equation can only be used if the asymmetry factor can be estimated. Therefore the valley between the peaks must not be greater than 9% of the height of the smaller peak. In practice, the asymmetry factor at the peak's half height is of more use. If the valley between the peaks is greater than 49%, integration with the perpendicular drop method is very prone to errors, because the second peaks sits on the first peak.

The plate number is dependent on the column length and the particle size. If one divides the column length by the plate number, the plate height H (µm) is obtained. Dividing H by the particle size d_p, one obtains the reduced plate height h, i.e. the number of particles due to one plate.

$$H = \frac{L}{N} \tag{1.14}$$

$$h = \frac{H}{d_p} \tag{1.15}$$

In theory, the minimum possible value for h is 2, but in practice, a value of 3 is usually acceptable in real chromatograms.

The three effects: flow inhomogeneity A, diffusion broadening B in competition with the flow and deterioration of the component exchange C with increasing flow are overlaid, to produce a Van Deemter Hu-curve. The essential statement is that there is a linear velocity u at which the plate height H has a minimum. A column has the highest separating performance at this flow, which is characterized by the linear velocity u_{opt}. In HPLC one usually works above the optimal flow, accepting a loss in the optimal separating performance because otherwise the analysis times would become too long. The corresponding broadening of the peak is particularly dependent on the component exchange term C, which causes the increasing slope of the Hu curve.

1.3
Chromatogram

A chromatogram is a graphical representation of all peaks eluting from the column superimposed on the baseline. The areas and heights of the peaks usually increase linearly in accordance with the amount of injected component. The integration systems enable an automated estimation of the area, height and other characteristic parameters.

The separation is usually carried out on a C18 column in HPLC and on a capillary column in GC. The separating ability of a column is characterized by the plate number, which determines the peak width relative to the retention time. A typical value for HPLC is 10 000 and for GC 90 000 plates.

$$\frac{s_p}{t_R} = \frac{1}{\sqrt{N}} \tag{1.16}$$

If the standard deviation of the peak is 1% of the retention time t_R, the plate number is 10 000.

1.4
Selectivity

The aim of chromatography is a separation, which is characterized by the different retention times of two peaks succeeding one another. A very efficient column with a high plate number is no guarantee of an efficient separation. The selectivity is characterized by the separation factor α, which is the quotient of the two retention factors.

$$\alpha = \frac{k_2}{k_1} \tag{1.17}$$

An α of 1.1 implies that the retention time t_s of the second peak is 10% longer than the retention time of the first peak. The resolution R, a measure of the quality of the separation between two peaks, depends not only on the distance between the two peaks but also on the peak width, which can be found graphically as the peak width at half height $w_{50\%}$.

$$R = 1.18 \frac{t_{R2} - t_{R1}}{w_{2_{50\%}} + w_{1_{50\%}}} \tag{1.18}$$

An R of 1.5 is known as baseline separation, although only $R = 2$ is really separation to the baseline.

If the retention times of two successive peaks lie sufficiently far away from each other then we have a separation down to the baseline. If not, the peaks merge together and the valley between them decreases until only a broadened peak can be seen. One does not have any chance of recognizing whether several components hide under the visible peak at (almost) the same retention time. This can be resolved only by variation of the separating conditions or with one or more specific detectors.

References

1 J. P. Foley, J. G. Dorsey, Equations for Calculation of Chromatographic Figures of Merit for Ideal and Skewed Peaks, *Anal. Chem. 55*, **1983**, 730–737.

2 S. L. Tamisier-Karolak, M. Tod, P. Bonnardel, M. Czok, P. Cardot, Daily validation procedure of chromatographic assay using gaussoexponential modelling, *J. Pharm. Biomed. Anal. 13*, **1995**, 959–970.

2
Integration Parameters

Daniel Stauffer and Hans-Joachim Kuss

The integration system determines the retention times, peak heights and peak areas as recorded as precisely as possible. The retention time is defined as the time at the maximum of the peak, when the greatest amount of the component is flowing through the detector. The peak height is measured at this time. A background signal is also detected when the analyte component being determined is not in the detector cell. Sometimes it is difficult to distinguish between this base signal and the signal corresponding to the component being analyzed. The integration system has, additionally, to determine the most probable trend of the baseline.

2.1
Peak Recognition Methods

2.1.1
The Classical Method

Using the classical method (Fig. 2.1) the difference (slope) between two successive data points is determined and the value compared with a threshold value that has been defined by the user or calculated by the system. If the threshold value is exceeded, the system marks the data point as the potential start of a peak and uses an algorithm to check whether this data point is definitely the start of a peak or if it is simply baseline drift or a spike (Fig. 2.2). The peak start is confirmed if further criteria (e.g. five subsequent data points show a positive slope) are fulfilled. After the peak start is confirmed, the apex of the peak must then be confirmed (Fig. 2.3). Each data point that is followed by a smaller signal value is marked as subsequent potential peak apex. The search for the peak end is performed in an analogous way (Fig. 2.4).

Quantification in LC and GC: A Practical Guide to Good Chromatographic Data
Edited by Hans-Joachim Kuss and Stavros Kromidas
Copyright © 2009 WILEY-VCH Verlag GmbH & Co. KGaA, Weinheim
ISBN: 978-3-527-32301-2

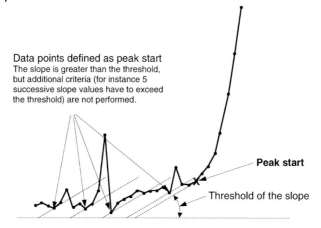

Data points defined as peak start
The slope is greater than the threshold,
but additional criteria (for instance 5
successive slope values have to exceed
the threshold) are not performed.

Peak start

Threshold of the slope

Figure 2.1 In the classical method of peak recognition, any data point
at which the slope to the next data point exceeds a defined threshold value
indicates the potential for the peak start. If further criteria are satisfied,
the point is confirmed as a peak start and the search for the peak end begins.
The slope of the signal and the regulation of the threshold value are defined
differently by the various manufacturers of the integration parameters.
The methods can, however, be explained by the description above.

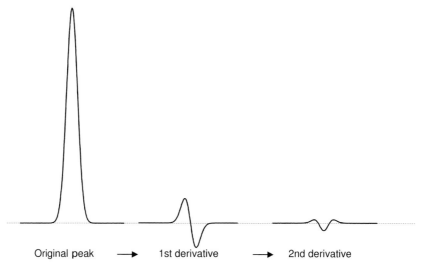

Original peak ⟶ 1st derivative ⟶ 2nd derivative

Figure 2.2 For the representation of the first derivative of the peak
the slope of the signal is recorded instead of the detector signal
at the respective time. The second derivative correspondingly shows
the slope of the curve of the first derivative. The first derivative
therefore corresponds to the slope, the second derivative to the
curvature of the peak. Minimum of the second derivative = peak apex.
Zero points in front of and after the minimum = ascending and
descending inflection points.

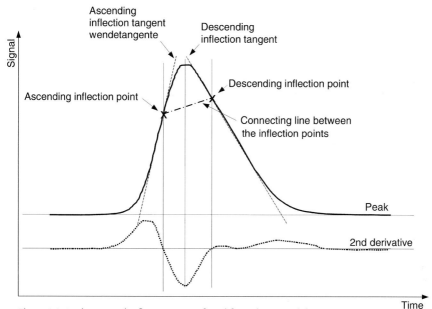

Figure 2.3 Peak apex and inflection points found from the second derivative.

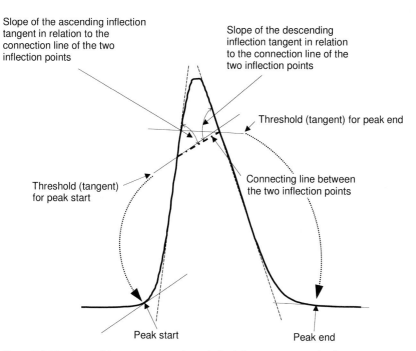

Figure 2.4 The slope of the two tangents through the inflection points with reference to the connecting line of the two inflection points is determined. A threshold value is derived (in %). At the peak apex, the tangent of every data point is compared with the respective threshold values in the direction of the peak start and end.

2.1.2
Alternative Method

The integration system known as "Apex Track"™, which is part of the integration program "Empower"™ (Waters), can be mentioned as an example of another method for the recognition of the peaks and for the baseline integration. Using the "Apex Track" the second derivative of the detector signal gives the peak apex (= retention time) and the inflection points are determined. Starting at the peak apex, at every data point in the direction of the start and end of the peak, a connecting line is drawn to the inflection point. This slope is compared with a threshold value, which is defined by the user or calculated by the system, to identify the peak start and end.

2.2
Integration and Integration Parameters

2.2.1
Data Acquisition and Integration with Empower 2

With Empower the peak recognition and determination of the course of the baseline can be carried out by two different methods:

- classical integration, peak start and end are investigated with the signal slope (first derivative);
- the Apex Track method, the top and the inflection point of the peak are determined by means of the second derivative and the peak start and end are identified by looking at the peak apex.

Data Acquisition
The manufacturer of Empower recommends adjusting the data collecting rate so that the narrowest peak is included (measured at 5% of the peak height [2–10]) with 15 data points within the peak width [1–7].

For regulation of the peak area all (unfiltered) raw data points are used, no matter which sampling rate or peak width are used for the peak recognition parameters.

Peak Recognition
If a higher data collecting rate is set, leading to more than 15 data points per peak width, then a bunching rate is calculated from the peak width and the sampling rate. Thus many more data points are averaged so that 15 data points result within the peak width.

$$\text{number of summarized data points} = \frac{\text{peak width} \times \text{data sampling rate}}{15}$$

The signal is smoothed by averaging the data. This smoothed signal is only used for the recognition of the peak start and end. The calculation of the peak area is then carried out with the raw data.

Peak Apex (Retention Time, Peak Height)

The maximum signal value within the peak plus two data points before and two data points after this value are used to calculate the maximum signal height by means of a parabolic fit. The time at the maximum signal value corresponds to the retention time.

Baseline

The baseline is drawn from the peak start to the peak end; in the case of an unresolved group of peaks then it is drawn to the end of the last peak.

Two neighboring peaks are considered to be resolved completely, if the peak width of the broader peak is not larger than three times the distance between the end of peak 1 and the start of peak 2. If the ratio is less than three, the peak pair is identified as being completely resolved, otherwise it is identified as a peak group.

In the example (Fig. 2.5) the peak pair 1/2 is identified as a group, because the width of peak 2 (w_2) is greater than three times the distance EB1-2. The peak pair 2/3 is completely resolved, because w_3 is only twice the distance EB2-3. A baseline point is set at the end of peak 2 and the start of peak 3.

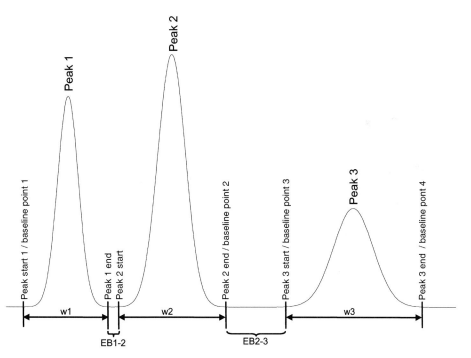

Figure 2.5 No baseline points are put at the end of peak 1 and the start of peak 2 because the width of the wider peak of the peak pair 1/2 is greater than three times the distance EB1-2. However, the width of peak 3 is less than three times the distance EB2-3, hence, the end of peak 2 and the start of peak 3 baseline points are set.

2.2.2
Data Acquisition and Integration with Chromeleon

Data Acquisition
The manufacturer of Chromeleon recommends that the data acquisition parameters are adjusted so that the narrowest peak is measured with 20 data points. For small peaks with a low signal/noise ratio the number of data points should be increased to approximately 40 points per peak.

The data acquisition for detectors delivering digital signal data is controlled by the data collection rate. For detectors with analog signal output, which must first be digitized by an AD-converter, the sampling rate determines the data acquisition.

In addition to the data collection rate or sampling rate, the time step between two data points t_{DP} must be defined. This is not necessarily the reciprocal value of the sampling rate. Using a sampling rate of 100 Hz, for example, the standard value of t_{DP} is 0.01 min. Increasing the step to $t_{DP} = 0.1$ min and Average = on, only the average of 10 data points, respectively, is stored. This data bunching causes a smoothing of the signal.

The manufacturer advises caution when adapting the step t_{DP} automatically as it depends upon the slope. Only for fast chromatograms is there a clear advantage to reducing the great amount of data. A constant step t_{DP} is recommended for a precise and reproducible analysis.

Peak Recognition
The peak recognition is determined by the two values peak-slice and sensitivity (Fig. 2.6). Beginning with the first measured data point, a rectangle is formed, with sensitivity on the time axis and peak-slice on the signal axis. By reflection of the sensitivity in the direction of the negative signal values the area of the rectangle is doubled. A second rectangle is formed using the last data point in the first rectangle as the zero point for this rectangle. Provided all data points are in this rectangle, the signal change is interpreted as noise. If a point is found outside this rectangle, a peak recognition algorithm is initiated to identify the peak start. After the recognition of the peak start and the peak apex, Chromeleon establishes the peak end in a similar way.

2.2.3
Data Acquisition and Integration with EZChrom Elite

Data Acquisition
The manufacturer of EZChrom elite also recommends the use of 20 data points over the narrowest peak. The number of data points to be taken can be adjusted using either the sampling frequency (in Hz) or the reciprocal value sampling period (in ms). For smoothing and reduction of the number of data points the width is adjusted to summarize many data points (data bunching), so that 20 data points are sampled in the time interval given by the width.

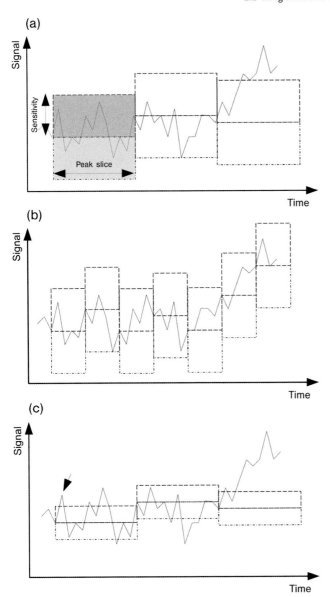

Figure 2.6 (a) The sensitivity of the peak recognition using Chromeleon is influenced by the two parameters sensitivity and peak-slice.
(b) Increasing the peak slice makes the peak recognition less sensitive. The peak identified later, or maybe not at all, the slope of the signal is interpreted as baseline drift. (c) Reducing the sensitivity makes the peak recognition more sensitive.

Peak Recognition

The peak recognition is adjusted through the threshold value. It is possible to find the threshold value by marking the baseline between two peaks. The greatest value of the first derivative is used for the calculation, which is hence the greatest signal slope in the section being considered.

2.2.4
Data Recording and Integration with ChemStation

Data Collection

The analog detector signal is sampled with a high, constant frequency. The data collection rate is adjusted to establish how many data points are averaged and stored as raw data in the computer. As an alternative to the data collection rate the half width of the narrowest peak (minimum peak width) can also be entered. The data collecting rate is then chosen so that during the half width of the peak, 12 data points are stored. As Fig. 2.7 shows, the noise is smoothed by this data bunching.

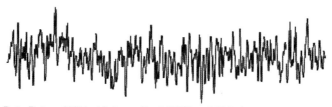

Data Rate = 100Hz, Minimum Peak Width = 0.002min

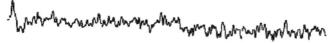

Data Rate = 10Hz, Minimum Peak Width = 0.02min

Figure 2.7 Depending on the adjusted peak width, the detector signal is bunched to a greater or lesser extent to give the required number of averaged data points, leading to a smoothing of the signal recorded with a high frequency.

Peak Recognition

The peak recognition is controlled by five parameters: initial slope sensitivity, initial peak width, initial area (height) reject and shoulders (on/off).

Besides the parameters that steer the recording of the raw data, time-dependent integration parameters can be entered into the integration programs. These values influence the peak recognition, the course of the baseline and the peak mapping.

In particular, with respect to the parameters for data collection it is often almost impossible to find information in the manuals from the software manufacturers as to whether the sampling rate and peak width etc. have an effect on the raw data recording or the peak recognition.

3
Integration Errors

Hans-Joachim Kuss

When any new chromatographic method is developed, it is typically understood that optimization of the chromatography can be difficult and time consuming. In a regulated environment, much of this work is associated with the validation of the method. The integration and evaluation of the results is generally considered to be easy. If the integration system is installed according to the instructions no problems should be expected.

It would be naive to think that integration systems are perfect. Even a basic knowledge of chromatography indicates that excellent results can only be obtained if the chromatographic resolution is good. However, the contribution of the integration system to the total error is frequently assessed as being so low that it can be considered negligible. This is not necessarily the case in all instances, as will be pointed out in the following.

When computers are used we expect the precision to be high. If a result appears on the screen or is printed out, it is generally accepted that this result is definitely correct. Increased automation and the faster throughput of analyses also means there is less and less time to confirm the results of the analyses calculated concentrations are indeed correct. For routine analyses, calculation of the analysis results is often done by the chromatographic data system (CDS) without any human intervention, particularly when the control values are in a predefined range.

3.1
What Does the Literature Say on Integration Errors?

There is only one book exclusively devoted to the topic of chromatographic integration methods, written by Dyson (1998, 2nd edition) who gives an excellent summary of the possibilities of the integration. One of the first pages contains the statement: "No analytical report should be accepted unquestioningly". The book is full of many interesting older citations, for example from the 1969 publication by Westerberg, who described the insufficiencies of the perpendicular drop method for peaks merged together. He described a curve fitting procedure to separate shoul-

Quantification in LC and GC: A Practical Guide to Good Chromatographic Data
Edited by Hans-Joachim Kuss and Stavros Kromidas
Copyright © 2009 WILEY-VCH Verlag GmbH & Co. KGaA, Weinheim
ISBN: 978-3-527-32301-2

der peaks mathematically. Unfortunately, four decades later the perpendicular drop is still the most important method for the integration of merged peaks.

One part of Dyson's book refers to Dyson (1992), in which an MS DOS program together with a digital analog converter (DAC) is used for chromatogram simulation. The commercially available device "Calpeak" generates an arbitrary chromatogram with many peaks as an analog signal. It is the chromatograph in the coat pocket. The examples show that the relative areas of the peaks are known, but the absolute areas are not predictable.

Papas and Delaney (1986) wrote: "Unlike the rigorous quality assurance programs implemented for instrumentation, there is presently no effective means to conduct such a quality assurance program for these data systems". They then described how they generated, with a Pascal program on a VAX 11/750, chromatograms consisting of exponential modified Gaussian peaks (EMG) and changed them with a DAC into analog data. The integrations at signal to noise ratios (S/N) of 3, 6, 9 and 12 yielded deviations of more than 30%.

Papas and Tougas (1990) compared the integration results for peaks merged together using the perpendicular drop and the tangent method. After a multiple linear regression they introduced a model according to which one can decide between the two integration methods. None of the seven integrators used at that time contained such a complex algorithm, the same situation holds today.

Meyer (1994) used Excel 4.0 on a Macintosh LC II to simulate a Gaussian function. The time axis was in arbitrary units. The peak width was generally set to an arbitrary unit of 1. Retention time and area were used as "units" and the peak height H_{max} was calculated. The integration was done numerically in Excel. The Excel simulation could be extended to EMG peaks. With a simulated EMG peak pair, the second peak yields a distinct shoulder, a deviation of −11% was found for the first peak and a deviation of +22% for the second peak, which had half the area of the first peak.

Meyer (1995) showed, using Calpeak, that with merged peaks and a height relationship of 100 and 1000 the tangent method gave worse results than the drop method. Only if the larger peak is more than five times as wide as the smaller peak should the tangent method be preferred. The peak height is fundamentally less prone to error.

Felinger and Guiochon (2001) reported on earlier work with regard to the repeatability of HPLC data. They mentioned that the data formats and integration routines of the manufacturers of integration systems are not patented, however, they are treated confidentially. In this work they dedicated much effort to the correct simulation of noise. As a chromatogram, they used noise superimposed by one synthetic (Gauss, EMG or Langmuir) peak. The simulation was programmed in Fortran using the UNIX workstation of a university computer center. Areas and heights were calculated and compared with the integration result (Table 3.1) of the Chemstation.

Obviously one must expect integration errors below a signal to noise ratio S/N = 100 of more than 1% for the area and of more than 0.3% for the height. Nevertheless, the authors summarize: "The most important conclusion of this study is that the contribution of the random and systematic errors originating

Table 3.1 The values mentioned in the text by Fehlinger and Guiochon (2001) have been put into tabular form. The difference between integrated area and the calculated area divided by the calculated area is the accuracy of the area or height.

S/N	Accuracy of the areas (%)		Accuracy of the heights (%)	
	10	100	10	100
Gauss	3.0	0.3	1.5	0.1
EMG-peak	9.0	1.0	3.0	0.3
Langmuir	10.0	1.0	3.5	0.3

from the data analysis procedure is generally far smaller than the experimental or instrumental contributions arising from the factors that affect the repeatability of chromatographic measurements". This may be true for chromatograms with only one peak, in practice chromatograms are complex, consider for example Fig. 3.1.

Contrary results were found by Schepers and Ermer (2004). From reproducibility examinations they indirectly determined that the integration error is the predominant error between the limit of quantification LOQ and the fivefold LOQ.

Bicking (2006a, 2006b) showed measured chromatograms with two symmetrical peaks having a resolution of 4, 2, 1.5 and 1. He calculated the integration error using the drop, valley to valley, exponential skim and Gaussian skim method. The estimated areas at a resolution of 4 were considered to be the target values. The best integration, particularly for great peak height differences, yields a complex decision scheme, which can only be converted into practice with some difficulty.

In the published literature, warning signals can be found with respect to integration errors, but these are not particularly useful to the person actively using chromatography in a laboratory, because their problems consist of how to integrate the predominantly complex chromatograms, which could have many peaks, automatically. To judge the quality of the integration a generally accessible accuracy standard is required. In the best case scenario, the optimization of the integration is controlled by plausibility considerations. This is discussed in the following sections, to give support to the decisions made, using previously known target areas for the peaks, by simulation of chromatograms.

3.2
Integration in Routine Practice

The integration of peaks merged together is a reality of chromatography with which everybody working in a chromatographic laboratory is confronted. Unfortunately, the published considerations confine themselves almost exclusively to two peaks on a straight line and their influence on each other. Real chromatograms are usually substantially more complicated and are often on an uneven baseline.

We thank Carmen Hartmann, Grünenthal GmbH, Aachen and Gerhard Lang, Pari GmbH, Starnberg for providing some real chromatograms.

3.2.1
Integration, Simple and Automatically Feasible?

Figure 3.1a shows an isocratic chromatogram recorded with a data point interval t_{DP} of 200 ms. The chromatogram contains three main components (Peaks 4, 8 and 9) with some impurities. If one wants to quantify only the three main components, one would not expect any problems with the integration. Figure 3.1 b shows the same chromatogram, amplified by a factor of 32. Only now can one recognize how the baseline was set. The integration was done by LCSolution, using a width of 5 s, that is, the width at half height of the first peak. The slope test in the time window containing only a baseline between 9.2 and 9.8 min gave a slope of approximately 1000 µV min^{-1}. These are the conditions of Fig. 3.1 a and b. Using a slope of 200 µV min^{-1} the integration of Fig. 3.2 a looks very different. With a slope of 2000 µV min^{-1} a valley to valley integration for the larger peaks carried out (Fig. 3.2b).

> *The adopted baseline determines the integration.*
> *Always use sufficiently amplified chromatograms to observe*
> *the trend of the baseline.*

For every peak one can discuss which integration is better. The area and height of each peak is dependent on the integration of the neighboring peaks. The last peak standing free can be integrated in two ways, which, in turn, considerably influences the area of the second to last peak. The example shows clearly, that one can judge a CDS only if one knows how a partly merged group of peaks of different size is integrated. This is a typical problem with real chromatograms.

There is hope from Figs. 3.1 and 3.2 that the differences between the integration modes will be insignificant. Therefore the integrated areas are listed in Table 3.2 as the deviation from the base of the result (slope = 1000) of Fig. 3.1 b. Even for the tall peaks the deviations are considerable. The different integration shows a difference of 14% for the area of the last peak. The heights show substantially lower deviations.

> *The influence of integration parameters on your chromato-*
> *grams should not be checked using solely the predefined values.*

3.2.2
Comparison of Integration Systems with a Small Number of Tall Peaks

With very well separated tall peaks, CDS may have no difficulties with integration in measuring the areas and heights reproducibly and correctly. However, can one rely on different CDS measuring the same areas and the same coefficients of variance (CV) of the areas?

In a pharmaceutical company, in order to reach a decision when buying a new CDS, EZChrom and Empower were connected in parallel with one HPLC

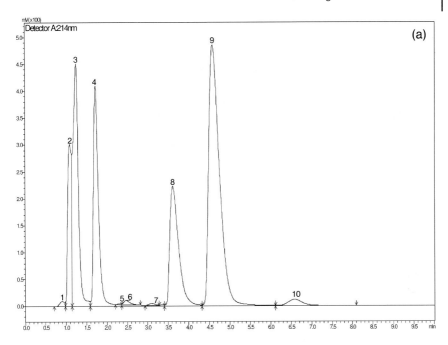

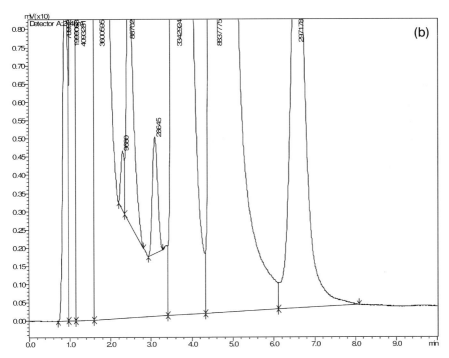

Figure 3.1 (a) Integration of an example chromatogram with LCSolutions and a slope of 1000 μV min^{-1}; (b) as in (a) but amplified 32 times.

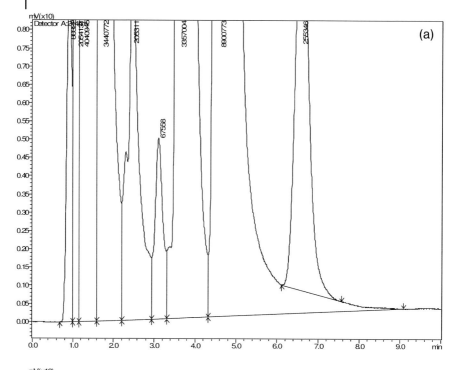

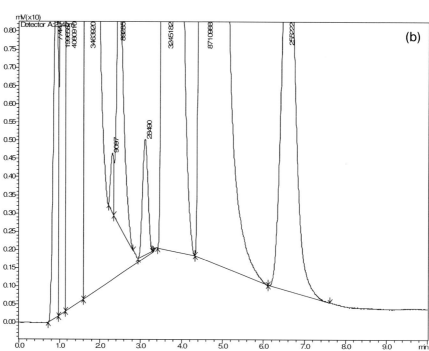

Figure 3.2 (a) As Fig. 3.1b but slope = 200 µV min^{-1}; (b) as Fig. 3.1b but slope = 2000 µV min^{-1}.

Table 3.2 The deviations of the areas and heights (in %) of the integration according to Fig. 3.2a and b relative to the integration according to Fig. 3.1b. The tall peaks are shaded in gray.

Slope	200	2000	200	2000
	Area		Height	
Peak 1	1.68	−1.96	0.02	−1.38
Peak 2	2.75	−0.12	−0.12	−0.09
Peak 3	−1.28	−0.30	−0.99	−0.08
Peak 4	−4.44	−3.80	−0.52	−0.17
Peak 5		−6.02		0.00
Peak 6	131.46	0.66	31.06	0.00
Peak 7	135.85	−0.54	54.51	−0.22
Peak 8	0.42	−2.92	−0.17	−0.83
Peak 9	0.71	−1.43	−0.08	−0.31
Peak 10	−14.08	−14.12	−4.51	−4.62

instrument. Seventeen samples with similar concentrations from a dissolution experiment and seven times a control sample with a slightly lower concentration were estimated in duplicate. The only aim of this test was to compare the results of the two CDS. A typical example of the chromatograms is shown in Fig. 3.3.

The integrated areas of the first peak are shown in the dot diagram (Fig. 3.4). The area values of EZChrom (μV s) were divided by 1 000 000 and those of Empower (nV s) by 1 000 000 000 to obtain comparable numbers. Practically identical results were obtained with the two CDS. For the dissolution tests (upper dot cloud) the CV of 1.86% for the areas (heights 1.77%) can be explained by the samples being taken at different times. The quotient of Empower area over EZChrom area amounts to 1002 ± 0.07% and the quotient of Empower height over EZChrom height to 1000.4 ± 0.02%.

The quotients from the area and height have a CV of 0.36%. The CV of the injection for the seven identical control samples of this series was calculated to 0.36% (lower dot cloud). In analogy with an internal standardization, reference to a different peak of the same chromatogram should be able to compensate to a large extent for the injection error. The area of the second peak divided by the area of the first peak yields a CV of 0.14%. Supposed that the injection error is eliminated, this CV serves as an estimate of the maximum integration error. The CVs are the same with EZChrom and Empower.

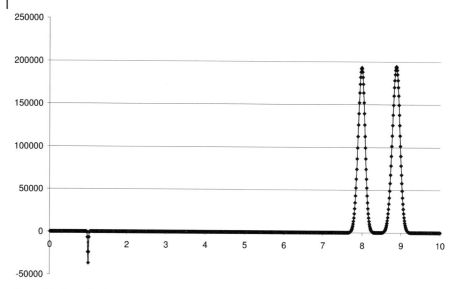

Figure 3.3 Example chromatogram as printout of exported Empower data read into Excel. Signal (y) axis in µV, time (x) axis in min.

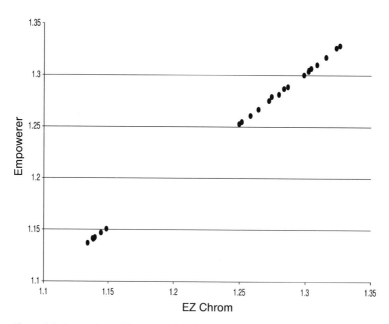

Figure 3.4 Comparison of the areas of 24 chromatograms with similar concentrations integrated with Empower and EZChrom in parallel at the same time. The control values are the group of seven slightly smaller values with the same concentration.

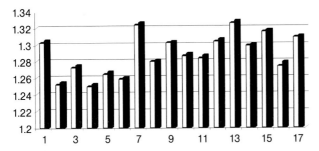

Figure 3.5 The 17 area values (upper point cloud in Fig. 3.4) integrated with EZChrom (white columns) and Empower (black columns).

The use of an internal standardization can fundamentally reduce the CV, particularly by overcoming injection (or sample preparation) errors. The difference between the two CDS is caused by a systematic deviation: The Empower value is insignificantly higher than the EZChrom value in every single case, as the bar chart shows (Fig. 3.5). All height values are also insignificantly higher in Empower. From a practical point of view: EZChrom and Empower give exactly the same areas and heights.

These values, obtained with a very simple and very good separation and excellent signal to noise ratio, provide a clue as to the realistic values which can presumably be reduced only with considerable effort:

Precision of the injection: 0.36% ($n = 7$)
Precision of the integration: less than 0.14% ($n = 24$)

Whether Empower or EZChrom has integrated "more correctly" does not have to be decided with this comparison because there is no accuracy standard. We are in the situation of a nearsighted precision marksman who cannot see the elements on his target because he does not have any glasses. He will never have the chance to improve.

It is not possible to resolve the problem of whether the systematic deviation can be explained by a difference in analog to digital conversion (sampling, integrating, delta sigma) or by the various integration algorithms. If one could resolve this, one would be able to work specifically on improvements to either the hardware or the software.

3.2.3
Comparison of Integration Systems for Numerous Small Peaks

In the context of the usual analytical procedures, a chromatogram was taken in a pharmaceutical laboratory with a C18 column and a flat phosphate buffer- acetonitrile gradient at 210 nm with the data system Empower. Figure 3.6 shows the Excel diagram made after exporting the Empower data into Excel. Three tall peaks and numerous small peaks can be seen in the concentration range of 0.1% to 1% of the tallest peak.

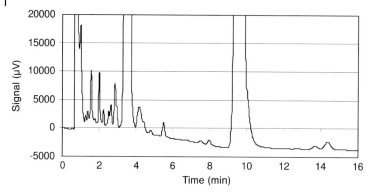

Figure 3.6 Chromatogram as an example of a gradient run using exported Empower data read into an Excel spreadsheet.

The same chromatogram was integrated 5 times. The first column of Table 3.3 lists the retention times. The other columns show the peak areas (µV s) obtained by: column 2, Empower in the traditional mode; column 3, using Empower in the traditional mode, where all the peaks were again integrated manually; column 4, Empower in the Apex mode; column 5, LCSolutions automatic integration with tangent method and perpendicular drop; column 6, LCSolutions exclusively with perpendicular drop.

If at least three area values were available, in the last column the CVs of the areas are listed. The three gray highlighted lines mark the three tall peaks. The number of peaks found is listed in the last line. The three single gray highlighted entries indicate peak shoulders, which none of the integration systems used could separate.

It can be seen that with the manual integration all areas decreased except for one. With Empower in Apex mode compared with the traditional mode eight additional peaks were found of which four, however, were below the quantification limit, so that 22 quantifiable peaks appeared.

To also integrate the data with LCSolution, the Excel data were copied into a text file, which was generated by LCSolution. The automatic integration used the tangent method for the first 10 peaks. Suppression of the tangent method gave higher areas for the first 10 peaks. For the small peaks high CVs were found. The areas of the last two peaks with Apex integration were not taken into account because the baseline had moved far too much and the areas came out far too large.

It is very unsatisfactory that the number of peaks and the areas obtained by the integration depend on which CDS is used. Is it useful to strictly regulate the calibration traceability and ignore the fact that the integration process is not under control? Without accurate standards for the areas and heights of the peaks one has to accept every reasonably plausible result. This apparent equivalence of the CDS has been accepted unchanged for almost 40 years and restricted further competive development of the integration systems.

The chromatogram simulation introduced in the following cannot improve the integration systems but can optimize their use. It is, virtually, the glasses for the

Table 3.3 Area values after integration of the chromatogram of Fig. 3.6 in
five different ways. The last row contains the number of the integrated peaks.

RT	Empower			LCSolutions		CV (%)
	Traditional	Manual	Apex	Tangent	Drop	
0.11			328			
0.58			453			
0.75	4285430	4264380	4284437	4510651	4292257	2.38
1.03	142777	138159	140290	30337	135780	41.52
1.3	16883	4471	15073	5517	16656	52.82
1.42	18286	6801	16392	8622	17939	40.18
1.6	84727	46191	69161	58943	90044	25.91
1.8	7820	1632	13822			78.57
2.03	64716	43652	86848	46630	63023	28.33
2.25	23143	12196	17457	12420	21604	29.16
2.45		898				
2.54	31169	14069	21535	17246	29924	33.24
2.66	32280	20224	25119	22005	29920	19.76
2.89	85482	47488	51958	66320	81544	25.58
2.98		15396	16566			5.18
3.68	8287311	8232904	8206488	8275794	8275794	0.42
4.21	100412	58719	68930	88023	88023	20.68
4.38		12211				
4.82	7262	3725	4463	4030	4030	30.94
5.1			425			
5.26			569			
5.51	21569	21569	23392	22394	22394	3.39
6.81			5397	2102	2102	59.44
7.55			26078	4644	4644	104.97
8.01	8331	8005	34428	8968	8968	84.22
9.76	10594747	10482034	10887257	10628447	10628447	1.40
10		123204				
12.75				1626	1626	
13.74		10159	900242	10793	10793	190.92
14.39	22542	27575	340776	29010	29010	156.31
30	18	23	26	21	21	22

precision marksman of the last section. He can improve his performance deliberately with the glasses, but he must do it himself – the glasses are only the aid.

3.3
Chromatogram Simulation

One usually has one CDS for every HPLC or GC instrument. It is thus not possible to check the integration error with real measured chromatograms because the other unavoidable errors cannot be separated.

Independent errors are combined into a measurable total error by adding up the individual variances (integration, injection, sample preparation):

$$Var_{tot} = Var_{Int} + Var_{Inj} + Var_{SP} \tag{3.1}$$

If one divides the variance by the square of the mean average value, an equation is given for the addition of the CVs

$$CV_{tot} = \sqrt{CV_{Int} + CV_{Inj} + CV_{SP}} \tag{3.2}$$

In order to acquire real chromatograms one must inject the sample(s) again and again and, in addition to the integration error, there is always an injection error, which lies in the region of 0.12% to 2.1% (Küppers et al. 2000). It is only possible to determine the dominant error. Table 3.4 shows the increase in the total error with increase in one of the other errors. For a five fold increase in one of the two errors the total error is only increased by 2% (not 20%).

It would be ideal if one could switch the peaks in a chromatogram on and off without having to re-inject. This could be used to determine the influence of an

Table 3.4 Two CVs add according to Eq. (3.2). The first column shows the ratio of the large to the small CV, the second shows the percentage that the smaller CV contributes to the total CV.

CV_{large}/CV_{small}	CV_{total}
2	11.8
3	5.4
4	3.1
5	2.0
6	1.4
7	1.0
8	0.8
9	0.6
10	0.5

additional peak on the integration of neighboring peaks. One could, for example, verify an internal standard. The internal standardization will work accurately only if the internal standard peak is completely independent of all other peaks and there is no influence from other peaks during the integration. But how can one prove this if the injection error conceals this influence?

Of course a good chromatographic separation is the basis of a good quantitative determination. It is worth the considerable effort, because the quantitative results are substantially improved. If the chromatogram consists of few peaks, it should generally be possible to optimize the resolution with minimum effort.

If one has a very crowded chromatogram (peak forest), the probability is low, that the resolution can be fundamentally improved. Improving the resolution of the worst peak pair may lead to peaks in another part of the chromatogram moving closer together. A point is reached where there is no other possibility and one accept the compromises. It is important to know the point at which one can no longer accept a compromise. Therefore, it is helpful to know what the integration error is for the problem in question and one should be able to determine the integration error separately. The only way to exclude all other possible interference factors is a digital chromatogram simulation.

- *Idea:* A chromatogram simulation should be developed, using Excel, which is so simple that everybody can use it. A lot of peaks will probably be needed and noise, complex baselines and tailing should be simulated. It must be possible to simulate each real chromatogram in a form that is identical to the original chromatogram. The data must be imported into the integration systems in the form of AIA files, so that different CDS can be compared using exactly the same data. The area and height must be predictable so that the accuracy can be checked.

- *Realization:* Excel, normal distribution, random number.

- *Aim:* (1) To test the correctness of the integration in order to check the CDS and optimize the integration parameters. (2) To test whether the usual evaluation using the peak area is better than an evaluation using the peak height.

- *Use:* To judge the integration performance of commercial CDS and, by recognizing integration errors, to reduce them.

3.3.1
Simulation of a Digital Chromatogram

A chromatogram simulation cannot improve the integration programs directly but can of course help to find the optimal parameters for the integration and to uncover the advantages and disadvantages of commercial CDS. All too often in trace analysis one must resort to the last possible means in order to improve the integration, this is, evaluating the peaks graphically by hand. This really is a user-dependent, subjective process and, therefore, not reproducible. An objective target value for the small peaks would be particularly helpful here.

In the following a chromatogram simulation is described and a step by step instruction guide is shown in the sections marked by a grey margin so that the reader can create his own Excel spreadsheet. The Excel-file Gaussian simulation.xls containing all the individual steps can be found on the CD. One's understanding and the possibility of carrying out one's own adaptations to individual problems is made easier if an effort is made to comprehend the single steps. This takes time, without question. If one does not have the time or would initially just like to obtain a summary reading the material in the sections marked by a grey margin can be skipped. A high speed reader can use the file 10peaks.xls on the CD.

3.3.1.1   One Peak
A Gaussian function describes the ideal shape of a peak:

$$f(x, \mu, \sigma) = \frac{1}{\sigma \cdot \sqrt{2\pi}} \, e^{-\frac{(x-\mu)^2}{2 \cdot \sigma^2}} \qquad (3.3)$$

This equation, standardized on the area of 1, is an essential basis of the statistics because it is the defining equation of the normal distribution upon which the (usual) parametric statistics are based. Therefore this function is included in Excel as Normdist $(x, \mu, \sigma,$ switching value$)$ – compare the Excel help function.

In a chromatogram x corresponds to the ongoing time t, μ corresponds to the retention time t_R and σ corresponds to the standard deviation of the peaks s_P (deviation of t_R). The switching value can be set to "0" or "1" only. As the function itself will be shown here, the switch is put on "0". The equation (with a variable area) could also be written as:

$$H = H_{max} \cdot e^{\frac{-(t-t_R)^2}{2 \cdot s_P^2}} \qquad (3.4)$$

where the signal value is H, i.e. the height measured at that time. Some readers may perhaps visualize this function better with an interactive simulation from the University of Basel: http://pharmtech.unibas.ch/modules/tinycontent6/content/Funktion/gauss.htm

Gas chromatographic theory originally took the idea of the plate number from an analogy with the theory of measuring the efficiency of distillation events and used it as a characteristic of the separating power. This is correct when the boiling-point of the components to be separated plays an essential role, as in GC. After the plate number had been established in GC for 20 years, the theory was also adopted for HPLC. The usefulness of the idea of the plate number is now accepted worldwide.

The usual symbol for the standard deviation of the Gaussian peaks, σ, is used in the general equations. This is not a usable term in Excel where the symbol s_P is used instead. The plate number N is defined by:

$$N = \left(\frac{t_R}{s_P}\right)^2 \qquad (3.5)$$

To simulate Gaussian peaks one only needs the retention time t_R and the standard deviation of the data points describing the peak, s_P, related to the retention time. For isocratic chromatograms s_P increases with the retention time, so that the quotient t_R/s_P remains almost constant. For chromatograms with a (linear) gradient s_P remains roughly constant. The plate number calculation is therefore not allowed, because N increases as the square of t_R.

Knowing the peak width at 50% of the peak height s_P is also known, because for Gaussian peaks $w_{50\%} = 2.35 \cdot s_P$. Another equation contains only the parameters that are the essentials in the report of every integration system: t_R, area A and height H.

$$N = 2\,\pi \cdot \left(\frac{t_R \cdot H}{A} \right)^2 \tag{3.6}$$

Another form of Eq. (3.5) solved for s_p is:

$$s_P = \frac{t_R}{\sqrt{N}} \tag{3.7}$$

A plate number of 8100 is just another way of saying that s_P is exactly one ninetieth of the retention time.

A basic decision for the data acquisition of a chromatogram is the distance between two data points t_{DP}. The generally accepted rule is that the smallest peak should be described by at least 20 data points.

If a 20-min chromatogram is taken with a data collecting rate of 5 Hz, i.e. 5 points per second ($t_{DP} = 200$ ms), this gives 300 points per minute and a total of 6000 data points. Each data point consists of the time from the start of the chromatogram and the corresponding signal value from the detector. For today's computer systems hundreds of such chromatograms would not require any storage requirements worth mentioning.

In cell B1 of an empty Excel spreadsheet type "Peak 1" and in cell C1 type "Peak 2". Mark both cells with the mouse and draw a black cross at the lower limitation to the right, expanding for example to "Peak 8".

Type in cell A10 "0" and in A11 "0.01". Draw the two cells to cell A1010 down. Then the time axis is established (10 min with 0.01 min means $t_{DP} = 0.01$ min $= 0.6$ s $= 600$ ms and leads to 1000 data points). Write "tR" in A2, "sP" in A3 and "PF" in A4 and, as preliminary data, "4" in B2, "=B2/90" in B3 and "1000" in B4.

We now turn to the most important value, the peak signal. Type in B10: "=$B$4*NORMDIST(A10;$B$2;$B$3;0)". The function in A10 is further dragged down to A11, A12 etc. to A1010 (indirect addressing). With the $ signs sitting in front this particular cell remains the same (direct addressing). After highlighting A10 to B1010 one gets, using the diagram assistant, a dot diagram, which corresponds to a chromatogram with one peak.

Instead of s_p one can enter the normally used plate number (root of (N)). However, because s_p and N are directly related, this is hardly an advantage. The disadvantage is that since the plate number increases linearly with the retention time with use of gradients, one should enter an arbitrary and variable size. However, s_p is simpler and can be entered as a constant value, because the peak width is roughly constant with linear gradients. With the peak factor, PF, one can vary the area (height) of the peak.

The peaks of an isocratic chromatogram do not have exactly the same plate numbers and the peaks of a gradient chromatogram do not have exactly the same width. However, in practice, these deviations do not influence the value of the chromatogram simulation as an integration test. It is always possible to type the widths measured into the spreadsheet.

3.3.1.2 Several Peaks

If one selects the parameters for t_R, s_p and PF correspondingly, then one can produce a second peak in just the same way for the next column. By addition of the values in columns B and C, a chromatogram with two peaks would be produced. This is expandable to many peaks.

> We change the cell B10 by adding "$" in front of A and deleting "$" before B, resulting in "=B$4*Normdist($A10;B$2;B$3;0)". Now we can, by simply drawing B10 to the right, produce as many peaks as we want. The error message "# number!" disappears by drawing the input cells B2 to B4 to the right.
>
> To put the 8 peaks of a chromatogram together, one must mark column B and type <insert> <columns>, to insert a new column. All columns to the right are shifted. Into the now empty cell B10 type "=sum(C10 : J10). After marking A10 to B1010 one gets once more a dot diagram, which corresponds to a chromatogram with 8 peaks.

With the previous table one can create a large number of peaks with arbitrary area and height and arbitrary retention times.

3.3.1.3 Noise

The noise is traditionally measured in a defined time range between the highest and the lowest data points. The peaks can be measured based on the lower and the upper noise limit or their mean. Sometimes it is argued that the noise, which has positive and negative deviations from the average, should only be accounted for by its half value. Therefore different values for the S/N are possible. The national or international terms of reference must be followed.

Since every detector has a noise, this must also be simulated. Without noise the integration systems cannot define a threshold (slope) value. This can be done simply with a number chosen randomly within a predefined range. One uses the function "Rand()". This produces a random number between 0 and 1 (which is calculated anew at every file start) which is multiplied by a factor to obtain the range. The maximum range is the value of the noise, indicated in μV. For the calculation of S/N the simulation uses the signal divided by the complete noise amplitude.

> One highlights column C, selects <insert> <columns> and types in C10 "=$C$4*Rand()" and enlarges to C1010. The value input in cell C4 is the (range of the) noise. The sum in column B is enlarged automatically so that all peaks and the noise are summed for the signal.

According to Papas (1986) one can also multiply two random numbers to give the so-called white noise.

The most realistic baseline can be drawn using a typical blank sample copied in Excel and overlaid by the peaks (Stauffer, personal communication 2007). This way, expensive unnecessary noise simulations can be avoided but an almost realistic situation can be created. With many CDS one can distribute the raw data in text format. With LCSolutions the noise of an SPD 10 AV UV detector was recorded at 214 nm and copied out into a text file for 10 min. The signal data were copied into an Excel file.

> Mark the content of the first column in "natural noise.xls" and copy it into column C of the simulation spreadsheet. The simulated chromatogram now has a real baseline.

Gradient chromatography is only possible when compared with to the chromatogram of a blank gradient run under the same conditions. This run works well as the background for the gradient simulation. However, one also can simulate the baseline course through one very wide peak, which works best if the retention time corresponds to the analysis time, and the width is set for example to approximately $t_R/3$. With the factor PF, the baseline course of the gradient can be emphasized more or less strongly.

3.3.1.4 Drift

A linear (or indeed non-linear) rising or falling baseline is termed drift. This can be simulated by one (or more combined) straight line equations. The intercept has the function of a positive or negative offset. Other simple functions can, in principle, be used or a combination of several possibilities during the time course of the analysis. A plate-shaped baseline is given by an underlying very wide negative peak.

> An offset and a drift can be simulated in an additional column with a straight line equation. One marks column D, selects <insert> <columns> and types into D10 "=$D$4*A10+$D$3" and enlarges to 1010. Cell D3 must contain the intercept and D4 the slope of the straight line.

To judge the influence of the drift on the integration one tall peak (Fig. 3.7) or two tall peaks (Fig. 3.8a) merged together were underlaid by a strong drift (drift1. cdf, drift2.cdf). The single peak can be clearly seen to be integrated incorrectly. With the two peaks the strong deviations of the area of –7.8% of the set point 2 100 000 for the first peak and of –12.9% for the second peak are evident. Both the end of the second peak and the perpendicular drop are set at the lowest signal value. Here the deepest point of the course is not the optimal point for the peak recognition. However, the integration algorithm of the program used here is too simple for this problem. (Figs. 3.7a and 3.8a). The result, amazingly, is even worse with EZChrom (Figs. 3.7b and 3.8b), although it looks as though EZChrom is also using the deepest signal point.

3.3.1.5 Peak Area

The integral of the normalized Gaussian function is exactly 1. The chromatogram simulation used µV to measure the signal and min for the time. Therefore the peak factor is the set point of the area in µV min. The area is expressed in µV s in most integration systems. This means an area of 60 µV s multiplied by the peak factor is used to obtain the expected area of the simulated peaks.

This information represents the essential breakthrough for the examination of the CDS. If the set point of the area is known exactly, every integrated area can be checked for accuracy, in spite of the peaks lying on an uneven baseline or being merged together. It is no longer necessary to optimize the integration just for plausibility, because the target value can be approximated with increasing confidence by changing the integration parameters. Several free-standing peaks are available in the later test chromatograms shown to prove the validity of the areas in the chromatogram simulation. In all cases the target values were found to be within a very good approximation.

Simulations provide you with peaks with a known area under the curve.

3.3.1.6 Peaks Merged Together

By variation of the retention times, the two peaks can merge together to a greater or lesser extent. Therefore it is possible to check the quality of the integration over a range of merged peaks. The more two peaks are merged together and the more different they are in size, the more strongly the areas deviate from the areas of the single peaks (compare with Meyer 1995). Using the chromatogram simulation, it is possible to merge more than two peaks and to construct a complicated baseline. As is described later, real chromatograms can be imitated in order to uncover the problems of the integration.

Of course peaks could be merged very simply. Type for example "5.2" into cell G2. A sufficient separation arises for quantification. If one reduces the peak factor to 100 and then to 10, one recognizes that at great concentration differences higher demands on the separation must be made, in spite of the same resolution value.

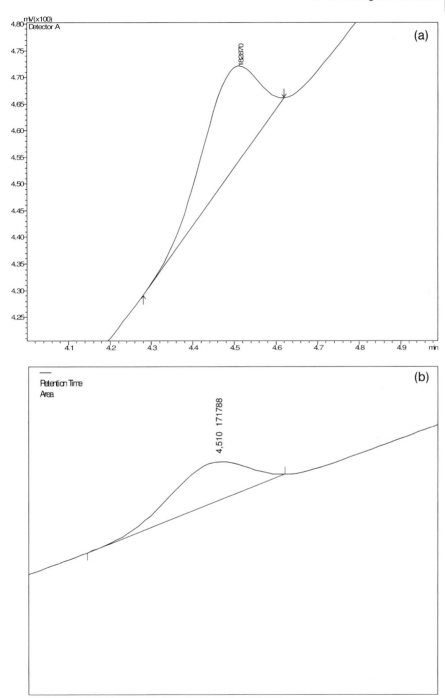

Figure 3.7 Extended part of a chromatogram with a peak on a 45° drift.
(a) Integration with LCSolutions, (b) integration with EZChrom.

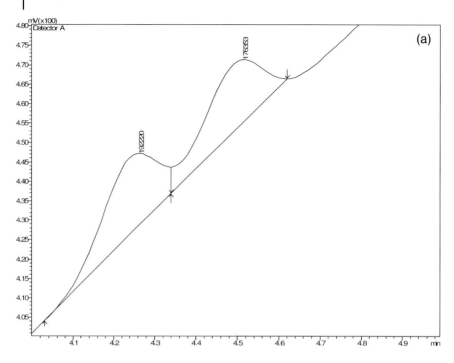

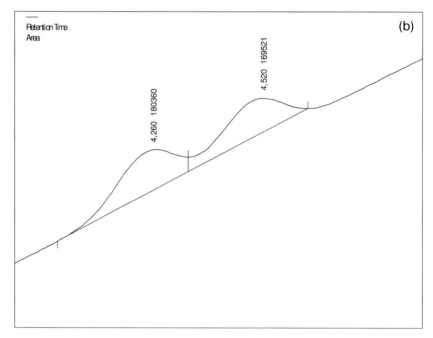

Figure 3.8 Extended part of a chromatogram of two peaks merged together on a 45° drift. (a) Integration with LCSolutions, (b) integration with EZChrom.

3.3.1.7 Unclear Baseline

At first the example in Fig. 3.9 looks very simple. Two peaks not particularly merged sit on a much broader peak taken as a baseline.

Every integration system is only able either to assume a baseline separation between the peaks or to connect the peak start of the first peak with the peak end of a following peak (Fig. 3.10a and b). In principle, only measured data points connected by a straight line can be used as support for the baseline. Both integration possibilities lead to considerable deviations in the areas of the first and second peaks.

The deviations of the peak height are lower by about a factor of 2. The integration systems do not have any chance of finding the right baseline in this example.

The only possibility to establish whether a valley point is also a baseline point is the resolution, which is indicated by LCSolutions (Fig. 3.10a) as being 1.2. Therefore the valley to valley result (Fig. 3.10b) must be excluded. Only a resolution of 2 or more must be interpreted as a baseline point in the valley between the peaks.

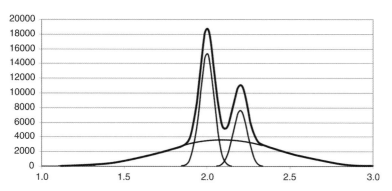

Figure 3.9 The target areas are 100 000 and 50 000 µV s, the target heights 15 344 and 7672 µV.

3.3.1.8 Tailing

A slight tailing may be simulated by the superimposition of two Gaussian peaks. The second peak with the same PF as the first peak has a slightly greater retention time ($t_{R2} = t_{R1} + F \cdot s_{P1}$) and is a little broader ($s_{P2} = s_{P1} + F \cdot s_{P1}$). The factor F can be varied between 0 and 1.2. With $F = 1$ a tailing arises with a tailing factor of about 1.55. The areas of the two peaks must be added together, to give the target area.

3.3.1.9 Data Point Interval

As a general rule, the manuals for the integration systems usually indicate that the narrowest peak should be described by 20 data points (DP). Depending on the range in which the retention factors lie, the last peak will contain 40 to 100 DP, because of the peak broadening in isocratic separations. For separations with gradients the peak width can vary by around a factor of 2 at most, which means the peaks are described by 20 to 40 DP.

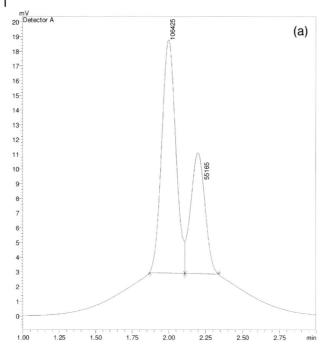

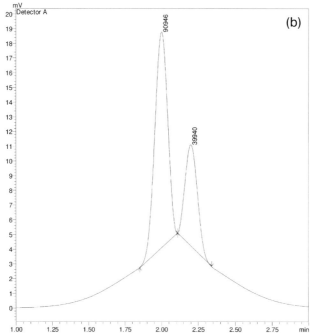

Figure 3.10 (a) Integration with LCSolutions by perpendicular drop (peak heights 15 763 and 6827 µV), (b) valley to valley (peak heights 14 599 and 6 832 µV).

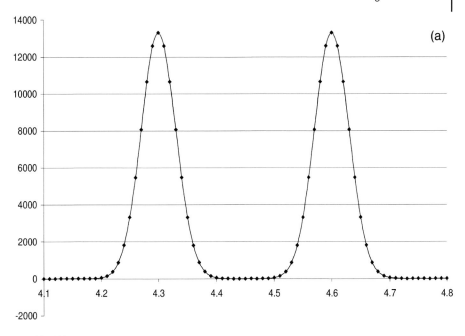

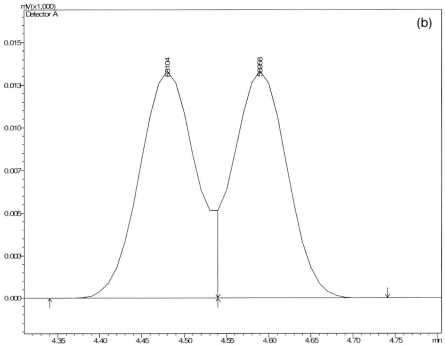

Figure 3.11 (a) With well separated peaks 20 DP suffice for precise integration.
(b) With peaks merged together 20 DP do not suffice for a precise integration.
5% of the area is assigned to one of the two peaks because t_{DS} is too large.

Two peaks having 20 DP each are shown in the Fig. 3.11a. With a peak height of about 13 000 µV and a noise of 10 µV, the S/N ratio is 1300.

For the integration with LCSolutions the areas of 60 047 and 60 082 µV s are very close to the target value of 60 000 µV s. They have a deviation of 0.08% and 0.14%. These deviations are so low that a possible improvement using more DP would be irrelevant, because the injection error will be higher.

The situation changes (Fig. 3.11b) if the two peaks are approximated to such a great extent that a resolution of about 1.1 appears, as happens frequently in reality. The integration system for the perpendicular drop has to decide between the right or the left valley point between the two peaks. Because the right DP with 5117 µV is slightly lower than the left DP with 5120 µV, the decision is predefined. A very small shift of these two values may switch the valley point to the other DP. Only if one looks at the result of the integration does it become clear that the area between the two DPs is about 5% of the total area. One arrives at a similar value by multiplying the above-mentioned signal of 5117 by the distance of the DP of 0.6 s. This inaccuracy is caused by the fact that the distance between the data points t_{DP} is far too wide. In this example the peak height delivers a correct result.

> *Twenty data points are sufficient for a singular peak;*
> *whereas merged peaks require 100 each.*

The peak height is largely insensitive to suboptimal adjustments of t_{DP}.

If one takes a 6 times shorter data point interval of 100 ms, the deviations of the areas are only 0.39% and −0.4%. It makes a very big difference for merged peaks whether the number of data points n_{DP} over the peak is 20 or 120. Twenty points are too few by far, only many more data points are required. Figure 3.12b shows the peak valley over an enlarged section. For accurate integration of merged peaks at least 100 data points per peak are required.

A visual comparison of the chromatograms in Figs. 3.11a and 3.13a shows no difference. The integration with LCSolutions leads to two-fold deviations compared with the integration with EZChrom. Obviously different integration algorithms are used, which the manufacturers do not reveal. The user is primarily interested in the influence on the integration. To check one's own CDS, the files 20 nDP.cdf and 120 nDP.cdf can be integrated with the individual integration systems.

If one has peaks merged together in the chromatogram, the t_{DP} should be adjusted to have approximately 100 DP around the peak width. Presumably the CDS manufacturers are not pleased about this because five-fold faster (and more expensive) ADCs could be necessary.

To adjust the desired data point interval t_{DP}, one must carry out an estimate of the peak width of the narrowest peak, which is, to a good approximation, the first relevant peak.

> *Never forget to optimize t_{DP} for every new problem.*

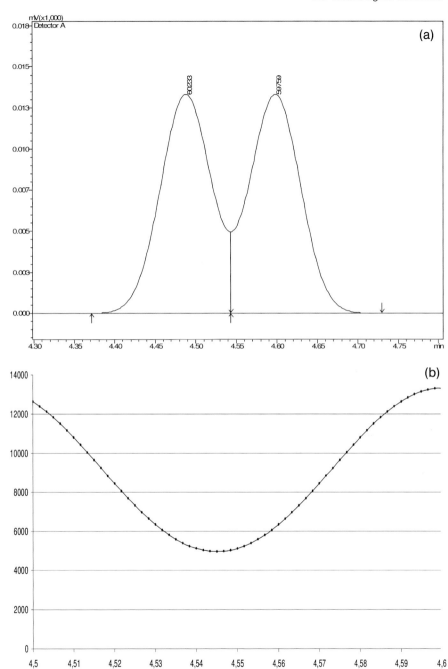

Figure 3.12 (a) Simulation of two peaks merged together, analogous to Fig. 3.11 b with a 6-fold lower t_{DP}. (b) The expanded part of Fig. 3.12a, this shows the much better adjustment of the data points due to the course of the curve than Fig. 3.11b.

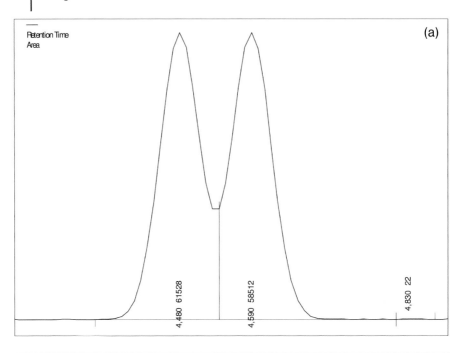

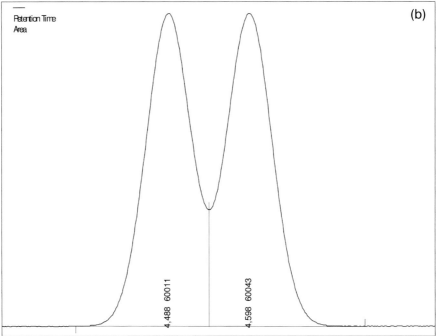

Figure 3.13 (a) Analogous to Fig. 3.11b, but integrated with EZChrom;
(b) analogous to Fig. 3.12a, but integrated with EZChrom.

How does one assess the peak width? If one has a printout of the integration system with the details of the peak start and end, the peak width is the difference. Perhaps this leads to an overestimation, which does not matter. One should be aware that the width sometimes means the width at the half height.

Another possibility is a graphical method. On a graph of the first peak, mark the baseline and draw tangents at the inflection points. The length of the baseline between the two intersection points with the tangents is an estimate for $4\ s_P$ and therefore the peak width in its traditional definition (compare Fig. 1.3). As a more realistic measure of the peak width, $6\ s_P$ can be used, which is in accordance with statistical considerations. If one knows the peak width, t_{DP} is selected such that the desired number of data points n_{DP} "fit into the peak width".

$$n_{DP} = \frac{6\ s_P}{t_{DP}} \tag{3.8}$$

$$w_{50\%} = 2.35\ s_P \tag{3.9}$$

The number of data points per peak is not provided for any CDS, but the half width can be shown. If just the 2.55-fold half width is divided by t_{DP} there are no problems in obtaining this number for every peak.

From the ratio of area to height an estimate of the half width of the first peak is obtained (Stauffer 2006, personal communication).

3.3.1.10 Other Characteristic Quantities

Using the Gaussian function, the height of a (Gaussian) peak can be calculated exactly:

$$H_{max} = \frac{A}{s_P\ \sqrt{2\ \pi}} \tag{3.10}$$

With the simulated (Gaussian) peaks t_R, A, H and N are known precisely. Therefore it is easy to check with what precision these values can be reproduced by the integration systems from the various manufacturers. Through the very simple method of simulation it should be possible for every chromatography user to model his own chromatograms (compare Section 3.3.6). Even if this is not exactly possible the parameters of the integration system used can be optimized because A and H are known. Until now, this has not been possible. Whether one uses the area or the height for the calibration and evaluation the quality of the results may change.

The optimization of the integration parameters is not possible without knowledge of the target value. The probability that the integration error will be compensated by comparing a control sample with the analysis sample is low. With an integration as "accurate" as possible one is more likely to get correct analysis results.

To be able to judge the simulated chromatogram as well as possible, one can calculate some additional useful characteristic quantities:

1. Target area in μV s.
2. Number of data points over the peak width.
3. Resolution (replace the width at baseline by 4 s_P).
4. Gaussian peak height.
5. Signal to noise ratio.

The peak factor PF indicates the target area in μV min. To obtain the usual μV s for the area, type in E5 "=60*E4" and expand until N5.

To calculate the number of data points for every peak n_{DP} one has to divide the real peak width (6 s_P) by t_{DP}. Type in cell D6 "=6*D3/($A$11-$A$10)" and expand until N6.

For the calculation of the resolution for Gaussian peaks, type in D7 "=(F2-E2)/(2*F3+2*E3)". Of course this is only possible after the second peak.

For the Gaussian height, type into cell E8 "=E5/(60*E3*root(2*PI()))". This is not valid for tailing peaks.

The signal to noise ratio can be easily calculated from the peak height and the noise. Type into cell E9 "=E8/$C$4". To get a warning sign for S/N below 10, cell E6 to N6 can be provided with <Format><Conditional Formatting> and [cell value is] [less than] [10] <Format> for example red on yellow ground or other enhancers.

In the same manner, one can list in F7 to N7 as a condition for the resolution [cell value is] [less than] [1] and as a condition for the number of data points into E5 to N5 [cell value is] [less than] [50].

3.3.1.11 **Gas Chromatogram**

The simulations described so far showed chromatograms typical for HPLC. There is no major difference between the HPLC and GC chromatograms but the separating performance of capillary gas chromatography is much higher than that of HPLC. Typical plate numbers are higher by about a factor of 10. This means narrower peaks and hence the possibility of separating considerably more substances from each other by around a factor of 3.

Since s_P was always 1/90 of t_R, a plate number of 8100 was used in the examples. This is a typical value for HPLC. If one wants to represent a GC chromatogram, one replaces, for example, the 90 by 300 (N = 90 000) in D3 and expands the calculation rule to I3. A GC temperature program can be visualized by typing "0.02" in D3 and drawing to the right.

Figure 3.14 shows the temperature program simulation of 70 narrow peaks ($s_P = 1$ s) with an analysis time of 15 min. Both the retention times and the peak factor were produced with the random function. The chromatogram looks similar to a real capillary gas chromatogram from the surrounding environment.

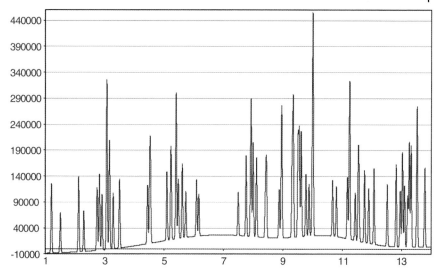

Figure 3.14 Simulation of a gas chromatogram with 70 peaks according to a random choice for the retention time and peak area.

The baseline is formed with two wide peaks. Using in this example "=(50+50*Rand ())*(50+50*Rand())" for the peak area, a (working) range of 150 000 to 600 000 μV s is predefined. Especially in GC chromatograms containing a lot of peaks, it can be advantageous to show not only the whole chromatogram, but also the single peaks in different colors, which are partly overlaid. The data of the single peaks are in the single columns and must only be highlighted.

A new chromatogram is produced at every call of "Gaschromatogram.xls", which covers 50 Mbyte. With a data acquisition time of 0.001 min, in 15 min 15 000 signal values occur and for 70 peaks more than 1 million cells in the Excel spreadsheet will be filled. Some computers are near to their limit with this.

A similar simulation for 10 peaks with 1020 data points in Excel was described by Meyer, developed about 12 years ago under the name "Zufalls-Chromatogramm. xls" and sent to interested people (Meyer 2006, personal communication).

Excel has a maximum table size of 256 columns. Therefore, in one table no more than 250 Gaussian peaks can be simulated. It is rare that one needs more peaks. The maximum number of lines is 65 536. At an acquisition rate of 200 data points per second, which is only necessary in fast gas chromatography, the maximum length of the simulated chromatograms is up to 54 min.

3.3.1.12 Applications of the Simulation

Some peak simulations are shown in the following, for which the most important applications can be checked with the chromatography:

- Calibration of a standard curve.
- Multiple measurements at the limit of determination.
- Isocratic analysis with peaks increasingly merged together.

- Gradient analysis with peaks decreasingly merged together.
- By-product analysis.
- Simulation of one's own chromatograms.

The problems in different laboratories vary considerably. However, it is possible to simulate one's own chromatograms and prove the accuracy of the integration.

The Excel simulation must be fed into the CDS. First one exports a text file, copies the Excel data matrix, reads the text file, stores it in the CDS format and exports as it a cdf file (AIA format), which should be readable by every CDS.

The procedure is described here for LCSolutions. The other CDS mentioned in this book are normally blocked at the point to import the text file into the CDS. The very similar software GCSolutions is not able to import a text file.

Change an arbitrary LCSolutions data (lcd) file into a text file with <file> <export data> <export data as ASCII>. Delete the data matrix and copy the data matrix created in Excel. Write down the right t_{DP} value on "Interval (msec)" and correct the value for "# of points" and "End Time (min)". Save the text file and open it in LCSolutions with <file> <open Data File> after you have selected the ASCII file (*.txt) from the list. The simulated chromatogram should be seen. You integrate and store in the LCSolutions data (lcd) format. Choose <view> <data Explorer> and click with the right mouse button on the before stored lcd-data file. Choose <file Conversion> and <LCSolutions Data File to AIA File>. It should be possible to read the change data format (cdf) file with every CDS.

For the transformation into an AIA file with LCSolutions only integer signal values (µV) are distributed in which the numbers after the decimal point are cut off. Therefore a noise of 10 is represented by the discrete numbers from 0 to 9 and the real noise is 9 µV, shown in Fig. 3.15. For the chromatogram simulation, therefore, values for noise of lower than 10 should be avoided. A noise of 1 would yield the signal value zero throughout the chromatogram. Once these limitations are known, they do not matter quite so much.

The advantage of the peak simulation described here lies in its simplicity and immediate availability provided that a computer with Excel is accessible.

The Excel spreadsheet is easily expandable. To prolong the analysis time, one marks all entries of the undermost line and draws down until the desired analysis time is reached. One increases the number of peaks by highlighting on the right cells and stretching to the right all entries in the column. The sum in column B must be re-sized accordingly.

If the spreadsheet contains too many columns: Set PF to zero for the peaks you do not need. This sets the height of the peak at zero and virtually turns it off. If the analysis time is too long, copy only the part of the chromatogram which is needed.

The integrations with LCSolutions were partly carried out by one of the editors (HJK), and partly by C. Zeidler, TB Shimadzu GmbH, 85375 Neufahrn. The

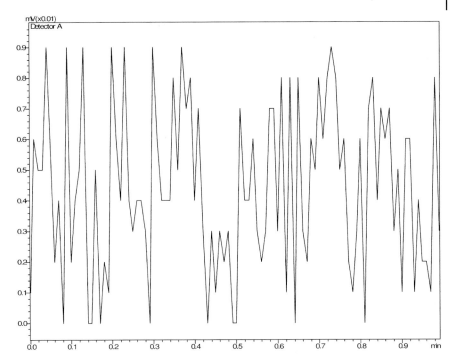

Figure 3.15 Excel simulation of noise with the random function multiplied by 10 and introduced in the LCSolutions format.

integrations with Cromeleon were done by M. Heep, Dionex GmbH, 65510 Idstein or P. Sauter, Dionex Softron GmbH, 82110 Germering. All integrations with Empower were carried out by B. Peuser or I. Prahl, Waters GmbH, 65760 Eschborn. The integrations with EZChrom elite (except Section 3.2.2) came from R. Fischer, Böhme KG, 82538 Geretsried.

All those involved are cordially thanked for their effort. Special thanks go to Mr Zeidler for his indication of how to use the text file for LCSolutions and to Mr Fischer who carried out the integrations alongside his regular professional activity.

3.3.2
Simulation of a Standard Curve

Calibration is the basis of any quantitative chromatographic determination, whether strict regulations must be fulfilled or not. There is the Mandel test which determines whether or not a low bend of the quadratic function yields a significantly lower residual standard deviation than the linear function. Furthermore, the variance homogeneity, the prognosis interval and the measuring uncertainty must be shown. A German standard (DIN 32645) deals with the limit of quantification according to the straight line calibration. In chromatography it is simpler and easier to determine the limit of quantification from the 10-fold signal to noise ratio.

With regard to how strongly the integration system contributes to the measuring inaccuracy, usually no information is available. For the measuring inaccuracy such as in the Ishikawa diagram (Meyer 2006) the integration is defined by different sub-structures in the listings of the parameters of influence. The difficulty consists in estimating the individual areas of uncertainty. If this is not individualized, the "Guide to the Expression of Uncertainty in Measurement" (GUM) remains a formalisted set of rules.

Quality control work is carried out way above the limit of quantitation (LOQ) and the integration error at the LOQ is unimportant. For the main component the requirements on precision and accuracy are particularly high. Often one has to measure very small concentrations of a by-product or maybe peaks have merged together so that the integration error can be considerable.

It is possible to produce a calibration curve for peaks beginning with the LOQ to much higher concentrations where the target areas are known from the chromatogram simulation. Then the deviation from the correct value can be found as a dependence on the concentration or the S/N, which gives the integration error only.

This was simulated for a simple chromatogram with a t_{DP} of 200 ms (5 Hz) and three neighbouring but well separated peaks within 4 min: Eich1, Eich2, Eich4, Eich8, Eich16, Eich32, Eich64, Eich128, Eich256, Eich512 and Eich1024, i.e. 11 calibration points with increasing concentrations of the first and third peaks of 1 ng up to 1024 ng were used (Table 3.5). Eich(ung) is the old German word for calibration.

The second peak is exactly the same in all chromatograms – at Eich128 all three peaks have approximately the same height – it can be considered to be an internal

Table 3.5 Target areas (μ s) of the three peaks in the 11 calibrating simulations.

Eich	Area 1	Area 2	Area 3
1	720	120000	1080
2	1440	120000	2160
4	2880	120000	4320
8	5760	120000	8640
16	11520	120000	17280
32	23040	120000	34560
64	46080	120000	69120
128	92160	120000	138240
256	184320	120000	276480
512	368640	120000	552960
1024	737280	120000	1105920

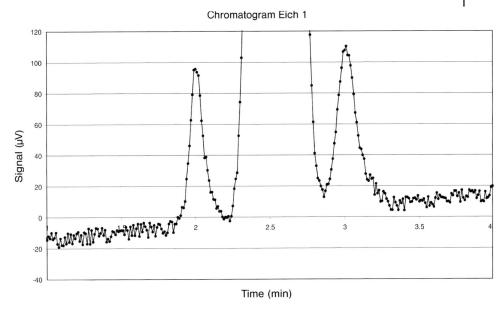

Figure 3.16 Simulation of Eich1 with S/N of 10.

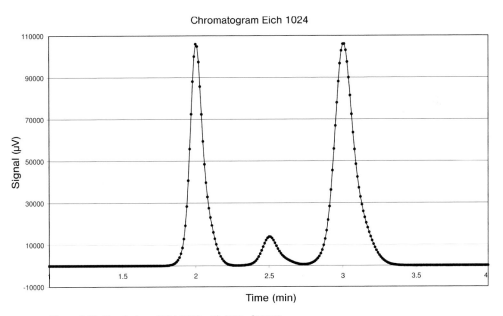

Figure 3.17 Simulation of Eich1024 with S/N of 10240.

standard. Peak tailing is seen for all peaks, with $A_f = 1.55$, which was produced by merging two Gaussian peaks as described above. The noise was set at 10 μV and the first and the third peaks in Eich1 have peak heights of approximately 100 μV

(Fig. 3.16). Therefore the signal to noise ratio is just 10 (compare Fig. 3,16). The peaks were described by about 50 data points. A gentle drift and a low negative offset were used. This is a particularly simple integration problem which represents a typical common situation in routine laboratory work.

Using Waters Empower the two independent integration modes "Apex Track" and "Traditional mode" were chosen for the integration test. With Chromeleon the integration was carried out independently by two employees of Dionex with different integration parameters. With EZChrom it was integrated twice with only the threshold being changed by around a factor of 2. In addition, LCSolutions from Shimadzu and Atlas from Thermo were also used. The double integration with Chromeleon and EZChrom shows the influence that different integration parameters can have. For a comparative judgement of the CDS, the better results must obviously be considered. The individual integrated areas and the evaluation of the accuracy are contained in the file "CalibrationResults.xls" on the CD.

The second peak with an S/N > 1000, which is treated as an internal standard, always contains the same data matrix. Only the noise is a slight modulator, which should not give a deviation of more than 0.1% . Therefore the same area should always be found, otherwise it would have to be assumed that the first and the third peaks have an influence on the integration of the second peak.

Table 3.6 shows that both the precision and the accuracy for the second peak lie below 0.1% in almost all instances and the deviations are irrelevant, because they lie below the inevitable injection error (at best 0.12%).

The result using the Apex-Track integration is very different: This gave a CV more than 10 times that with all other integration systems. Eich512 and Eich1024 in particular contribute to the poor result. In addition, all integrated areas are too low. On the other hand, the Traditional mode Waters integration is in good agreement with the other results.

In all the tables of results deviations of more than 1% are highlighted with light gray for a quicker overview, deviations over 5% are highlighted in dark gray and printed in bold.

The deviations of the two variable peaks before and after the constant peak are shown in Table 3.7a and b. In the lower range up to S/N = 80, the deviations are above 1%, at higher S/N ratios they are always below 1%. Again the exception is the Apex-Track, where they are higher and always negative. This means that, in principle, in all instances the baseline is drawn too high.

Only by changing the threshold from 50 to 100 does EZChrom find +5% at the third peak instead of –18% for the lowest "concentration" Eich1. The Chromeleon results also show that a different choice of integration parameters can give very different results. The deviations determined by Fehlinger and Guiochon (2001) with Chemstation for one peak are in accordance with the results shown here, where Chemstation was not examined.

If the second peak is used as an internal standard, no improvement in the results is possible because this peak is always the same in the simulation. In practice the internal standard varies with the injection volume and the sample preparation, and hence in a typical situation these errors will be reduced.

Table 3.6 Area values of the 11 calibrating simulations for 8 different integration modes. The lower 4 rows give a summary of precision and accuracy.

2. Peak Eich...	Shimadzu Lab-solutions	Waters Apex Track	Waters Tradi-tional	Dionex Chro-meleon	Dionex Chro-meleon	VWR EZ-Chrom	VWR EZ-Chrom	Thermo Atlas
1	119792	119171	120167	119820	120011	119792	120008	119920
2	119976	119068	120176	119700	120096	119976	119960	119914
4	119925	119008	120138	120120	120111	119925	119943	119958
8	119948	118791	120219	120000	120169	119948	119892	119998
16	120004	118448	120191	120060	120141	120004	119939	119985
32	120028	118019	120246	120120	120145	120028	119970	120037
64	120050	119370	120222	119580	120014	120050	119988	120008
128	120063	119044	120213	119280	120172	120063	120025	120084
256	120052	118464	120107	119760	120051	120052	119822	119875
512	119960	112965	120080	119820	119974	119960	119857	119817
1024	120163	110785	120095	119940	119981	120163	119893	119831
Mean	119996	117558	120169	119836	120079	119996	119936	119948
Std. dev.	94.8793	2877.3	56.49	254.69	75.181	94.959	63.735	84.915
Precision (%)	0.08	2.45	0.05	0.21	0.06	0.08	0.05	0.07
Accuracy (%)	0.00	−2.04	0.14	−0.14	0.07	0.00	−0.05	−0.04

The calibration data shown may lead to the assumption that the integration errors given in the tables are absolute deviations, which do not depend on the noise and the peak height. The integration depends, in principle, on the correct regulation of the peak start and peak end, between which one even (base) line is drawn. The uncertainty of this line is determined in practice only by the noise, and different positions of this line will lead to the same error in the area in absolute units for tall and small peaks, i.e. as a percentage this error disappears for tall peaks. It is clear from this that using the same concentration of the internal standard cannot, in principle, correct for the integration error, which is dependent on the concentration.

An internal standard is the best aid for reducing uncertainty in your measurements. However: It will not remove the integration error.

Table 3.7 Deviations of the areas (accuracy) (%) of the 11 calibrating simulations

(a) with 8 integration modes for the first peak.

1. Peak	Shimadzu	Waters	Waters	Dionex	Dionex	VWR	VWR	Thermo
Eich...	Lab-solutions	Apex Track	Tradi-tional	Chro-meleon	Chro-meleon	EZ-Chrom	EZ-Chrom	Atlas
1	−12.52	−12.08	14.17	−18.92	−6.39	−12.50	−17.78	−11.53
2	−4.75	−4.79	6.46	−2.92	4.51	−4.72	−13.19	−6.32
4	−4.15	−4.17	2.74	1.98	1.94	−4.13	−6.28	−2.88
8	−0.09	−2.00	2.92	−0.14	1.89	−0.09	−4.60	−1.41
16	−0.59	−1.44	0.95	−0.39	0.79	−0.59	−1.60	−0.40
32	−0.43	−1.55	0.62	0.35	0.35	−0.43	−0.63	−0.12
64	−0.14	−1.02	0.39	−0.62	−0.15	−0.14	−0.32	−0.13
128	−0.07	−0.78	0.17	−0.57	0.11	−0.07	−0.16	0.00
256	0.03	−0.78	0.08	−0.11	0.08	0.03	−0.15	−0.03
512	−0.03	−0.56	0.06	−0.06	0.02	−0.03	−0.06	−0.02
1024	0.02	−0.54	0.03	0.00	0.01	0.02	−0.02	0.00

(b) with 8 integration modes for the third peak.

3. Peak	Shimadzu	Waters	Waters	Dionex	Dionex	VWR	VWR	Thermo
Eich...	Lab-solutions	Apex Track	Tradi-tional	Chro-meleon	Chro-meleon	EZ-Chrom	EZ-Chrom	Atlas
1	−18.42	−11.94	10.09	−5.32	1.02	−18.43	5.00	0.19
2	−1.87	−12.36	4.40	−11.07	−0.60	−1.85	−0.46	−2.87
4	−1.86	−9.00	−0.37	−0.77	−0.76	−1.85	−1.46	−1.27
8	−1.43	−7.28	1.59	−1.86	0.73	−1.42	−0.22	−0.10
16	−0.91	−5.77	0.42	−0.21	−0.02	−0.91	0.03	0.20
32	−0.22	−5.67	0.39	−0.01	0.00	−0.22	−0.19	0.09
64	−0.26	−1.11	0.12	−0.49	−0.20	−0.26	−0.10	0.01
128	−0.24	−0.90	0.03	−0.32	−0.01	−0.24	−0.03	0.01
256	−0.05	−0.79	0.08	−0.10	0.04	−0.05	0.02	0.06
512	−0.04	−1.98	0.03	−0.04	−0.01	−0.04	0.01	0.02
1024	−0.01	−1.60	0.02	0.00	0.00	−0.01	0.00	0.02

Table 3.8 If the noise range is 10% of the peak height (S/N = 10), then the maximum deviation (Max%) of the areas is just 10%. Because the peak foot widens, the expected deviation in the height is only half the deviation of the area.

S/N	N/S	Max (%)
10	0.1	10
100	0.01	1
1000	0.001	0.10

As an approximation one can assess the integration error of the area from the ratio of noise to signal, i.e. from the reciprocal value of the S/N. A noise of 10% (S/N = 10) provides the possibility of increasing or decreasing the peak height by up to 5%. The area changes approximately twice as much because the peaks widen substantially at the base of the peak (see Table 3.8).

Using the optimal integration parameters the integration error may be smaller, but with less optimal parameters the integration error can be greater than those given in the table. Dong (1999) wrote: "Theoretical considerations indicate that peak area precision is inversely proportional to signal-to-noise and the number of sampling points, and is proportional to peak width". At an S/N = 11, he found a CV of 5.5%.

The calculation of the concentrations of the components in the samples being analysed is always done relative to the control samples. It could thus be conceivable that deviations could be tolerated because the calculation compensates for them. In this instance the integration system would then be wrong, but reproducibly wrong. Considering linear systematic errors, this argument is correct, but in spite of this, the systematic errors are consistently not compensated for. As shown previously the integration errors are constant systematic errors.

The calibration is usually carried out with reference to a straight line graph. The intercept (that is: y axis value at $x = 0$) is, in most instances, not significant in chromatography. From the statistical point of view, it can therefore be neglected. Chromatographically, a positive intercept indicates an interfering substance, a negative intercept means a negative peak at this retention time. Both situations are failures of the chromatography, which must be corrected. Mathematically the intercept represents a correction factor for the analysis values:

$$y = \frac{Y}{b} - \frac{a}{b} \tag{3.11}$$

The correction factor is the x axis value at $y = 0$, which can be neglected for large signal values y. Using a large working range at small y values and a positive intercept the correction may be so large that negative concentrations are calculated.

To assess the extent of the false results, one can convert the signal values for the calibration into concentration values using the above equation (which can be done automatically by every CDS). By comparison with the predefined concentration values one can easily calculate the deviations. More than 50% difference between the known and calculated concentrations cannot be tolerated at any concentration point.

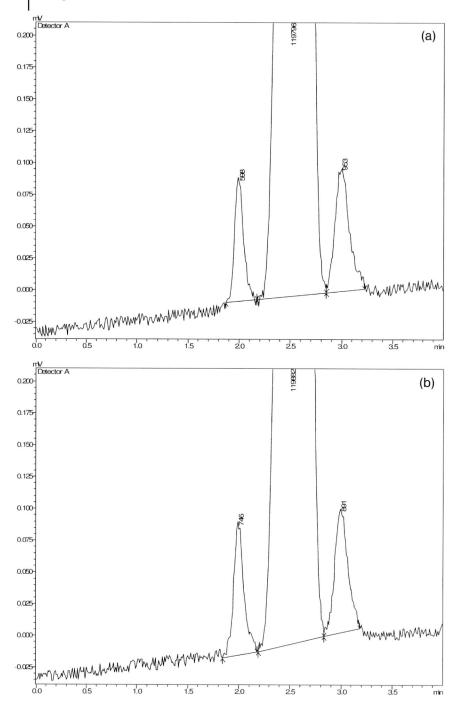

Figure 3.18 (a) LCSolutions integration of the simulation Eich1-2, where the smallest area is found for the first peak. (b) LCSolutions integration of the simulation Eich1-10, where the largest area is found for the first peak.

3.3.2.1   Tenfold Simulation at the Limit of Quantification

Due to the high noise with Eich1, the integration is most critical. To obtain more information at this "concentration" this file was called up ten times, and the noise calculated afresh every time and stored as Eich1-1 to Eich1-10. These files were integrated with LCSolutions and Chromeleon.

The simulation delivers, as a comparison of Fig. 3.18 (a) and (b) shows, slightly different peaks, because the noise overlays the whole chromatogram including the peaks. As the integration of the area is averaged over all the data points of the peaks, most of this deviation is compensated for.

The individual values in Table 3.9a (integrated with LCSolutions) and Table 3.9b (integrated with Chromeleon) are different. If one compares the values of the respective AIA files Eich1-1 to 10, there is absolutely no correlation. It seems that the two integration algorithms of LCSolutions and Chromeleon are so different, because of the way they handle the noise, that the integrated areas do not correlate at all. This should be taken into account when transferring methods and using different CDS.

The difference of 20% between the two areas in Fig. 3.18 (a) and (b) can be explained graphically. In one instance the baseline follows the upper limits of the noise and in the other it follows along the lower limits of the noise. Only measured points can be used as data for the baseline.

The peaks used for the calibration tests have tailing and are merged together slightly. To judge whether better results can be found when the peaks are not merged together, the presupposed plate number was increased from 3600 to 10 000. The peaks became narrower and were separated to the baseline. As expected, the result with LCSolutions improved considerably (Table 3.10a and Fig. 3.19). However, Chromeleon had a better precision with 3600 plates than LCSolutions, which did not improve even with 10 000 plates (Table 3.10b).

3.3.3
Simulation of an Isocratic Chromatogram

The simulation of a 20-min isocratic (isothermal) chromatogram contains 22 Gaussian peaks, which have a plate number of 10 000. With a t_{DP} of 200 ms there are 6000 data points. The first peak has a retention time of 2 min, the distances to the subsequent peaks are either 0.5 or 1 min. The peak pairs merge more and more as the peak width increases in parallel with the retention time, however, between the peak pairs the separation always reaches the baseline. Peaks 21 and 22 are at a distance of 1.5 min two free-standing peaks at the end of the chromatogram. The baseline was constructed by a very broad negative peak with retention time 8 min, which give a plate-shaped baseline. The peak factors are either 16 666.67 or 1666.67, i.e. the target areas are 1 000 000 and 100 000 µV s. Therefore the deviations from the target area can be "seen". If the value at the top of the peak begins with 100 or 99, the deviation must be less than 1%.

The number of data points per peak is 36 to 342, i.e. the information density increases. The noise is set to 100 µV and the range for the S/N is 380 to 3325 for

Table 3.9 The simulation of Eich1 (S/N = 10) was called up 10 times and only the noise was calculated afresh every time. After integration with (a) LCSolutions and (b) Chromeleon the precision and accuracy were recorded.

(a) LC Solutions

N = 3600	Eich1-1	Eich1-2	Eich1-3	Eich1-4	Eich1-5	Eich1-6	Eich1-7	Eich1-8	Eich1-9	Eich1-10	Precision (%)
LCSolutions											
1. Peak	744	598	702	704	726	728	739	743	654	746	6.97
2. Peak	120026	119795	119851	119884	120058	119966	119978	119888	119939	119882	0.07
3. Peak	1009	953	850	876	1080	986	1040	868	937	891	8.52
											Accuracy (%)
Dev 1 (%)	3.33	−16.94	−2.50	−2.22	0.83	1.11	2.64	3.19	−9.17	3.61	−1.61
Dev 2 (%)	0.02	−0.17	−0.12	−0.10	0.05	−0.03	−0.02	−0.09	−0.05	−0.10	−0.06
Dev 3 (%)	−6.57	−11.76	−21.30	−18.89	0.00	−8.70	−3.70	−19.63	−13.24	−17.50	−12.13

(b) Chromeleon

N = 3600	Eich1-1	Eich1-2	Eich1-3	Eich1-4	Eich1-5	Eich1-6	Eich1-7	Eich1-8	Eich1-9	Eich1-10	Precision (%)
Chromeleon											
1. Peak	761	679	769	781	783	785	738	694	782	707	5.73
2. Peak	120052	119947	120003	119985	120078	120095	119973	119866	120096	119882	0.07
3. Peak	1038	972	1025	1006	1066	1113	1042	948	1054	1004	5.15
											Accuracy (%)
Dev 1 (%)	5.69	−5.69	6.81	8.47	8.75	9.03	2.50	−3.61	8.61	−1.81	3.88
Dev 2 (%)	0.04	−0.04	0.00	−0.01	0.07	0.08	−0.02	−0.11	0.08	−0.10	0.00
Dev 3 (%)	−3.89	−10.00	−5.09	−6.85	−1.30	3.06	−3.52	−12.22	−2.41	−7.04	−4.93

Table 3.10 (Similar to Table 3.9a and b. The simulation of Eich1 (S/N = 10) was changed. The plate number was increased from 3600 to 10 000.

(a) LCSolutions

N = 10 000	Eich1_1	Eich1_2	Eich1_3	Eich1_4	Eich1_5	Eich1_6	Eich1_7	Eich1_8	Eich1_9	Eich1_10	
LCSolutions											Precision (%)
1. Peak	709	674	749	703	723	728	683	753	724	703	3.59
2. Peak	119961	120069	119988	120061	119998	120057	120073	120006	120047	120085	0.04
3. Peak	1077	1068	1073	1087	1109	1063	1062	1132	1070	1067	2.11
											Accuracy (%)
Dev 1 (%)	-1.53	-6.39	4.03	-2.36	0.42	1.11	-5.14	4.58	0.56	-2.36	-0.71
Dev 2 (%)	-0.03	0.06	-0.01	0.05	0.00	0.05	0.06	0.01	0.04	0.07	0.03
Dev 3 (%)	-0.28	-1.11	-0.65	0.65	2.69	-1.57	-1.67	4.81	-0.93	-1.20	0.07

(b) Chromeleon

N = 10 000	Eich1_1	Eich1_2	Eich1_3	Eich1_4	Eich1_5	Eich1_6	Eich1_7	Eich1_8	Eich1_9	Eich1_10	
Chromeleon											Precision (%)
1. Peak	769	777	769	710	724	768	787	740	754	772	3.27
2. Peak	120046	120061	120068	120066	120083	120082	120085	120073	120076	120012	0.02
3. Peak	1125	1179	1123	1072	1126	1128	1080	1055	1103	1030	3.94
											Accuracy (%)
Dev 1 (%)	6.81	7.92	6.81	-1.39	0.56	6.67	9.31	2.78	4.72	7.22	5.14
Dev 2 (%)	0.04	0.05	0.06	0.06	0.07	0.07	0.07	0.06	0.06	0.01	0.05
Dev 3 (%)	4.17	9.17	3.98	-0.74	4.26	4.44	0.00	-2.31	2.13	-4.63	2.05

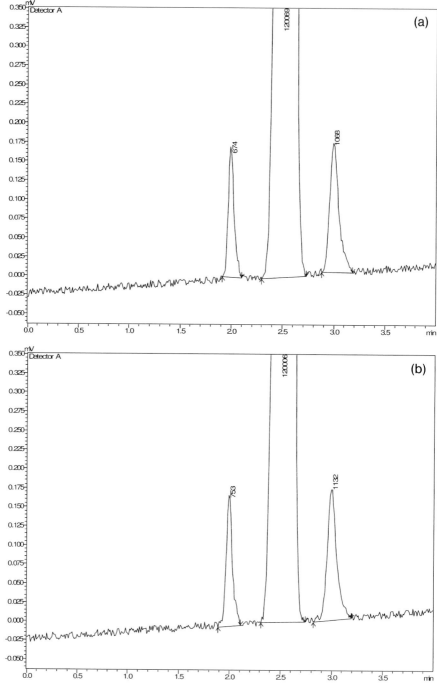

Figure 3.19 (a) Integration of the simulation Eich1_2 with LCSolutions.
This sample had the smallest area for the first peak in Table 3.10a.
(b) Integration of the simulation Eich1_8 with LCSolutions.
This sample had the largest area for the first peak in Table 3.10a.

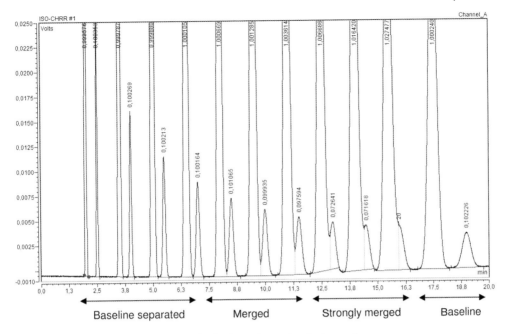

Figure 3.20 Integration of Iso-Chr.cdf with Chromeleon using the integration parameters shown in the following table.

Retention T	Name	Value	Channel
0.000 min	Fronting Sensitivity Factor	2.0	All Channels
0.000 min	Tailing Sensitivity Factor	3.0	All Channels
0.000 min	Minimum Area	1.0E-5 "[Signal]*min"	All Channels
0.000 min	Peak Slice	5.00 s	All Channels
7.500 min	Sensitivity	0.55E-3 "[Signal]"	All Channels
7.500 min	Peak Slice	20.00 s	All Channels
7.500 min	Maximum Rider Ratio	0.00 %	All Channels
12.000 min	Peak Shoulder Threshold	2.00	All Channels

the larger peak and 35 to 266 for the smaller peak. The last peaks are wider than the first peaks by a factor of approximately 8. Their S/N is correspondingly less. One could summarize nine data points (bunching) and reduce the noise for the last peaks by a factor of about 3. There is, nevertheless, a disadvantage if the last peaks have the same area. The integration conditions should be adapted for the widening of the peaks by a factor of almost 10. This is not required for chromatograms with gradients (GC: temperature program).

Figure 3.20 shows the 11 peak pairs, of these the first four have baseline separation (valley = 0%), so here the integration may not cause any problems. The next three peak pairs are not expected to show any particularly large deviations because the valley between the peaks is less than 50%. The next three peak pairs are separated so badly that they should be used for quantitative purposes with extreme caution. Large deviations are to be expected here. At the end are two free-standing peaks, which serve for comparison and confirmation. In principle, the target values should be found, however, their width is almost 10 times greater than the width of the first two peaks.

For the integration with Chromeleon, eight parameters had to be adjusted (Fig. 3.20). The deviations are below 1% up to the sixth peak pair. The smaller peak of the seventh peak pair, with a valley of about 50%, loses 2.4% of its area. This very large, unacceptable deviations arise, because the peak end is visibly too high. A shoulder peak (20) is recognized, however the area is useless with a deviation of –32%. The integration does not look optimal in the latter part of the chromatogram. It is important not to represent a chromatogram simply by standardizing on the tallest peak, otherwise the quality of the integration cannot ever be checked for plausibility.

For the integration with Empower Apex Track (Fig. 3.21) the deviations are similar to those with Chromeleon but continuous over the course of the chromatogram. The area of the smaller peak (14) at a valley of 50% has a deviation of –4.2%, whereas the height has a deviation of only –0.2%. The same peak has an area deviation of 7.97% and a height deviation of 1.1% using the Gaussian

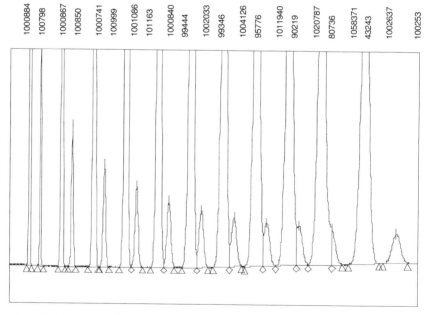

Figure 3.21 Integration of Iso-Chr.cdf with Empower Apex Track using the following integration parameters:

1	Iso-Chr		Iso_Chr_Apex			1061	
Algorithm	Peak Width	Threshold	Min. Height	Liftoff %	Touch-down %	Int. Event	Int. Event
Apex Track	20	90	200	0.000	0.000		

skim for the first peak. With Empower in the Traditional Mode, then −1.94% for the area and +0.73% for the height are found. The results depend on the degree of merging. Even for merged Gaussian peaks and a valley of 50% the results are quite different and dependent on the integration mode. The Gaussian skim has the worst result for Gaussian peaks (!). The shoulder peak is recognized but the partitioning line can be is seen visually as being drawn too late, giving the large deviation of the area of −57% (height −2.2%).

Using EZChrom (Fig. 3.22) peak 14 has an area deviation of −3.7% and a height deviation of 0.6%. These are values which are in accord with the Empower results. The shoulder peak 20 was not divided off and therefore the area of peak 19 is 10% too high. The area loss of 6.5% with the last free-standing peak is striking. The integration conditions are obviously anything but optimum. It is astonishing that the choice of the individual parameters influences the integration result so much. The differences that can be obtained through these parameters are just as high as the differences between the integration systems.

The last peaks, which are merged together in some instances, are primarily integrated with the tangent method. This yields areas that are considerably less than the areas obtained with the perpendicular drop, that are already generally too small. The tangent method yields totally nonsensical areas (and heights) in this instance, in spite of appearing plausible.

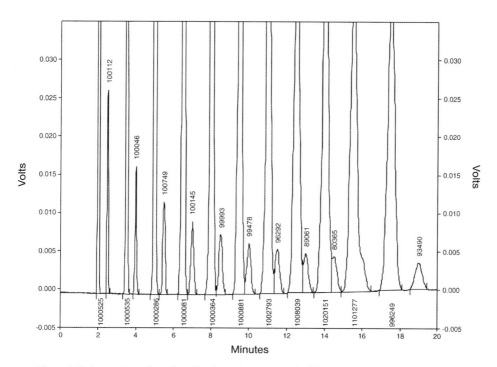

Figure 3.22 Integration of Iso-Chr.cdf with EZChrom using the following integration parameters: Chromatogram ISO CHR, width 0.2, threshold 200, minimum area 300.

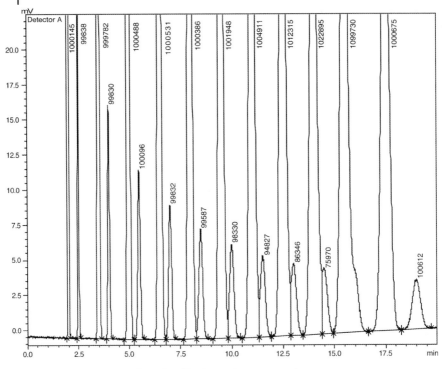

Figure 3.23 Integration of Iso-Chr.cdf with LCSolutions using the integration parameters shown in the following table:

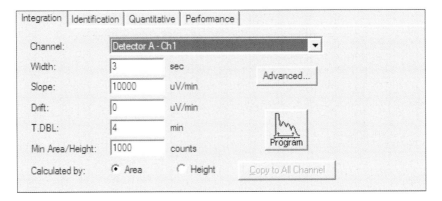

With LCSolutions (Fig. 3.23) the deviations appear similar to those obtained with the other integration systems. The heights show substantially lower deviations than the areas. It is important to adjust the width at half height, if in doubt, lower values should be preferred. The slope can be adjusted with the slope test. Furthermore, the correct value for the doubling time T.DBL, the time after which the width has doubled in size, is necessary.

*Check, contrary to your previous procedure, if the peak heights
rather than the areas yield better results*

With the free-standing small Peak 22 (the last peak) integration with Chromeleon gives an area 2.2% too high and integration with EZChrom gives an area 7.5% too low (Table 3.11). In Fig. 3.22 it can be seen that EZChrom with a width = 0.2 and threshold = 200 sets the peak start too late. Obviously the integration conditions are not optimized with either Chromeleon or EZChrom. The plate-shaped baseline can be recognized only by magnification.

For Peak 20 (shoulder peak) a separation is only indicated by Chromeleon and Empower (Table 3.11), but the area is clearly erroneous. The separation point has visibly moved unfavorably with Empower. With the four different integration systems peaks 18 and 16 are integrated amazingly, although the minimum point between Peak 15/16 and 17/18 is clearly defined. The differences can hardly be explained by a different recognition of the baseline. Obviously merged Gaussian peaks are not only wrong because the valley is more than 50% but also they have been wrongly integrated in different ways. All deviations from the accurate value are so high that accurate quantification is impossible.

With a valley of less than 50% the integration of Peak 14 gives a nearly acceptable area. A valley of 50% should be the absolute limit for quantification using the area. If the valley is higher, the perpendicular drop method no longer suffices. The peak height, however, yields considerably lower deviations (isoresults.xls).

Table 3.11 Areas of the proximate peaks in the last part of the chromatogram – comparison of CDS. The target area is 100 000 µVs.

Area (µV s)	Peak 14	Peak 16	Peak 18	Peak 20	Peak 22
Chromeleon	97 594	72 641	71 618	68 245	102 226
Empower	95 776	90 219	80 736	43 243	100 253
EZChrom	96 292	89 061	80 365	n.a.	93 490
LCSolutions	94 827	86 346	75 970	n.a.	100 612

3.3.4
Simulation of a Gradient Chromatogram

The simulation of a separation with gradients (GC: temperature program) was carried out (Grad-Chr.cdf) with Gaussian peaks of the same peak width. It contains 36 Gaussian peaks that have a standard deviation of the peak of 0.05 min = 3 s (Fig. 3.24). The uneven peak numbers have the retention times 2, 3, 4 up to 19 min and have a peak (size) factor (PF) of 16 666.67 µV min, the even peak numbers have the retention times $(2.14 + 0) \times 0.015$ min, $(3.14 + 1) \times 0.015$ min, $(4.14 + 2) \times 0.015$ min up to $(19.14 + 17) \times 0.015$ min up to 19 min and a PF of 166.67 µV s. This means 18 peak pairs which are merged significantly and can

Table 3.12 Deviations of the areas (%) and heights (%) of the chromatograms in Figs. 3.24 to 3.29.

Empower Apex		Apex Gauss		LCSolutions		EZChrom elite		Chromeleon	
Area	Height	Area	Height	Area	Height	Area	Height	Area	Height
4.70	0.18	4.62	0.18	10.03	0.22	9.97	0.19	4.41	0.19
−47.56	−14.84	−46.81	−30.08					−44.30	
3.21	0.06	3.80	0.06	9.85	0.09	9.94	0.07	2.60	0.07
−32.70	−6.97	−38.64	−22.23					−26.32	
2.71	0.01	2.59	0.01	2.73	−0.02	2.72	0.01	2.74	0.02
−27.79	3.85	−26.62	−11.53	−29.95	3.57	−28.02	3.77	−27.74	3.92
1.53	−0.01	1.94	−0.01	1.48	0.01	1.31	−0.01	1.31	−0.01
−16.06	1.02	−20.23	−7.30	−14.80	1.17	−11.95	0.96	−13.82	1.01
0.84	−0.02	1.43	−0.02	0.95	0.00	0.85	−0.02	0.87	−0.01
−9.20	0.19	−15.06	−4.50	−10.08	0.33	−9.44	0.10	−9.16	0.23
0.50	−0.02	0.76	−0.02	0.42	−0.05	0.51	−0.02	0.51	−0.01
−5.82	−0.09	−8.44	−1.32	−6.01	−0.62	−6.05	−0.27	−5.83	−0.17
0.26	−0.02	0.50	−0.02	0.27	−0.03	0.27	−0.01	0.27	−0.01
−3.36	−0.13	−5.75	−0.68	−3.24	−0.15	−3.64	−0.25	−3.52	−0.20
0.09	−0.02	0.29	−0.02	0.27	0.02	0.10	−0.02	0.11	−0.02
−1.86	−0.16	−3.92	−0.40	−2.74	−0.18	−2.26	−0.35	−1.83	−0.17
0.06	−0.02	0.12	−0.02	0.01	0.00	0.07	−0.02	0.07	−0.01
−1.30	−0.11	−1.89	−0.14	−2.64	−0.77	−1.68	−0.25	−1.38	−0.13
0.03	−0.01	0.06	−0.01	−0.02	−0.01	0.18	0.04	0.02	−0.01
−0.90	−0.07	−1.18	−0.08	−2.25	−0.73	1.22	0.68	−0.46	−0.08
−0.01	−0.01	−0.01	−0.01	−0.12	−0.09	0.84	0.17	0.00	−0.01
−0.43	−0.05	−0.43	−0.05	−1.72	−0.59	4.50	2.10	−0.52	−0.07
0.01	0.00	0.01	0.00	0.01	−0.03	1.58	0.29	−0.02	−0.01
−0.15	0.03	−0.15	0.03	−0.64	−0.45	7.29	3.11	−0.49	−0.17
0.01	0.00	0.01	0.00	−0.03	−0.02	1.84	0.32	0.03	0.01
0.02	0.06	0.02	0.06	−1.19	−0.28	7.23	2.93	0.05	0.01
0.02	0.01	0.02	0.01	−0.04	0.01	1.11	0.16	−0.36	−0.12
0.35	0.16	0.35	0.16	−0.34	−0.17	1.01	0.41	−2.97	−1.22
−0.23	−0.07	−0.23	−0.07	0.07	0.05	0.40	0.12	−0.20	−0.06
−1.81	−0.65	−1.81	−0.65	0.06	−0.01	4.71	1.78	−1.90	−0.69
−0.15	−0.05	−0.15	−0.05	−0.14	−0.04	0.87	0.15	−0.12	−0.04
−0.95	−0.35	−0.95	−0.35	−1.68	−0.62	1.10	0.38	−1.00	−0.41
−0.12	−0.04	−0.12	−0.04	0.03	−0.01	0.17	0.05	−0.04	−0.01
−0.55	−0.17	−0.55	−0.17	−0.07	−0.01	0.80	0.29	−0.56	−0.19
−0.08	−0.02	−0.08	−0.02	0.04	0.00	0.18	0.05	0.01	0.00
−0.19	−0.02	−0.19	−0.02	0.21	0.22	1.12	0.39	−0.19	−0.06

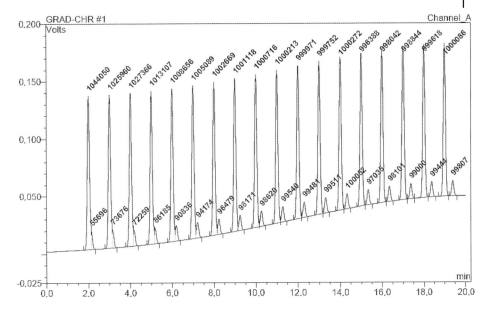

Figure 3.24 Simulation of a gradient chromatogram.
Integration with Chromeleon using the following conditions:

Retention Time	Name	Value	Channel
0.000 min	Peak Shoulder Threshold	1.00	All Channels
0.000 min	Rider Threshold	100.00 %	All Channels
0.000 min	Maximum Rider Ratio	0.00 %	All Channels
0.000 min	Minimum Area	1.0E-6 "[Signal]*min"	All Channels
0.000 min	Lock Baseline	On	All Channels
15.000 min	Lock Baseline	Off	All Channels
18.000 min	Valley to Valley	On	All Channels

be much better separated in the following. To simulate the gradient baseline rise, a very broad peak with a retention time of 20 min was used, i.e. only the increasing part of the peak. The areas of the peak are either 1 000 000 or 100 000 µV s. Therefore the deviations of the target area can be "seen" in the figures. If the value at the top of the peak begins with 100 or 99, the deviation is less than 1%. The deviation in the areas and heights are shown in Table 3.12.

To simulate the gradient chromatogram, the same peak couple with a size ratio of 10 to 1 was moved 18 times relative to each other in increments of 0.9 s from the end to the start of the chromatogram. The peak pair is, at the beginning, merged very significantly and by the end of the chromatogram is baseline separated. Therefore high deviations are to be expected at the beginning and low deviations at the end. If the tall peak gains 1% in area, then the small peak loses 10% in area. The simulated baseline increase might not cause any difficulties. To test this, the file "Grad-noBL.cdf" can be integrated with a straight base line in addition to the file "Grad-Chr.cdf".

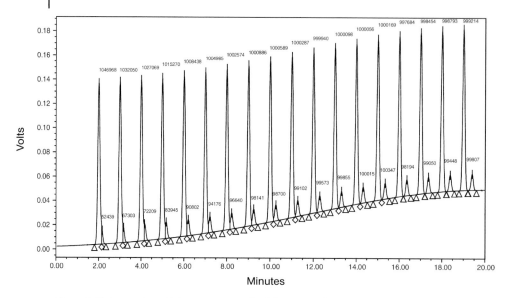

Figure 3.25 Integration with Empower using the following conditions:

7	Grad-Chr		Grad_Chr_Apex			1051	
Algorithm	Peak Width	Threshold	Min. Height	Liftoff %	Touch-down %	Int. Event	Int. Event
Apex Track	AUTO [15.50]	AUTO [12.00]	50	0.100	0.300	**Detect Shoulders**	

The integration with Chromeleon gives the expected results (Fig. 3.24), which look very similar to the results with Empower Apex Track (Fig. 3.25). It can be seen in Table 3.12 that the heights yield considerably lower deviations. The area deviations fall below 1% at the 10th peak pair, whereas the height deviations are already below 1% at the 5th peak pair.

The simulation contains exclusively Gaussian peaks. Therefore a clear improvement using the Gaussian skim of the taller peak should be found. It has to be assumed that the smaller peak hardly influences the taller peak. Therefore it should be favorable to continue the Gaussian course (Fig. 3.26) of the tall peak here. Unfortunately, exactly the opposite is the case. Table 3.12 shows that the deviations of the areas and the heights are greater using the Gaussian skim rather than the normal Apex integration. It is unclear in which cases the Gaussian skim actually leads to a better integration.

The integration with EZChrom shows incredibly large deviations for the last peak pairs, which are only slightly merged. If one integrates under the same conditions, but without "Reset baseline", a straight line is drawn between the end of the seventh peak pair and the end of the last peak pair. Obviously "Reset baseline" should be used even more frequently. However, this is not recognized in the usual

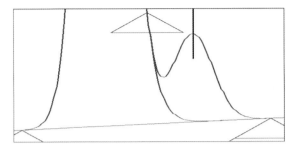

Figure 3.26 Peak pair at 8 min integrated under the following conditions:

8	Grad-Chr		Grad_Chr_Apex_Gauss			1053	
Algorithm	Peak Width	Threshold	Min. Height	Liftoff %	Touch-down %	Int. Event	Int. Event
Apex Track	AUTO [15.50]	AUTO [12.00]	50	0.100	0.300	Detect Shoulders	Gaussian Skim

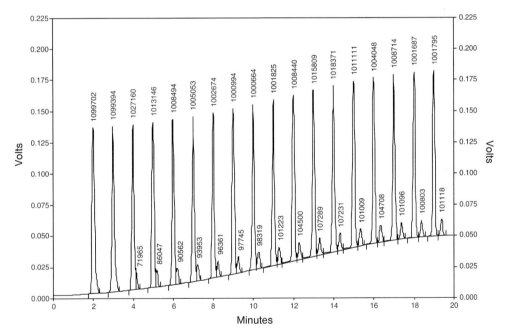

Figure 3.27 Integration with EZChrom using the following conditions:
Width 0.2, threshold 50, 3 x reset baseline 15.71, 17.65 and 18.67 min.

representation of the chromatogram (Fig. 3.27) but only with the expansion of the signal axis (Fig. 3.28). By increasing the distance between the peaks in a peak pair one can establish whether instructing "Reset baseline" is always necessary.

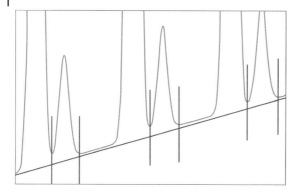

Figure 3.28 The chromatogram in the time range 13–15 min from Fig. 3.27 is expanded. No correct baseline is found.

The results with LCSolutions (Fig. 3.29) correspond to those of Chromeleon and Empower. The first two shoulders are not separated. Therefore the averaged area is indicated for the first two tall peaks. It does not make practical sense to indicate a (wrong) area for shoulder peaks either. The adjustment of the integration parameters is very simple: With a peak width at half height of 3 s, a slightly lower value of 2 s is used and the slope is found using the slope test and rounded up or down.

3.3.5
By-Product Analysis

One generally has a very tall peak and several much smaller peaks in by-product analysis. Some of the small peaks may be situated on the descending portion of the main product peak, that is, on the tailing of this peak, and can only be recognized at 100-fold magnification. Therefore a realistic simulation of the main peak is essential, along with an exact copy of the tailing. The tailing of the by-product peaks is much less significant, compared to this.

To carry out a simulation that is as realistic as possible one could use the EMG function, which is decribed in the next two sections. Another suitable simple function with four parameters was described by Losef (1994):

$$y = \frac{A}{\exp\left[-a\,(x - c)\right] + \exp\left[b\,(x - c)\right]} \tag{3.12}$$

According to an actual chromatogram using the following factors $a = 15$, $b = 4$, $c = 1.930\ 433\ 9$ and $A = 836\ 530$ as input data, a peak was produced with a height of 500 000 µV, resulting in tailing being measured at various peak heights (Table 3.13).

Using the function of Losef (Fig. 3.30), six symmetrical peaks were positioned on the tailing. The standard deviation, s_p, of the small peaks was increased from 0.12 min in steps of 0.02 min to 0.28 min. The by-product peaks have an area of

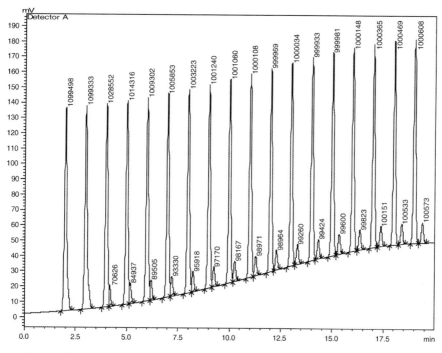

Figure 3.29 Integration with LCSolutions using the following conditions:

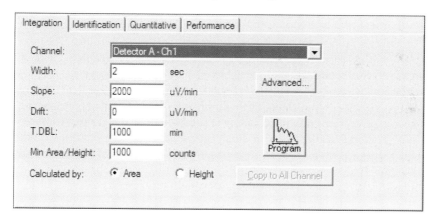

Table 3.13 The asymmetry of the Losef function.

Peak height (%)	Asymmetry factor	Tailing factor
0.01	3.30	2.15
0.10	3.15	2.07
1.00	2.88	1.94
5.00	2.67	1.83
10.00	2.53	1.77

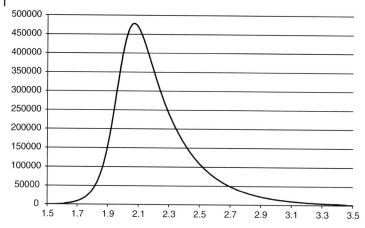

Figure 3.30 Simulation of a peak according to a real example with the function of Losef.

1% or, in a second series, of 0.1% of the area of the main peak. The retention time of the main peak is 2 min. The retention time of the by-products was increased in steps of 4 s_p (see Table 3.14), to give a resolution of 1 between the by-product peaks. Every second peak has double the area and height. Only some chromatograms are shown here. All chromatograms are included in the file Losef.ppt on the CD.

All six peaks have a resolution of 1, i.e. there is no base line separation between any of the peaks. This represents a common integration problem. If the by-product peaks are few and well separated, no integration problem exists. Depending on the peak width, at least one, or at most three, peaks disappear in the main peak. This effect is stronger for the 0.1% peak area than for the 1% peak area. The target area of the 1% peaks is alternating between 64 800 and 129 600 µV s, the target area for the 0.1% peaks is alternating between 6480 and 12 960 µV s.

The first two peaks are completely hidden by the main peak (Fig. 3.31). A separation is possible only by changing the chromatographic conditions considerably. However, every reduction in the tailing of the main peak would already improve the separation from the following small peaks. The third peak sits as a shoulder with a recognizable maximum on the main peak. The original peak is strongly distorted. The envelope of the fourth to sixth peak imitates the original peaks almost unchanged, i.e. it is no longer influenced by the main peak.

Table 3.14 Calculation of the retention times of six by-product peaks overlaying the mean peak according to Losef.

$= 2 + 4\,s_p$
$= 2 + 8\,s_p$
$= 2 + 12\,s_p$
$= 2 + 16\,s_p$
$= 2 + 20\,s_p$
$= 2 + 24\,s_p$

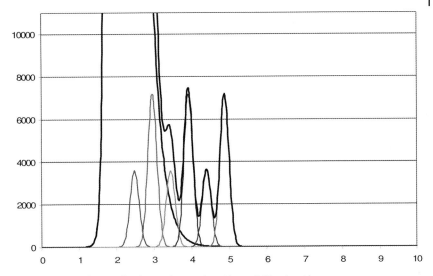

Figure 3.31 Simulation of six by-product peaks with $s_p = 0.12$ min with a retention time delay according to Table 3.14 are overlaid on the Losef peak at 2 min.

To examine only the effect of the peak fusion a very low noise of 5 µV was underlaid and integrated with LCSolutions and the conditions width = 10 and slope = 200. Depending on the quality of the detector, the noise could be higher than that assumed here, which will make the results even worse.

LCSolutions always choose the tangent method (Fig. 3.32) automatically. The perpendicular drop was enforced (Fig. 3.33) with the instruction "tangent off" at 2.5 min. The third peak is still just recognizable but no longer separated. For the fourth by-product peak the correct area must be between the values from Figs. 3.32 and 3.33. For the following peaks the tangent method obviously loses too much area.

There are no investigations concerning the tailing at 1% or 0.1% of the peak height. Small differences in the peak fit parameters can lead to large differences, possibly in the lowest concentration area in the last decrease of the tall peak. It is conceivable that in reality the exponential decay is an overlapping of several exponential functions with different time constants. Dolan (2002a) postulates exaggerated tailing at the base of the peak, which can be recognized only at great magnification, without giving any experimental evidence for this.

Real chromatograms can be very complex and there are borderline cases in which a decision is made between the tangent and the perpendicular drop methods. In individual cases the Exponential or Gaussian Skim method can lead to a good approach towards reality. However, in general, the Skim methods are not recommended as suitable integration methods (cf. Bicking 2006 B).

Figures 3.34 and 3.35 were additionally magnified 10 times. The foot of the peak becomes visibly broader and the series of six peaks overlap more. A first shoulder is found with the fourth peak.

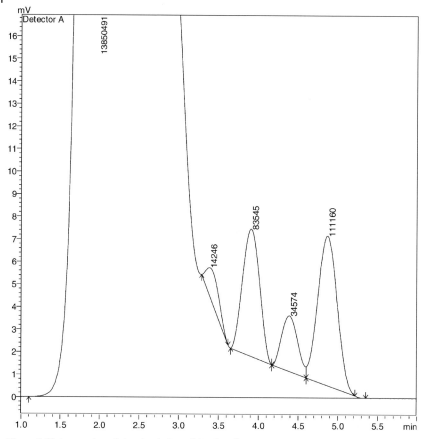

Figure 3.32 Integration of the simulation of the data from Fig. 3.31
with LCSolutions using default conditions.

The demand to prove the presence of by-products at only 0.05% of the main
component exhausts the analytical possibilities. A main peak with a peak height
of 2 V produces a minimum (0.05%) by-product peak of 1 mV. For an S/N = 10
the noise may not exceed 100 µV. Using a UV-detector at low wavelengths this
sometimes cannot be achieved.

The simulations in Figs. 3.36 and 3.37 show by-product peaks in the 0.1% range.
Then the foot of the main peak appears broader because one must increase the
chromatogram by a factor of around 10 once again so as to present the small
peaks. This has the consequence that an additional peak is more likely to virtually
"hide" in the foot of the peak. These possibly "hidden" by-products are invisible,
but they could be biologically active. Components that are chemically very similar
to the main product presumably also only have slightly different retention times.
According to a suggestion by Dolan (2002) one should think about capturing the
tailing of the main peak (heart cut) and injecting it once again with as small an
amount of the main substance.

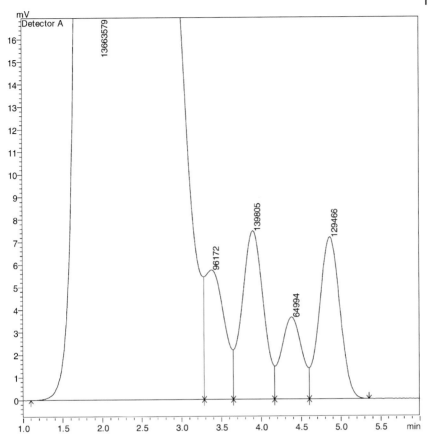

Figure 3.33 Integration of the simulation of the data from Fig. 3.31 with LCSolutions with suppression of the tangent method.

The area for all peaks directly following the main peak is severly underestimated with the tangent method. This can be desirable for the by-product analysis but is nevertheless wrong. Using the perpendicular drop method the area is assessed too high but to a substantially lesser extent. Furthermore, the positive deviations of the integrated areas are still considerable for the first by-product peaks, but not for all further peaks. Except for only a few exceptions the peak height has smaller deviations from the target value.

Table 3.15 The proportional deviations of the target value integrated with the tangent and perpendicular drop methods for the by-product peaks in the 1% area for Figure 3.34.

Tangent		Perpendicular drop	
Area (%)	Height (%)	Area (%)	Height (%)
−63.16	−45.82	10.08	11.03
−28.39	−16.95	2.37	0.32
−34.52	−22.76	−2.41	0.12
−10.22	−5.68	0.26	−0.07

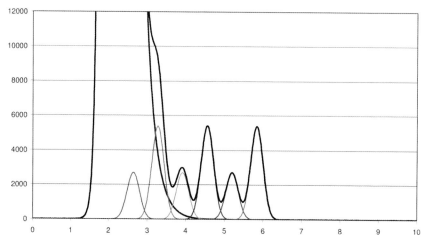

Figure 3.34 Simulation of six by-product peaks with $S_p = 0.16$ min and a retention time delay according to Table 3.14 superimposed on the Losef peak. PF varies in the range 1080 µV min to 2160 µV min.

Table 3.16 The proportional deviations of the target value integrated with
the tangent and the perpendicular drop method for the by-product peaks
in the 1% yrea for Figure 3.35.

Tangent		Perpendicular drop	
Area (%)	Height (%)	Area (%)	Height (%)
−45.64	−30.14	5.60	2.68
−50.57	−33.91	−2.94	0.14
−21.34	−12.76	1.20	−0.08
−26.39	−16.56	−3.07	0.08
−6.90	−4.08	0.64	−0.08

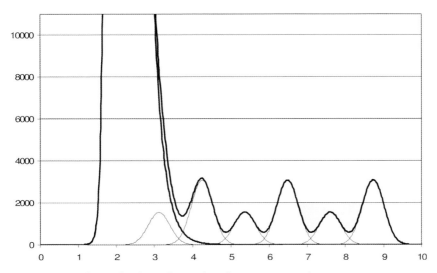

Figure 3.35 Simulation of six by-product peaks with $s_p = 0.28$ min and a
retention time delay according to Table 3.14 superimposed on the Losef peak.
PF varies in the range 1080 µV min to 2160 µV min.

Table 3.17 The proportional deviations of the target value integrated with the tangent and the perpendicular drop methods for the by-product peaks in the 0.1% area for Figure 3.36.

Tangent		Perpendicular drop	
Area (%)	Height (%)	Area (%)	Height (%)
−74.49	−59.15	11.23	20.23
−29.53	−16.91	2.11	−0.20
−37.35	−23.41	−6.37	−2.06
−10.68	−5.77	−2.65	−0.66

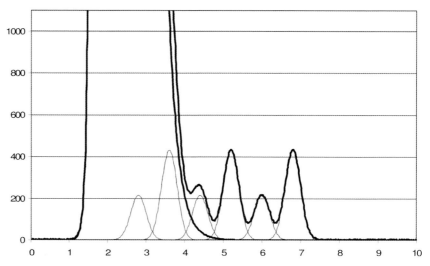

Figure 3.36 Simulation of six by-product peaks with $s_p = 0.2$ min and a retention time delay due to Table 3.14 superimposed on the Losef peak. PF varies in the range 108 µV min to 216 µV min.

Table 3.18 The proportional deviations of the target value integrated with the tangent and the perpendicular drop methods for the by-product peaks in the 0.1% area for Figure 3.37.

Tangent		Perpendicular drop	
Area (%)	Height (%)	Area (%)	Height (%)
−53.64	−35.01	−6.23	0.08
−23.66	−13.57	1.20	−0.57
−29.29	−18.77	−8.10	−1.87
−7.93	−4.47	0.02	−0.25

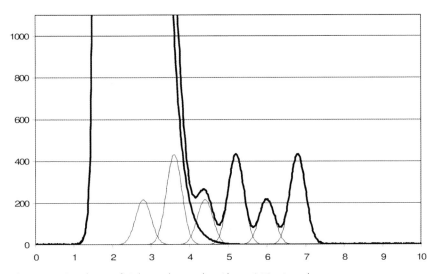

Figure 3.37 Simulation of six by-product peaks with $s_p = 0.28$ min and a retention time delay according to Table 3.14 superimposed on the Losef peak. PF varies in the range 108 µV min to 216 µV min.

Table 3.19 Westerberg's report translated into input data (grey) for chromatogram simulation.

t_R	A	H_P	t_R	s_P	PF
16.77	56841	8114	0.77	0.04658	947
16.98	92643	14182	0.98	0.04343	1544
17.28	60942	9086	1.28	0.04460	1016
17.7	1376500	196580	1.7	0.04656	22942
18.18	23722	3307	2.18	0.04770	395
18.8	7964	827	2.8	0.06405	133
18.97	37273	4520	2.97	0.05483	621
19.28	29864	2952	3.28	0.06728	498
19.68	767730	104790	3.68	0.04871	12796
19.83	181180	19143	3.83	0.06293	3020
20.03	52884	6251	4.03	0.05625	881
20.25	215030	26505	4.25	0.05394	3584
20.5	487850	53074	4.5	0.06112	8131
21.47	14712	1739	5.47	0.05626	245
21.82	14180	1514	5.82	0.06227	236
22.13	62416	5906	6.13	0.07027	1040
24.18	88048	8584	8.18	0.06820	1467
26.13	22065	2044	10.13	0.07179	368

3.3.6
Post-Simulation of Real Chromatograms

In 1969 Westerberg published a chromatogram with 18 peaks with two shoulder peaks included. Using a suitable calculation method all 18 peaks were integrated. None of the commercial integration systems used today are able to integrate this chromatogram (Westerberg.cdf) as well as was done by Westerberg 40 years ago. Today's integration systems still work almost exclusively with the perpendicular drop method, developed at a time when one actually cut out peaks from the chart recorder print-out and weighed them.

The report from the Westerberg publication is printed on the left-hand side of Table 3.19. As the retention time of the first peak is at 16.77 min, 16 min were subtracted from all times to reduce the "analysis time". The areas were multiplied by 1 000 000 (results in μV s) and divided by 60 (to give PF in μV min). The standard deviation s_P as a measure of the peak width was calculated according to Eq. (3.7).

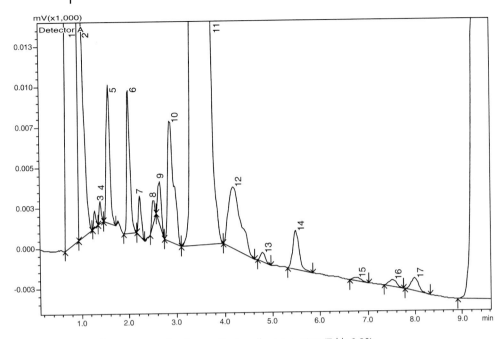

Figure 3.38 Chromatogram showing a valley to valley integration (Table 3.20).

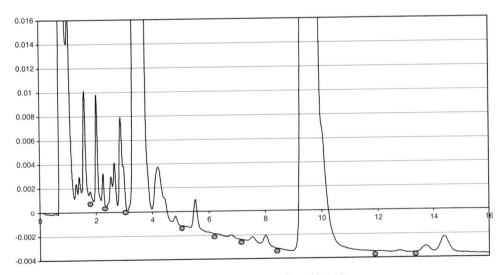

Figure 3.39 For all valleys characterized by a resolution of 2 (Table 3.20) a baseline point was set.

Table 3.20 Integration report for Figure 3.38.

Peak	t_R	Area	Height	w50%	Tf	Resolution
1	0.726	4285467	1057484	0.057	1.680	–
2	1.005	118345	17564	0.000	–	0.955
3	1.275	3.693	1055	0.056	1.278	0.924
4	1.391	4635	1372	0.054	1.342	1.062
5	1.574	43078	8564	0.078	1.162	1.434
6	2.001	41170	8812	0.070	1.231	2.914
7	2.230	10075	2458	0.068	1.240	1.711
8	2.516	6254	1447	0.064	1.042	2.283
9	2.641	11202	2500	0.078	1.309	0.955
10	2.865	62930	7425	0.116	1.649	1.172
11	3.664	8233312	638527	0.204	0.627	2.517
12	4.180	68421	3808	0.266	1.522	1.201
13	4.788	4030	565	0.119	1.219	1.773
14	5.485	22394	2426	0.141	1.191	2.834
15	6.752	2102	203	0.173	1.293	4.459
16	7.509	4644	424	0.179	1.146	2.433
17	7.977	8968	805	0.173	1.155	1.450
18	9.736	10628447	572318	0.277	0.676	4.276
19	12.744	1626	120	0.222	1.019	6.615
20	13.678	8781	483	0.296	0.985	2.040
21	14.348	25801	1253	0.323	1.135	1.247

The results printed in the three columns on the right-hand side of Table 3.19 are taken from the worksheet "report" of the Excel file "Westerberg.xls". If one copies this part of the table with <Edit> <Paste Special> <Values ><Transpose> into the worksheet "data" a chromatogram appears under "C-gram", which resembles that in Westerberg's publication with no visible differences.

With a prepared Excel worksheet the simulation of a chromatogram is very simple with (arbitrarily many) Gaussian peaks.

In the chromatogram (Fig. 3.6) in Section 3.2.3 three tall peaks can be seen. At an HPLC meeting in 2007, 27 scientists drew the baseline on this chromatogram manually. After the second tall peak the baseline was always the same. Between the first and second tall peaks 18 participants drew the baseline as a connection between the peak end of the first and the peak start of the second tall peak. Nine

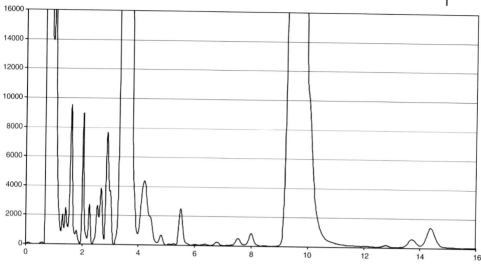

Figure 3.40 The chromatogram of Fig. 3.39 with the baseline subtracted.

participants drew the line (perhaps assuming a matrix effect) considerably higher. One participant wrote that without additional information no clear solution was possible. Of course one has to use all possible additional information: blank sample, blank gradient, comparison of the baseline of several similar samples, estimation of the probability of a matrix effect.

Let us assume that no additional information is available. One can then judge the resolution of the small peaks between the first and second tall peaks. A resolution of 2 is a clear indication of a baseline between two peaks if the concentration difference between these two peaks is not too large. To obtain only the information relevant to the peaks a valley to valley integration is shown in Fig. 3.38 and Table 3.20 shows the corresponding chromatogram report.

Everywhere the resolution exceeds the value 2, in Fig. 3.39 a baseline point is marked. These points are then connected to each other. The connecting lines can, with straight line equations that are valid in the respective area, be modeled as a baseline. The assessed baseline can then be subtracted for every data point.

After subtracting the baseline, the difference chromatogram (Fig. 3.40) is easier to integrate. This is similar to the subtraction of an individual blank gradient run. Since the peak areas have not been changed, the final result is given with its inaccuracy. Of course the follow-up simulation of a complex baseline is arduous and only applicable for very important chromatograms. However, it can be the basis for making important decisions in unclear integration conditions.

References

1 Bicking, M., Integration Errors in Chromatographic Analysis, Part I: Peaks of approximately equal size. LCGC Int. 24/4 (2006a) 402–414.

2 Bicking, M., Integration Errors in Chromatographic Analysis, Part II: Large Peak Size Ratios, LCGC Int. 24/6 (2006b).

3 Doerffel, K., Statistik in der analytischen Chemie, Deutscher Verlag für Grundstoffindustrie, Leipzig 1990.

4 Dolan, J. W., Peak Tailing and Resolution, LCGC Int. 20 (2002a) 430–436.

5 Dolan, J. W., Resolving Minor Peaks, LCGC Europe 15/9 (2002b) 578–580.

6 Dong, M., Enhancing HPLC Precision and Ruggedness, Perkin-Elmer Newsletters: Views (Fall 1994) 7–8.

7 Dyson, N., The Validation of Integrators and Computers for Chromatographic Measurements, Int. Lab. 22 (June 1992) 38–46.

8 Dyson, N., Chromatographic Integration Methods, 2nd ed., The Royal Society of Chemistry, Hertfordshire 1998.

9 Fehlinger, A., Guiochon, G., Validation of a chromatography data analysis software, J. Chromatogr. A 913 (2001) 221.

10 Foley, J. P., Dorsey, J. G., Equations for Calculation of Chromatographic Figures of Merit for Ideal and Skewed Peaks, Anal. Chem. 55 (1983) 730.

11 Johnson, E. L., Reynolds, D. L., Wright, D. S., Pachla, L. A., J. Chromatogr. Sci. 26 (1988) 372.

12 Kipiniak, W., A Basic Problem – The Measurement of Height and Area, J. Chrom. Sci. 19 (1981) 332.

13 Küppers, S., Renger, B., Meyer, V. R., Autosamplers – A Major Uncertainty Factor in HPLC Analysis Precision, LCGC Europe 13/2 (2000).

14 Kuss, H. J., Gewichtete Lineare Regression, in: Handbuch Validierung in der Analytik (Hrsg.: S. Kromidas) Wiley-VCH, Weinheim 2000.

15 Kuss, H. J., Weighted Least-Squares Regression in Practice: Selection of the Weighting Exponent, LCGC Europe 16/12 (2003a).

16 Kuss, H. J., Quantifizierung in der Chromatographie; in: HPLC-Tipps II (Hrsg.: S. Kromidas), Hoppenstedt Bonnier Zeitschriften GmbH 2003b.

17 Losef, A., On a Model Line Shape for Asymmetric Spectral Peaks, Appl. Spectrosc. 48 (1994) 1289–1290.

18 Miller, J. N., Miller, J. C., Statistics and Chemometrics for Analytical Chemistry, Pearson Education Limited 2000.

19 Meyer, V. R., Fallstricke und Fehlerquellen der HPLC in Bildern, Wiley-VCH, 3. Aufl., Weinheim 2006.

20 Meyer, V. R., Generation and Investigation of Chromatographic Peaks Using a Spreadsheet Program, LCGC Int. 7/10 (1994) 590.

21 Meyer, V. R., Chromatographic Integration Errors: A Closer Look at a Smaller Peak, LCGC Europe 13/3 (1995) 252–260.

22 Papas, A. N., Delaney, M. F., Evaluation of Chromatographic Integrators and Data Systems, Anal. Chem. 59 (1987) 54A.

23 Papas, A. N., Tougas, T. P., Accuracy of Peak Deconvolution Algorithms within Chromatographic Integrators, Anal. Chem. 62 (1990) 234–239.

24 Schepers, U., Ermer, J., Preu, L., Wätzig, H., Wide concentration range investigation of recovery, precision and error structure in liquid chromatography, J. Chromatogr. (B) 810 (2004) 111–118.

25 Westerberg, A. W., Detection and Resolution of Overlapped Peaks for an On-Line Computer System for Gas Chromatographs, Anal. Chem. 41 (1969) 1770.

4
Simulation of Chromatograms

Uwe D. Neue

4.1
Introduction

The theories of bandspreading and retention in chromatography are now well developed. Since the mid-1980s, computer-assisted method development software has been available that helps the chromatographer in the development of new HPLC methods. Recently, programs have become available that specifically predict the performance of HPLC columns for a range of practical operating conditions (ACQUITY Column Calculator, developed by Uwe Neue, Waters Corporation, USA, and Dimitri Ginnis; www.unige.ch/sciences/pharm/fanal/lcap/divers/telechargements.php, developed by Davy Guillarme with contributions from Dao Nguyen, Serge Rudaz, Jean-Luc Veuthey at the University of Geneva, Switzerland). The primary reason for the interest in such software tools is the complexity of the parameters that influence the performance of a separation. Column length, particle size, mobile phase composition, temperature, analyte molecular weight etc., all influence the performance of a separation. The complexity of these relationships makes it rather difficult for the practitioner to understand the optimization parameters. Even for a well-versed theoretician, it is often not easy to see through the complex interrelationships between the various parameters.

However, performance optimization requires a reasonable knowledge of the details of the underlying parameters, and computer modeling can give us an advantage when we try to gain an understanding of the overall optimization. Commercial software is available (for example Drylab, developed by Lloyd Snyder and coworkers, or ChromSword, developed by Sergey Galushko), but the literature dedicated to the details of such simulations is limited. In the following, I would like to give an overview of such details. Ultimately, the reader will be able to simulate realistic chromatograms in a spreadsheet. The target here is not to describe the complete optimization software, but a significant portion of the underlying basic chromatographic theory is included in order to create simulations with features identical with real chromatograms.

Quantification in LC and GC: A Practical Guide to Good Chromatographic Data
Edited by Hans-Joachim Kuss and Stavros Kromidas
Copyright © 2009 WILEY-VCH Verlag GmbH & Co. KGaA, Weinheim
ISBN: 978-3-527-32301-2

I will start with the simulation of peaks and noise, and then move on to the generation of real chromatograms. In the final sections, I will combine aspects of performance and retention modeling with everything that has been discussed previously in order to generate gradient chromatograms based on a complete understanding of retention and performance. At the end, it is fun to see a realistic chromatogram emerge from a spreadsheet. Plus, I think that the reader will learn a lot along the way ...

4.2
Peak Simulation

4.2.1
Symmetrical Peaks

The simulation of a chromatogram relies primarily on the generation of peaks based on the equation for a Gaussian distribution. In a spreadsheet that simulates a chromatogram, time t is represented on the x-axis and a Gaussian peak with the mean retention time t_R can be generated with the following formula:

$$S = S_{max} \cdot \exp\left[-\left(\frac{t}{t_R} - 1\right)^2 \cdot \frac{N}{2}\right] \tag{4.1a}$$

S is the signal at the time t, S_{max} is the proportionality factor that determines the height of the signal, and N is the column plate count. This equation is identical with the standard formulation of a Gaussian peak (as shown in other parts of this book), but uses the plate count, which may be more convenient for the chromatographer. The standard version is formulated with the time-based standard deviation σ of the peak

$$S = S_{max} \cdot \exp\left[-\frac{(t - t_R)^2}{2 \cdot \sigma^2}\right] \tag{4.1b}$$

With just the simple equation (4.1a), it is already possible to generate reasonable chromatograms. An example is shown in Fig. 4.1.

In some cases, it is desirable to generate chromatograms from combinations of different columns. A typical example is size-exclusion chromatography on columns with different pore sizes, or the combination of columns with different retention or plate count properties in the same mobile phase. Examples of the latter could be the combination of a C_8 and a C_{18} column, or the combination of a 3 μm column with a 10 μm guard column. In these cases, the retention times are additive. However, we cannot simply add the plate counts of the various columns under these circumstances. A chromatogram of the combined columns must be generated by summing the retention times and the variances of the two columns. For our simulation, this can thus be handled in the following way:

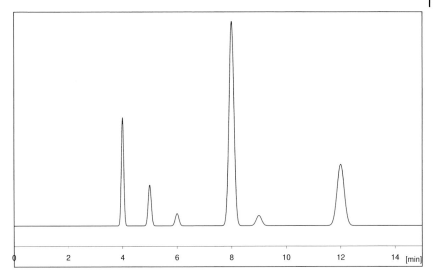

Figure 4.1 Simple simulated chromatogram with 8500 plates.

$$S = S_{max} \cdot \exp\left[-\frac{(t - t_R)^2}{2 \cdot \left(\dfrac{t_{R,1}^2}{N_1} + \dfrac{t_{R,2}^2}{N_2}\right)}\right] \tag{4.2}$$

Here $t_{R,1}$ and $t_{R,2}$ are the retention times and N_1 and N_2 the plate counts in each column, and, of course, t_R is the sum of the two retention times $t_{R,1}$ and $t_{R,2}$. $t_{R,1}^2/N_1$ and $t_{R,2}^2/N_2$ are the variances of the first and the second column, which are additive.

4.2.2
Peak Tailing

We can proceed with making the chromatogram more realistic by adding some tailing to the peaks. For this purpose we create exponentially modified Gaussian peaks. This is the peak profile most commonly used by theoreticians to describe tailing peaks. The reason for this is that the exponentially modified Gaussian peak has a few properties that are useful for the theoretician. The total variance of the peak is the sum of the variance of the parent Gaussian peak σ^2 and the square of the characteristic parameter τ of the exponential decay. The square of the peak asymmetry can be measured at 4.4% of the height of the peak and is therefore approximately:

$$A_s^2 \approx \frac{\sigma^2 + \tau^2}{\sigma^2} \tag{4.3}$$

The equation for the exponentially modified Gaussian distribution is as follows:

$$S = \frac{A}{\tau} \cdot \exp\left[\frac{1}{2} \cdot \left(\frac{\sigma}{\tau}\right)^2 - \frac{t - t_G}{\tau}\right] \cdot \int_{-\infty}^{z} \frac{\exp\left(-x^2/2\right)}{\sqrt{2 \cdot \pi}} \cdot dx \qquad (4.4a)$$

$$z = \frac{t - t_G}{\sigma} - \frac{\sigma}{\tau} \qquad (4.4b)$$

where t_G is the retention time of the unmodified Gaussian peak. The exponentially modified Gaussian function is, therefore, a combination of the exponential function with the error function.

However, to generate exponentially modified Gaussian peaks in a spreadsheet, attempting to write complicated equations is not necessary. It is simpler to think about how exponentially modified Gaussian peaks can arise in a chromatogram. The classical example is the modification of a Gaussian peak profile generated by a perfect column in combination with the time constant τ of a detector.

To understand this, let us see how we can create an exponential decay in a spreadsheet. If we have a signal of value 1 in the first cell of the spreadsheet, and the decay from cell to cell is $d = \frac{1}{2}$, we find a signal of $\frac{1}{2}$ in the second cell, a signal of $\frac{1}{4}$ in the third cell, a value of $\frac{1}{8}$ in the fourth cell and so on. In other words, the value of the current cell is generated by multiplying the value in the previous cell by a constant factor, here $\frac{1}{2}$. This constant factor can be any value that we like to choose, provided it is smaller than 1.

Let us assume that we are creating a spreadsheet function with the following formula in column B (shown here for cell B4):

$$EXP(-(\$A4^2/\$B\$1^2)/2) \qquad (4.5)$$

The value in $B$1 is the standard deviation of the Gaussian function, and the value in $A4 is the x-value. In order to create a value that will result in an exponentially modified Gaussian peak in column C, we use the exact same principle as given in the previous paragraph: we subject the signal from the previous cell to an exponential decay, and add this value to the newly generated Gaussian value in the new line. Therefore in cell C4 we write:

$$(1 - C\$2)*EXP(-(\$A4^2/\$B\$1^2)/2) + C\$2*C3 \qquad (4.6)$$

The value in C$2 is the value that creates the exponential decay from cell to cell. Here it is the decay from cell C3. We also need to subtract the value in C$2 from the newly generated exponential function in order to maintain a normalized sum over all cells in column C, i.e. a constant peak area.

In our example above, the value that created this exponential decay was $\frac{1}{2}$. We now need to specify the exponential decay d in general terms:

$$d = \exp\left(-\frac{\Delta t}{\tau}\right) \qquad (4.7)$$

where Δt is the time interval between sequential data points. This is the value that is applied in the spreadsheet, and is stored in the location C$2 in the example

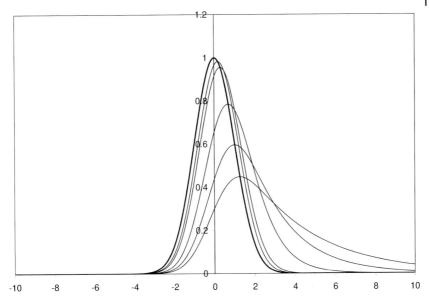

Figure 4.2 Exponentially modified Gaussian peaks.
$d = 0$ (Gaussian), $d = 0.950, 0.970, 0.990, 0.995, 0.997$.

above. In order to generate an exponentially modified Gaussian peak in a chromatogram with a reasonable time increment of the decay (or sampling rate), the values of d that are useful in a spreadsheet simulation are between 0.97 and 0.997. A value of 0 stored in C$2 for d means that there is no exponential decay, and the peak is Gaussian.

Let us now consider an example of exponentially modified Gaussian peaks generated with the spreadsheet function just described (Fig. 4.2)! The thick line represents the pure Gaussian peak. As we increase the value of d, the peak becomes more and more unsymmetrical, but the peak area remains constant. Interestingly, the maximum of the exponentially modified Gaussian peak is always on the downward slope of the original Gaussian peak.

A fixed value of d for the entire chromatogram will act on the peaks in the same way as a detector time constant: the early-eluting peaks will tail more than the peaks retained for a longer time. If we want to maintain a constant peak tailing throughout the chromatogram, the ratio of the standard deviation of the peak σ to the value of the time constant τ needs to remain constant:

$$\frac{\sigma}{\tau} \equiv c \tag{4.8a}$$

The standard deviation of the peak increases with its retention time t_R and is inversely proportional to the square root of the plate count N:

$$\sigma = \frac{t_R}{\sqrt{N}} \tag{4.8b}$$

In other words, we need to change the value of d in Eq. (4.5) for each peak.

$$d = \exp\left(-c \cdot \frac{\Delta t}{t_R} \cdot \sqrt{N}\right) \tag{4.9}$$

The value of c generates the desired σ/τ ratio, i.e. the desired tailing.

The peak height of a tailing peak can not readily be calculated from fundamental equations. The following polynomic approximation gives an accurate peak height up to a $\tau:\sigma$ ratio of $5:1$, which appears to be sufficient for most practical purposes:

Height $= \exp(-5.9124 \cdot 10^{-4}\, x^6 + 1.0122 \cdot 10^{-2}\, x^5 - 6.8070 \cdot 10^{-2}\, x^4 + 0.22510\, x^3 - 0.35125\, x^2 - 5.6178 \cdot 10^{-2}\, x)$

where x is the $\tau:\sigma$ ratio. With other words, it can be used until the height of the tailing peak becomes as small as 34% of the symmetrical parent peak. The predicted value is always better than 0.5%, mostly better than 0.2%.

4.3
The Baseline

Baseline noise can be generated simply by using the RAND() function of the spreadsheet. In order to maintain a baseline value of 0, a value of 0.5 is subtracted from the RAND() value, and the magnitude of the noise is scaled by the value $B$1:

$$\$B\$1*(\text{RAND}() - 0.5) \tag{4.10}$$

If the nature of this noise is not satisfactory, a more realistic noise can be generated using

$$\$C\$1*(\text{RAND}() - 0.5)*\text{RAND}() \tag{4.11}$$

With this function, the maximum noise is still ±0.5, but the population density increases around the baseline value of 0. Of course other, more sophisticated, tools can also be used to generate the noise. This has already been discussed in other parts of this book. A simple way to create a slightly more sophisticated noise is to sum the values generated by Eq. (4.10) or (4.11) over several adjacent cells.

Another simple issue is the generation of baseline drift. Any suitable function can be employed here, even a random drift by accumulating the RAND() function over the execution time of the chromatogram (the standard RAND() function will always generate a positive value, therefore an accumulation will always generate a positive drift). Another highly practical approach to the generation of real baselines for our artificial chromatograms has also been described in earlier sections of this book: we can record a real baseline, with all the usual bumps and drift, and we can superimpose our artificial peaks onto the recorded baseline. Overall this gives the best and most realistic test cases for integration software: we have a real baseline, and the properties of the peaks (area and height) are known *a priori*.

4.4
The Chromatogram

Now we can proceed to assemble realistic looking chromatograms. We will take
the chromatogram from Fig. 4.1, and add peak tailing and random noise. In

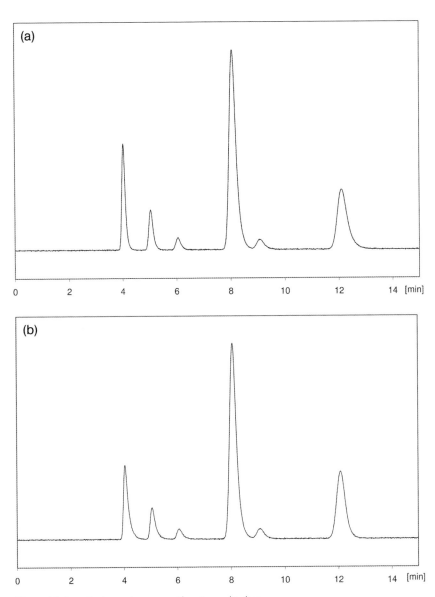

Figure 4.3 Sample chromatograms with noise and tailing.
(a) Constant peak tailing, (b) peak tailing decreases with retention time
due to a constant exponential decay function.

Fig. 4.3a, the tailing is the same for all peaks, and in Fig. 4.3b it is caused by a detector time constant or other extra-column effects, i.e. it decreases with increasing retention time. These example chromatograms already look quite realistic. In the cases shown here, the plate count of the parent peak was kept the same over the entire chromatogram. The peak tailing algorithm was also constant for all peaks in Fig. 4.3a (constant σ/τ ratio). In real chromatograms, variations in plate count and peak tailing can occur. The possibilities are manifold and are left to the imagination of the user.

There are cases of peak distortions that cannot be simulated by a simple exponentially modified Gaussian distribution. One case found in a real chromatogram was a strong tailing at the very bottom of the peak, in addition to peak deformations similar to the cases discussed above. This situation could be simulated readily by assuming that a second independent tailing phenomenon was taking place. The peak shape observed in the real chromatogram could then be generated as the sum of two exponentially modified Gaussian peaks with significantly different exponential contributions. An example of this is shown in Fig. 4.4. The two peak functions with significantly different tailing characteristics form the complete peak.

We can see that exponentially modified Gaussian peaks are quite flexible and can simulate many different situations. In general, plausible peak distortions can be created readily by summing peaks with the same or slightly different retention times and the same or different tailing characteristics. Thus without too much difficulty, one can visualize non-uniform flow through the column or peak distortions due to the design of column inlet and outlet fittings.

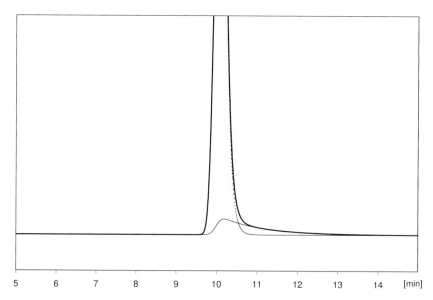

Figure 4.4 Example of peak tailing caused by two independent phenomena.

4.5
Simple Retention Modeling and Real Chromatograms

In the following sections, we will explore the foundations of retention modeling in reversed-phase chromatography. We will start with an isocratic chromatogram, learn a few simple rules of thumb of reversed-phase retention, then proceed to gradient chromatography. Ultimately, we will be able to generate gradient chromatograms that incorporate all the relevant features of a real set of analytes.

4.5.1
Isocratic Chromatography

In our isocratic chromatogram, we want to

- simulate the retention and the peak shape of several peaks;
- change the retention factor with solvent composition;
- change plate counts, peak shapes and response factors for each peak.

I will include detailed instructions on how to construct a spreadsheet that allows us to do all this. In order to make the instructions more readable, I have opted to name the cells involved and present most of the formulas with the cell names.

We will begin by putting a time scale on our chromatogram. Let us take a column dead time of 0.75 min as t_0. We write "t_0" into cell A1 of our spreadsheet and 0.75 into cell B1. Next, we assign the name "t0" to cell B1 by writing this name into the upper left corner of the spreadsheet, where it says "B1" before our entry. All other procedures for naming the cells proceed in a similar way.

Next, we define the "time increment" (cell A2) in cell B2: 0.0066, and name cell B2 "TI". Now we can construct the "time" axis (cell A14), which starts with 0 in cell A15, and continues with A15 + TI in cell A16. We pull the value in cell A16 down to cell A1530 to a retention time of 10 min, calling the array "Time". We have now constructed the time axis of our chromatogram: 10 minutes, with an increment of 0.0066 min (~ 0.4 s).

Next, let us briefly review the theoretical equation that shows how the retention factor k changes with the solvent composition Δc. This can readily be found in many textbooks:

$$\ln(k) = k_0 - B \cdot \Delta c \quad (4.12)$$

The numerical value of B tends to be around 10 for small molecules (such as typical pharmaceuticals) in an acetonitrile/water mobile phase. It is a bit lower if methanol is the organic modifier, and the value tends to increase with increasing molecular weight of the analyte.

We start by defining the change in solvent composition ("delta c" in A3, name the cell B3 "dc"). If the value is 0, the retention factors in the chromatogram will be the ones that we will define in the next line. If the value is larger, for example 0.15, the retention factors in the chromatogram will be lower.

In the cells B4 to G4, we define the retention factors using the starting conditions ("k_0" in cell A4). Let us select six values between 0 and 15, and put them into cells B4 to G4. Next, we select the array B4:G4 and name it "k0" in the upper left corner of the spreadsheet. We will change these retention factors with the solvent composition "delta c" (A3) in cell B3, named "dc". For the time being, let us put a 0 into cell B3. The change in the retention factor with solvent composition will be calculated using Eq. (4.10). We select *B*-values for each peak between 8 and 12, write them into cells B5 to G5, and name them "Exponent" in cell A5, and name the array B5:G5 "B" as done above for the previous array.

In the next line, we define the "Plate Count" (cell A6) for each peak in cells B6 to G6, which we call "Plates". We can select plate count values between 5000 and 10 000 in row 6. In row 7, we define the "τ/σ ratio" (cell A7). Good values to start with are values between 0 and 2 in cells B7 to G7. The only thing that is missing is the "Peak Area" (cell A8) for each peak. We select values between 500 and 5000 in cells B8 to G8 and give this array the name "Area".

Thus we have defined all input parameters necessary for our chromatogram. We will now use these values in the following cells. First, we will define the retention factors "k" (cell A10) in the cells B10 through G10. In order to do this we write in cell B10

$$=k0*EXP(-B*dc) \tag{4.13}$$

and fill in the neighboring cells to cell G10 with this value. This calculates the retention factor according to Eq. (4.10). In our case, the value will be identical to the value in cells B4 to G4, as we still have a value of 0 in cell B3. The retention times "t_R" (cell A11) are calculated in cells B11 to G11, by writing, in the classical spreadsheet notation, into cell B11

$$=(B10 + 1)*\$B\$1 \tag{4.14}$$

and filling this value into the neighboring cells and naming the group "tr".

We now calculate the peak height for a symmetrical peak from the peak area in line 8. In A12 we name it "Response", and fill the following formula into cells B12 to G12:

$$=Area/60/(SQRT(2*PI())*(tr/SQRT(Plates))) \tag{4.15}$$

and call these cells "Response". This peak height will be used as the basis for the peak height of the tailing peaks generated later, in agreement with the model shown in Fig. 4.2.

As a final step, we define the value for the "Tailing d" (A13) based on the τ/σ ratio in line 7 using Eq. (4.9). As the τ/σ ratio can be 0, we need to define two cases with an IF statement in B13:

$$=IF(B7 = 0,0,EXP(-(1/B\$7)*TI/tr*SQRT(Plates))) \tag{4.16}$$

The value is either 0, if the τ/σ ratio is 0, or the calculated value, if the τ/σ ratio is not 0 (see Eq. (4.9) and the surrounding text for explanation, and Fig. 4.2 for examples). The statement in cell B13 is used to fill in the cells sideways through to cell G13, and call the cells "d".

Now we are ready to create the "peaks 1 to 6" (cells B14 to G14). We start the chromatogram in cell B15:

$$=EXP(-((Time/tr - 1)\char`^2)*Plates/2)*Response \qquad (4.17)$$

by writing the value for a Gaussian peak with the retention time "tr" from line 11, the plate count from line 6 and peak height or "Response" from line 12. Again, this value is used to fill in the cells sideways through to cell G15.

In the next line, in B16, we add the value of the exponential tail that we had created in line 13, following the principle described in Eqs. (4.5) and (4.6), and demonstrated in Fig. 4.2:

$$= (1 - d)*EXP(-((Time/tr - 1)\char`^2)*Plates/2)*Response + B15*d \quad (4.18)$$

We now fill all cells from B16 to G1530 with this expression. In H14, we write "Noise" and define the noise of the chromatogram in the cells below. A random noise can be generated as described in Section 4.3, or we can import noise from a real chromatogram, after suitable scaling. In column I, we copy the cells from A17 to A1530, and in column J we sum the values for each peak to generate the entire "Chromatogram" (J14) by writing into cell J15:

$$= SUM(B15:H15) \qquad (4.19)$$

Finally, we fill in with this expression down to cell J1530. The combination of columns I and J gives us the desired chromatogram, which we can plot. We can also change the actual observed retention by changing the solvent composition "delta c" in position B3.

A slightly modified version of this spreadsheet has been used to demonstrate peak overlap and the improvement in resolution that can be gained from changes to the strength of the mobile phase. Here, a chromatogram has been created from nine different components. When plotted (see Fig. 4.5a), only seven peaks can be seen. What is happening? There are two peak overlaps in the chromatogram, but a slight adjustment of the solvent composition (in B3) makes it possible to resolve all nine peaks (Fig. 4.5b).

We are now able to generate as many chromatograms as we wish. However, a few more points need to be added. In order to produce a chromatographic peak without undue distortion, we want to have at least 40 data points over the (baseline) width of the narrowest peak. It is actually a worthwhile exercise to select fewer data points and observe what peak distortions result. At the same time, we do not want to make the spreadsheet unduly large, so we want to select the minimum number of data points that will still give us a suitable simulation. This is very

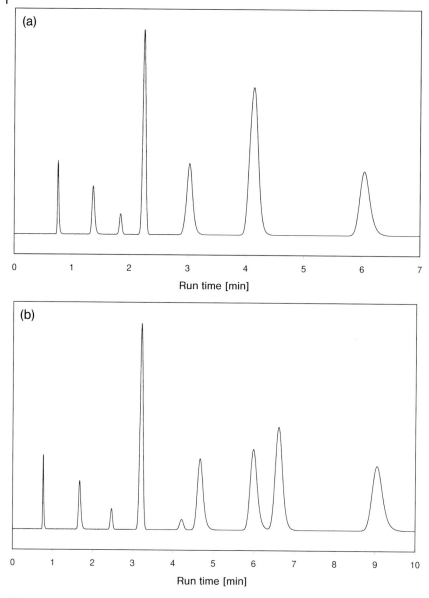

Figure 4.5 Two isocratic chromatograms with the same number of peaks (9).
(a) Two peak overlaps, (b) both peak pairs have been resolved by simply
decreasing the elution strength of the mobile phase.

similar to the case of data collection in a practical situation. The optimization of
the sampling rate is an important decision.

More comments are necessary with respect to the expansion of the chromatogram with the nomenclature used here. The time axis of the spreadsheet cannot

be simply expanded with a "fill down" or "fill left" procedure. However, we can fill empty cells into the spreadsheet (i.e. for a downward fill before the last row of the spreadsheet), followed by a "fill down" or "fill left" procedure over the empty cells up to the last cell. This "trick" permits an expansion of the time axis or the addition of more peaks into our chromatogram.

4.5.2
Gradient Chromatography

For the generation of gradient chromatograms, we want to accomplish the same thing as we did above for isocratic chromatography. We must

- simulate the retention and the peak shape of several peaks;
- change the gradient retention factors (i.e. the peak position in the chromatogram) through the gradient execution;
- change plate counts, peak shapes and response factors for each peak independently.

In gradient chromatography, the peak widths are typically rather constant over the entire chromatogram. In order to simulate a gradient chromatogram, we need to understand the relationship between peak width and retention time as a function of the analyte properties. Detailed explanations of the theory are not given in the following discussions. These can be found in publications or textbooks [1–3]. Here, simply the equations needed, followed by detailed instructions on the implementation in a spreadsheet will be shown.

The retention factor of a peak in gradient chromatography k_g depends on two parameters: the retention factor at the beginning of the gradient k_0, and the gradient itself, specifically the gradient slope G.

$$k_g = \frac{1}{G} \cdot \ln(G \cdot k_0 + 1) \tag{4.20}$$

As in isocratic chromatography, the gradient retention factor is calculated from the retention time t_r and the column dead time t_0:

$$k_g = \frac{t_r - t_0}{t_0} \tag{4.21}$$

The gradient slope G depends on the gradient run time t_g, the column dead time and the difference in solvent composition from the beginning to the end of the gradient Δc,

$$G = B \cdot \Delta c \frac{t_0}{t_g} \tag{4.22}$$

the relationship between the retention factor and the solvent composition (shown in Eq. (4.10), and the value of the slope of the logarithmic retention factor and the solvent composition B described previously in Eq. (4.12).

The true retention factor at the point of elution k_e, which determines the peak width, depends on the gradient execution:

$$k_e = \frac{k_0}{G \cdot k_0 + 1} \tag{4.23}$$

The peak width at the point of elution is a function of this retention factor at the point of elution, the plate count N, and the band compression factor C_g:

$$w = C_g \cdot \sqrt{N} \cdot t_0 \cdot (k_e + 1) \tag{4.24}$$

This band compression factor has been derived in the literature as

$$C_g = \frac{\sqrt{1 + p + p^2/3}}{1 + p} \tag{4.25}$$

The factor p is closely related to the gradient steepness parameter G.

$$p = G \cdot \frac{k_0}{k_0 + 1} \tag{4.26}$$

Now we have all the equations that are needed to simulate realistic gradient chromatograms, including the dependence of retention and peak width on the gradient execution. Let us now assemble the spreadsheet for the "Simulation of a Gradient Chromatogram", which we write into cell A1. The resulting plot is shown in Fig. 4.6.

First, we will define the gradient parameters. It is a reversed-phase gradient with a run time ("$t_g =$" in cell A2) of 30 min (cell B2). As in Section 4.5.1, we will name the cells for an easy understanding of the formulas used. Therefore we give the name "tg" to cell B2 by writing "tg" into the upper left corner of the spreadsheet. The column dead time ("$t_0 =$" in cell A3) is 1 min (cell B3, named "t0"). The "time increment" (cell A4) with which we will construct the time axis of the chromatogram is entered in cell B4 (named "TI"). The difference in solvent composition from the beginning to the end of the gradient ("delta $c =$" in cell A5) is 0.55 in cell B5 (named "dc"), and the value of the solvent composition at the beginning of the gradient ("$c_0 =$" in cell A6) is 0.25 in cell B6 (named "c0"). This means that we run a gradient from 25% to 25% + 55% = 80% of the strong solvent. The gradient slope ("slope =" in cell A7) is calculated in cell B7 from the values just entered:

$$= B4*B3/B2 \tag{4.27}$$

This defines the gradient. Now we will declare the retention properties for all analytes that we would like to elute in the gradient. We will start with the retention times for all analytes in 100% water ("$k_{00} =$" in cell A9). These values should

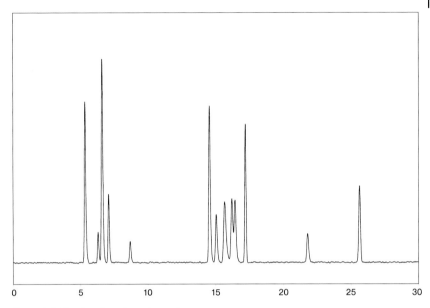

Figure 4.6 Simulated gradient chromatogram.

be between 50 and 10 000 for our gradient. The values that have been selected for the chromatogram in Fig. 4.6 are shown in Table 4.1. In row 9, we enter these values for all of the peaks that we want to see in the chromatogram. We select the cells B9:N9 and name this group "k00". Next, we enter the values for the slope of plot of the natural logarithm of the retention time versus the solvent composition ("B =" in A10) into row 10 (naming the cells B10:N10 "B"). Typical values are around 10 for analytes with a molecular weight around 250. Next, we enter the peak areas ("Peak Area =" in A11) for all analytes into row 11 (named "area" as above), and the plate counts ("N =" in A12) into row 12 (named "N" as above). The last item that needs to be defined is the tailing, i.e. the τ/σ ratio ("τ/σ =" in A13), in row 13. If the peaks do not tail, we enter 0. For slightly tailing peaks, a value of 1 is suitable.

It needs to be pointed out clearly that we are using the isocratic plate count in row 12. This is the true plate count, defined as the ratio of the column length to theoretical plate height. The fundamental definition of the theoretical plate height is in length units inside the column, and thus the plate count is defined independently of the gradient or whether we are using isocratic chromatography or gradient chromatography. Therefore we can use typical isocratic plate count numbers for our modeling of the analyte behavior in gradients.

We have now defined everything that we need to simulate the gradient chromatogram and for executing all the necessary calculations.

Before we proceed, a few words are needed to comment on the values in the first two columns in Table 4.1 (and the first two rows in the spreadsheet). The entries are ordered by the retention factor found in 100% water. However, this does not

Table 4.1 Retention properties of analytes for the example gradient chromatogram.

k_{00}	B	Peak area	N	τ/σ
80	10	160 000	10 000	1.2
200	12	30 000	10 000	0
300	13	196 000	8 000	1
350	13	66 000	10 000	0.9
400	12	24 000	9 000	1
900	10	100 000	8 500	1.5
1 000	10	94 000	10 000	1.1
1 900	12	60 000	10 000	1
2 100	12.5	190 000	11 000	1
2 600	12	74 000	8 500	1
5 000	13	130 000	10 000	1
5 000	11	40 000	10 000	1
6 000	10	90 000	10 000	1

mean that this is the order of elution in the gradient chromatogram. The elution order is determined by three factors: the retention factor in water, the slope B of the relationship of the logarithm of the retention factor with the solvent composition, and the gradient execution, i.e. the starting composition and the slope of the gradient. Once the spreadsheet is complete, the user can manipulate the gradient parameters to observe the changes in the resolution and the elution order.

Let us now proceed with the calculations! In row 15, we calculate the retention factor for each compound at the beginning of the gradient (A15 = "k_0 ="), starting with B15:

$$=k00*EXP(-B*c0) \tag{4.28}$$

This is the same equation as shown above (Eq. 4.11) for isocratic separations. c0 is the solvent composition at the beginning of the gradient, and the other two parameters are the compound-specific values of the retention factor in 100% water (k00) and the compound-specific slope B of the relationship of the logarithm of the retention factor with the solvent composition. We will need this value for further calculations, so we select cells B15:N15 and name them "k0".

Next, we calculate the gradient slope G for each compound in row 16 (A16 = "G =" and name B16:N16 "G"). Note that this value will vary for each compound, as it depends on the slope of the relationship of the logarithm of the retention factor with the solvent composition. In B16 we write:

$$=B*dc*t0/tg \tag{4.29}$$

and fill in this value sideways to cell N16.

In row 17, we calculate the gradient compression factor for each peak, in accordance with Eqs. (4.25) and (4.26). The entry in A17 is "CF=", and the one in B17 is

$$=(SQRT(1 + G*k0/(k0 + 1))$$

$$+ (G*k0/(k0 + 1))\wedge 2/3))/(1 + G*k0/(k0 + 1)) \tag{4.30}$$

As above, we fill in this equation sideways to N17 and name this group of cells "CF".

We now calculate the gradient retention factor and retention factor at the point of elution in the next 2 rows (A18 = "k_g ="; A19 = "k_e =") by writing in cell B18 and B19, respectively:

$$=(LN(G*k0 + 1))/G \tag{4.31}$$

$$=k0/(G*k0 + 1) \tag{4.32}$$

which are implementations of Eqs. (4.20) and (4.23). The cells B18:N18 and B19:N19 are filled with the same equations and are called "kg" and "ke", respectively.

In row 20, position B20, we simply calculate the gradient retention time (A20 = "t_g") from the gradient retention factor "kg" and the value for "t0" defined in cell B3:

$$=(kg + 1)*t0 \tag{4.33}$$

followed by the peak width (A21 = "sigma") in B21:

$$=t0*(ke + 1)/SQRT(N)*CF \tag{4.34}$$

After filling these values to column N, we name the cells B20:N20 "tg", and the cells B21:N21 "sigma". This calculation uses the isocratic plate count from row 12. Note that the peak width is narrowed by the peak compression factor.

The peak height ("Response" in line 22) is calculated from the peak area in row 11. It is a function of the retention factor at the point of elution and the plate count. The larger this retention factor, the lower the peak height. It is also influenced by the peak compression, calculated in row 17. This has already been incorporated into the calculation of the "sigma" value discussed in the last paragraph. In cell B22 we therefore write:

$$=Area/60/(SQRT(2*PI())*sigma) \tag{4.35}$$

fill this value to column N, and call this assembly of cells "Response". This subtlety changes the peak height as a function of the gradient execution, taking all influences into account.

In the final line of the set-up, line 23, we calculate the value of the parameter that is used to create the peak tailing from the τ/σ ratio in row 13, the peak width "sigma" from row 21, and the time spacing of the cells in the spreadsheet defined as the time increment "TI" in cell B4

$$=IF(B\$13 = 0,0,EXP(-(1/(B\$13*sigma)*TI))) \tag{4.36}$$

If the peak is not tailing (B13 = 0), the value is 0, otherwise, it is the calculated value in accordance with Eq. (4.9).

We have now accomplished all the preliminary work to set up the spreadsheet to create the chromatogram. The retention times, the peak width and peak shape values for all peaks have been defined based on the gradient parameters and the properties of our analytes. Next, we need to generate the response as a function of the gradient run time for all peaks, followed by summing all peaks to create the final chromatogram.

In A25, we describe the column of values below by "Time". In cell A26, we enter the value 0. In line A27, we add the value of the time increment "TI":

$$=A26 + TI \tag{4.37}$$

We fill in the value of cell A27 down to position A6026 up to a time of 60 min.

Next, we write in cell B26 the response expected for the peak defined in the column above:

$$=Response*EXP(-((\$A26 - tr)^2)/(2*sigma^2)) \tag{4.38}$$

This is the Gaussian function for this peak. We fill this value into all the cells in row 26 up to column N. In the next line, specifically in cell B27, we define the value for the peak response that includes any tailing, as described in Eq. (4.6):

$$=(1 - d)*Response*EXP(-((t - tr)^2)/(2*sigma^2)) + d*B26 \tag{4.39}$$

This formula is now filled into all cells of the spreadsheet from cell B27 to cell N6026.

We have now created all the peaks in the chromatogram. There are two things left to do: one is to create the baseline noise in a similar way to that shown earlier, and the final thing is to assemble the chromatogram by summing all the values for all the peaks and the noise.

We write "Noise" into cell O24, and a value for the noise into cell O25. For our current simulation, a value of 30 is suggested. In cell O26, we write:

$$=(RAND() - 0.5)*O\$24 \tag{4.40}$$

We fill in this value down to cell O6035. The reason for filling it beyond the reach of the rows used before becomes apparent in cell P26, where we accumulate the noise to make it more realistic:

$$=SUM(O25:O34) \tag{4.41}$$

However, the details of the generation of noise (and drift) are entirely up to the user. We have mentioned earlier that it may be better to use real chromatographic noise to form the baseline for our artificial chromatogram.

We copy the values from column A into column Q (Q26: = A26), and we fill this value down to Q6026. The final task is to add the individual peaks generated and the noise in column P by writing into R26:

$$=SUM(B26:N26) + P26 \tag{4.42}$$

and filling this value down to row R6026.

Finally, we can generate the chromatogram by plotting the values in column R (the "detector response") versus the values in column Q (the retention time). All peaks "elute" in less than 30 min, so we can limit the time axis to a maximum value of 30, and we should have a plot that is – after some fine tuning of the layout of the axes – very similar to Fig. 4.6. We started with 13 "compounds", and indeed we observe 13 peaks, although some are not perfectly resolved.

We can manipulate the gradient parameters to observe what is happening in the separation. This is an interesting and very useful exercise, even for the experienced chromatographer. The beauty of any spreadsheet exercise is that one can get the results of the "experiment" instantaneously. One does not have to wait for 30 to 45 min until the results are in!!! A further optimization of the gradient may result in the chromatogram shown in Fig. 4.7. Peaks 9 and 10 are now completely resolved, the resolution between peaks 2 and 3 has improved further, and the cost in analysis time is small. Still better solutions may be achievable but I leave it to the interested user to explore such possibilities. In order to do that, one would change the gradient run time (in cell B2), the difference in solvent composition from the beginning to the end of the gradient (cell B5), or the solvent composition at the start of the gradient (cell B6). Of course, in order to change the gradient steepness, we could also manipulate the column dead time t_0. However, this would also change the plate counts, and in this simple version of the gradient simulation, the changes in plate count are not incorporated. Therefore, using the column dead time as an optimization parameter is not recommended in this simulated chromatogram.

As described in Section 4.5.1, we can add additional "peaks" by inserting empty columns into the "peak" section of the spreadsheet, and then using the "fill right" command to fill the empty cells. Then we can change the characteristic properties of the new peaks. In a similar way, the time frame can be expanded, if needed.

Now it is time to have fun with the simulations.

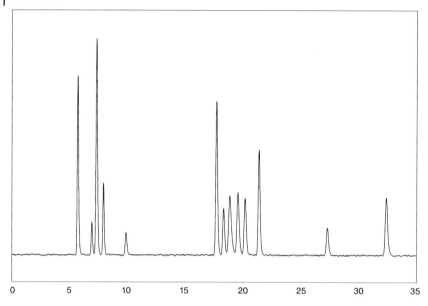

Figure 4.7 Optimized chromatogram of the peaks shown in Fig. 4.6.

4.6
Outlook

In this chapter, I have described in depth how artificial chromatograms can be produced. I have shown how peak tailing can be generated, and how gradient retention times can be estimated based on well established chromatographic theory. The reader is encouraged to play with the spreadsheets. I am convinced that even the experienced chromatographer can learn something, as I did when I developed the spreadsheets.

The spreadsheets can be expanded to include a prediction of the plate count as a function of the operating conditions of the column. This has been done in the ACQUITY Column Calculator mentioned in the Introduction. Also, the relationships between retention and solvent composition are not necessarily linear [4], and this feature can be included in more sophisticated models. An entirely different area is the retention of ionizable compounds as a function of the mobile phase pH. The theory of this condition is well developed, and the necessary background information is available in the literature [5, 6]. One can also incorporate extra-column bandspreading. Before long, we could have an artificial chromatograph in our hands ...

The pleasure of including additional ideas is left to the users of my spreadsheets.

References

1 *High-Performance Gradient Elution*, Lloyd R. Snyder, John W. Dolan, Wiley Interscience, 2006.

2 U. D. Neue, *Gradient Elution Theory*, published on-line at Chromedia.nl, April 2007.

3 U. D. Neue, D. H. Marchand, L. R. Snyder, Peak compression in reversed-phase gradient elution, *J. Chromatogr. A 1111* (2006), 32–39.

4 U. D. Neue, Non-linear retention relationships in reversed-phase chromatography, *Chromatographia 63* (2006), S45–S53.

5 X. Subirats, E. Bosch, M. Rosés, Retention of ionisable compounds on high-performance liquid chromatography XVI: Estimation of retention with acetonitrile/water mobile phases from aqueous buffer pH and analyte pK_a, *J. Chromatogr. A 1121* (2006), 170.

6 X. Subirats, E. Bosch, M. Rosés, Retention of ionisable compounds on high-performance liquid chromatography XVII: Estimation of the pH variation of aqueous buffers with the change of the methanol fraction of the mobile phase, *J. Chromatogr. A 1138* (2007), 203.

5
Integration of Asymmetric Peaks

Hans-Joachim Kuss

Anyone who has ever worked in a chromatography laboratory will have come into contact with the problem of the tailing of peaks. One just has to live with it but tailing should be kept as low as possible. Any reduction in the tailing is progress as far as the chromatography is concerned and permits a better quantification. The following discussion shows how tailing has an adverse effect on the integration of peaks merged together.

The simulations of EMG peaks are based on the Excel spreadsheets of Uwe Neue, explained in Chapter 4.

Figure 5.1 shows two equal peaks with clear tailing, merged together and with a valley of a little less than 50% between the peaks. From Fig. 5.1 (a) and (b) the asymmetry factor as the quotient $A_f = b/a$ or the tailing factor as the quotient $T_f = (a + b)/2\,a$ can be calculated. The sum of a and b yields the peak width w at the corresponding height. Labsolutions, for example, can provide $T_{f5\%}$, $T_{f10\%}$, $w_{5\%}$, $w_{10\%}$ and $w_{50\%}$ in its report. A_f and T_f are interconvertible provided that one "remains at the same height level".

$$T_f = \frac{a + b}{2\,a} = \frac{1}{2}\left(1 + \frac{b}{a}\right) = \frac{1}{2}(1 + A_f) \tag{5.1}$$

$$A_f = 2\,T_f - 1 \tag{5.2}$$

The deeper a and b are measured, i.e the further from the maximum peak height at which they are measured, the more exactly the tailing can be represented. However, only for very well separated peaks can values at 5% of the height be determined. If the peaks in Fig. 5.1 move even closer together, the characteristic quantities a and b can no longer be determined at 50% of the height.

As is shown in the following, it is not advisable to quantify tailing peaks with a higher valley because the integration error increases too much. In reality, merged peaks are found in many chromatograms, thus the characteristic quantities at 50% of the height should be preferred. Finally, it clearly cannot be accepted, to have no characteristic values, especially for the most critical peaks in the chromatogram.

Quantification in LC and GC: A Practical Guide to Good Chromatographic Data
Edited by Hans-Joachim Kuss and Stavros Kromidas
Copyright © 2009 WILEY-VCH Verlag GmbH & Co. KGaA, Weinheim
ISBN: 978-3-527-32301-2

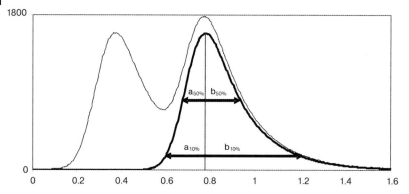

Figure 5.1 Two simulated tailing peaks merged together with a valley a little less than 50% of the height. The two peaks are identical, as can be seen by the bold line.

Integrating with the perpendicular drop the height of the first peak is correct, but the area is not; the height of the second peak as well as the area are increased.

Peaks with tailing can be described by the model of the Gaussian function modified exponentially (EMG). The exponential decay has the time constant τ. The extent of the tailing is represented as the relative value τ/σ. The usual symbol, σ, is used for the standard deviation of the Gaussian peaks in the general equations. This is not a usable term in Excel, so instead the symbol s_p is used.

The Gaussian function with a superimposed e-function leads to a slight widening of the increasing arm of the peak (the *a* section increases easily) and a much stronger widening of the decreasing peak arm (the *b* section increases considerably). There are other descriptive models for tailing peaks which in practice are of no consequence.

The EMG function is a model and only one approach towards the consideration of real peaks, developed by Foley and Dorsey [1]. However, no other useful alternative has been proposed in the 25 years since its introduction. Hence it is better to use the EMG function so as to maintain the present belief that real chromatograms consist of ideal Gaussian peaks. Almost everyone working in a chromatography laboratory would prefer a separation with little tailing to a similar separation but with more tailing. In the usual plate number calculations the worst effect of tailing is neglected almost completely. In their last sentence, Foley and Dorsey [1] strongly recommend the use of their equation (see Eq. (1.7) in Chapter 1) for the plate number calculation. All international sets of rules in use today settle for flexible calculations, which leads one to believe in a high efficiency.

For the full description of an EMG peak, only the following quantities are necessary: t_R, *a* and *b* at a defined height and the peak area as a size parameter. With $a = b$ the special case of a Gaussian peak is fulfilled.

How does one get from the experimental quantities *a* and *b* to the theoretical parameters τ and σ? Many equations can be found in the original publication for the characteristic quantities at 10, 30 and 50% of the height. An exemple equation is shown here for the "Gaussian part" of a tailed peak:

$$\sigma = \frac{w_{50\%}}{2.5 \, A_{f50\%}} \qquad\qquad (5.3)$$

From the asymmetry factor at 10 or 50% of the height, the τ/σ relationship can be read from the table in the publication by Tamisier-Karolak et al. [2]. The two sizes σ and τ/σ that are necessary for the simulation of tailing peaks are thus available. A table for the calculations with the empirical equations shown in the same work can be found on the CD.

Only the retention time t_R and the peak width at half height $w_{50\%}$, which delivers the standard deviation s_p, is necessary for the simulation of Gaussian peaks. From the area and the height s_p can also be readily calculated. In addition for the simulation of EMG peaks only the asymmetry factor $A_{f10\%}$ is required, which yields the input value τ/σ.

In practice, using the Excel spreadsheet "Simulation with Tailing.xls" the perfect simulation of real peaks is almost possible. In addition to the parameters needeed for Gaussian peaks, i.e. retention time, area and height, the asymmetry factor is necessary. This can only be determined if the valley between two peaks

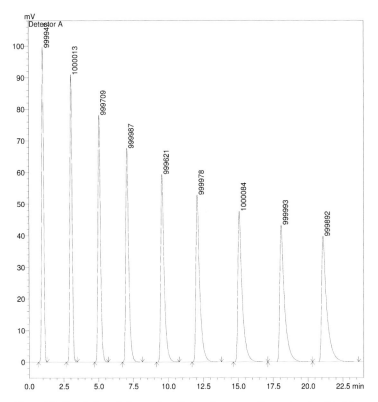

Figure 5.2 Simulation of nine peaks with baseline separation and a τ/σ of 0, 0.5, 1, 1.5, 2, 2.5, 3, 3.5 and 4. The peak ends appear plausible also at 100-fold magnification. The target value of the area is 1 000 000 µV·s.

is at less than 10% of the height of the smaller peak. Of course peaks in complex chromatograms are merged more significantly and often it is only the integration of these peaks that is particularly interesting. If the CDS permits it, one could use the tailing factor at 50% of the height. If the peaks are merged even more strongly then the integration becomes doubtful anyway.

Figure 5.2 shows 9 peaks separated to the baseline with a τ/σ of 0, 0.5, 1, 1.5, 2, 2.5, 3, 3.5 and 4. However, the integration yields the set point of the area. Higher requirements are needed for the integration of tailing peaks than for Gaussian peaks, because the peaks increase more steeply than they decrease. However, this does not seem to be an integration problem.

5.1
The Valley Between Merged Peaks

Figure 5.3 shows two equal Gaussian peaks merged together. The valley between them has a height of about 27%. The minimum point is exactly in the middle of the two retention times and with the complete symmetry there is complete error compensation. When the second peak has its area reduced to 10% of the original, the separation looks substantially worse. The valley point is now 68%, which means there is a dramatic decrease in the separation quality. The resolution is the same in the two cases because the concentration difference is not taken into account in Eq. (1.18). The minimum between the peaks is moved to the right. On integrating with the perpendicular drop method the tall peak therefore gains more area than it loses. In this example the area of the taller peak is too high by 1%, but the area of the smaller peak is too small by 9.4%. The deviation in the height of both peaks is well below 1%.

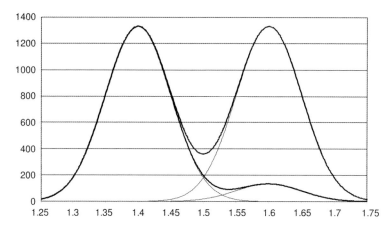

Figure 5.3 Two Gaussian peaks merged together. The upper bold line shows two peaks of equal size. The lower bold line characterizes the case where the second peak has its area reduced to 10% of the original.

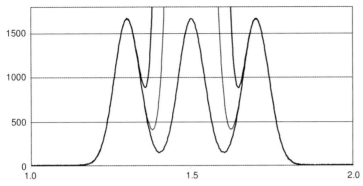

Figure 5.4 Three similar Gaussian peaks were slightly merged.
The valley between the peaks is at approximately 10%.
The second peak was then increased 10-fold and 100-fold.

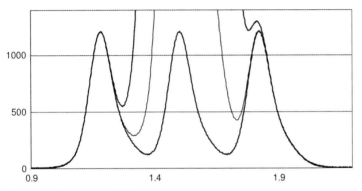

Figure 5.5 Three similar EMG peaks were slightly merged.
The valley between the peaks is at approximately 10%.
The second peak was then increased 10-fold and 100-fold.

Three equal Gaussian peaks were merged together so that the valley between them was nearly 10% of the peak heights. Then the middle peak increased by a factor of 10 or 100 (Fig. 5.4). The separation became visibly worse: the valley increased to 30 or 60%. As the characteristic peak parameters remain constant, in all three cases the resolution was found to be 1.23. By extending the middle peak the valley moved to the peak maxima of the two outer peaks. Therefore, on integration with the drop method, the area became smaller. The 10-fold (100-fold) taller peak takes 1 to 2% (6 to 7%) of the area away from the two outer peaks. An additional peak would further change the integration of the three peaks being considered. Significantly different peak heights worsen the integration of the area of the small peaks considerably. The peak height is not influenced by the minimum point between the peaks.

*Large differences in peak height lead to large integration errors
in the area (not in the height) even with ideal Gaussian peaks.*

The influence of the tailing was then examined. All peaks (Fig. 5.5) had an asymmetry factor of 1.55. The peaks became broader and the retention times were adapted so that a valley of about 10% again resulted. The scale of the illustration was adjusted accordingly. The 10-fold (100-fold) taller peak takes 2 to 4% (10 to 14%) of the area away from the two outer peaks. Adding tailing the integration error is doubled. A more significant tailing leads to an even bigger error. The peak height of the first peak is not influenced, and the peak height of the third peak is only slightly affected. The integration system LCSolutions as others automatically integrate the last peak using the tangent method. As the drop method already leads to a loss in area, here the tangent method is insufficient.

5.2
Small Peak Between Larger Ones

The illustration 3.47 in the book by Dyson [3] shows a chromatogram with a small peak between two tall symmetrical peaks. The illustration shows a tan-

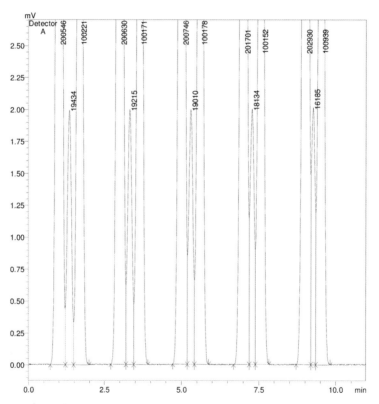

Figure 5.6 Three Gaussian peaks were merged together slightly and then merged more and more. The first peak has a target area of 200 000 μV s, the second peak 20 000 μV s and the third peak 100 000 μV s.

gential integration for the middle peak. The conclusion of Dyson is: "There is no integrator solution to this problem; the only solution is improved peak resolution."

The simulated chromatogram "5mal3Peaks.cdf" clearly shows that after integration using LCSolutions (Fig. 5.6), the tangent method would be definitely wrong here. The perpendicular drop method alone leads to reduced areas because the two minima are pushed toward the middle peak, jamming it between them. Therefore, a clear statement can be made concerning the above problem: The tangent method delivers areas far too small. The heights of all peaks in all triplets are correct.

With the analogous simulation "5mal3PeaksT" and limited tailing of the peaks the result is better because an error compensation takes place (Fig. 5.7). Again the area is reduced, but the middle peaks cut off more of the tailing from the first peak but loose less area by reducing its own tailing. The result depends on the extent of the tailing and one cannot rely on this error compensation. When required, the follow-up simulation of real chromatograms gives an indication of how they can be integrated better.

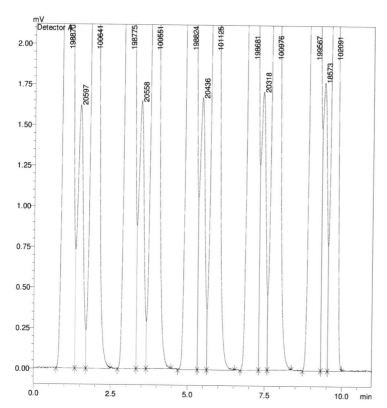

Figure 5.7 Three EMG peaks ($\tau/\sigma = 2$) were merged together slightly and then merged more and more. The first peak has a target area of 200 000 μV s, the second peak 20 000 μV s and the third peak 100 000 μV s.

5.3
Peak Pairs

Figure 5.8 shows the same chromatogram as in Fig. 5.2 but with peak pairs, which have an inner distance of 30 s at an s_p of 4 s. Because of the increased tailing the valley is increased from 0.2 to 48%. At $\tau/\sigma = 1$ the second peak is already taking more than 1% of the area from the first peak because the tailing is cut off. At $\tau/\sigma = 2$ the amount of area transferred is 7%, at $\tau/\sigma = 3$ it is 16% and at $\tau/\sigma = 4$ it is as much as 25%.

Tailing worsens the integration. But is it the tailing itself or raising the height of the valley? The chromatogram in Fig. 5.8 was modified. Now all peak pairs have a valley at 10% (Fig. 5.9). Comparison of the areas in Figs. 5.8 and 5.9 clearly shows that the height of the valley is the decisive factor. The extent of the tailing is nevertheless important because otherwise no area would be transferred from peak 1 to peak 2.

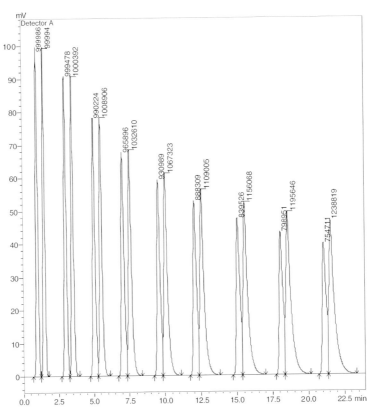

Figure 5.8 The first peaks are the same as in Fig. 5.2. A second similar peak was added. The increased tailing leads to an increase in the height of the valley between the peak pairs.

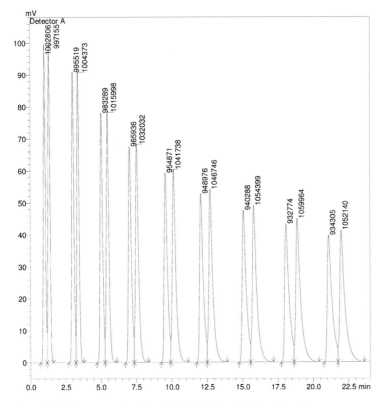

Figure 5.9 The same peak pairs as in Fig. 5.8. The retention time difference was adjusted, resulting in a valley of approximately 10% for all peak pairs.

5.4
Simulation of a Calibration

The simulation of a chromatogram with six peaks is discussed in the following. It is based on a real chromatogram of a tricyclic antidepressant. This separation was carried out with a C18 column and an acidic buffer mixed with acetonitrile. Tailing could not be avoided completely. The two outer peaks stand free. The third and fourth peaks are merged. The valley between them is slightly deeper than 50%. Such chromatograms, which are not particularly complex, are often used to carry out a concrete analysis. Figure 5.10 shows the lowest calibrating concentration for which the last peak has an S/N of 10.

From Fig. 5.10 it can be seen that integration by perpendicular drop cuts off the last part of the tailing of peak 3 and that this area is then added to the area of the fourth peak. In real analyses, if peaks 3 and 4 never occur together, the calibration for substances 3 and 4 must be carried out separately in two parts. The erroneous integration of the areas cannot occur in this case.

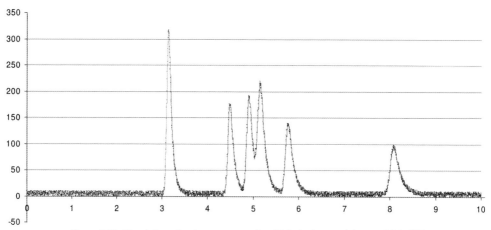

Figure 5.10 Simulation of a chromatogram in which the last peak has an S/N of 10. Because of the tailing the middle peaks are integrated insufficiently.

In gas chromatograms with several peaks merged together then one calibration solution will contain one substance from each of the pairs of peaks, and the other solution will contain each of the other substances from the pair. It is clear from such a calibration and a chromatogram with all peaks, that the integration error directly influences the analysis result. If the integration of the analysis is wrong, there will be no benefit to the accuracy of the calibration. In the simulated example, the third peak influences the area of the fourth peak and vice versa.

Unfortunately, it is to be expected that with real analyses, all peaks could be present. Should one then (partly) calibrate with individual substances or with the complete solution? Of course all substances together in one solution require the lowest amount of effort for the calibration.

The calibration was carried out with the file "Analysis Simulation.xls" for the concentration stages 1, 2, 4, 8, 16 and 32. For the AIA chromatograms Kalib1.cdf to Kalib32.cdf, the factor was set in cell B6 on this number. For the six "analysis samples" the areas of the first and fourth peak were set to a middle value, the area of the other peaks received 97% of the random function by "=Rand()*97% + 3%" to simulate a working range with a concentration quotient of 32. The area of the calibration tests varied around a maximum of ±1. The retention times span a window of ±0.25%. The chromatogram of Analyse3.cdf is shown in Fig. 5.11.

With method coefficients of variance (VVK; cp. eg. 7.10) of less than 1%, the calibration for peaks 1, 5 and 6 looks excellent (on the CD as Peak1(5; 6).xls). The VVK is also 0.57% at peak 2 but all analysis values have a systematic deviation of approximately +1.8%.

A small method coefficient of variance is proof of good precision of the calibration tests but not of the accuracy. A systematic deviation will not be recorded and, even worse, it thus cannot be recognized at all. Multiple measurements on the same sample does not help here either. The strategy used for the simulation

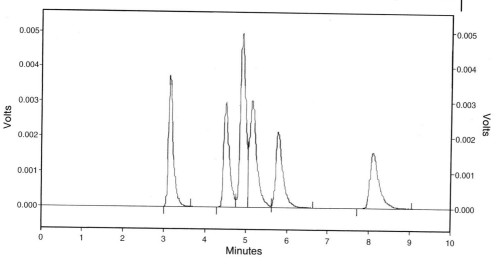

Figure 5.11 Chromatogram of Analyse3.
The peak areas 2, 3, 5 and 6 were determined with the random function.

is also possible in real samples, namely to hold one concentration constant and vary the others.

>   *A wrong integration is not also mirrored in the calibration.*

For peaks 3 and 4 in particular, this can be seen, where the straight line calibrations are not visually different from the other straight lines (Fig. 5.12). The set concentrations, the concentrations calculated for the straight line calibration and the proportional deviation are shown in Table 5.1. The error observed is complicated but systematic. The size of the area of peak 3 is mainly responsible for the error in both peak areas. At best, the area that is cut off could be given back to the third peak but the fourth peak already contains the correction due to the additional area and can have a positive or negative deviation.

>   *The calibration does not make good an integration error.*

If the analysis samples contain all peaks and – as is seldom the case – the height relationships of the peaks remain the same within narrow limits, then calibration can be performed with a solution containing all of the substances. There must be clear guidelines to the limits within which this is allowed.

Unfortunately, the addition method cannot make good the integration error either. The simulated tests ADD0, ADD1 and ADD2 yield a value for the area of the fourth peak that is 15% too high.

In conclusion, all possible approaches to calibration lead to significant errors due to the insufficient separation of tailing peaks. Chromatograms with peaks similarly merged, with tailing, are presumably evaluated all over the world every

Table 5.1 Six analysis simulations using the simulated calibration.

	Peak 3			Peak 4		
	Target value	"Measured"	%-Deviation	Target value	"Measured"	%-Deviation
Analyse 1	6.08	6.97	14.6	13.51	13.33	−1.3
Analyse 2	13.41	15.70	17.1	13.51	13.87	2.6
Analyse 3	24.23	28.59	18.0	13.51	14.35	6.2
Analyse 4	2.69	2.93	8.9	13.51	12.63	−6.6
Analyse 5	24.11	28.45	18.0	13.51	15.20	12.4
Analyse 6	1.28	1.25	−2.4	13.51	12.27	−9.2

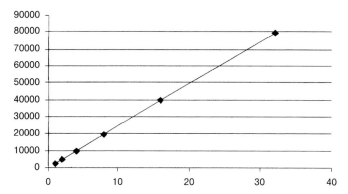

Figure 5.12 Straight line calibration for peak3 with a VVK of 0.58%. The "wrong" slope of 2486 instead of 2220 µV per concentration unit does not lead to a deviation of the linear function.

day even though the errors are too high to be tolereated. The only possibility for improvement is better chromatographic separation or better integratoin of the peaks. The present integration systems that use the perpendicular drop method are not sufficient to solve the everyday problem of the integration of tailing peaks merged together with a valley of 35%.

> *Integration systems that only work with the perpendicular drop method, are badly suited to the integration of merged tailing peaks.*

5.5
Exponential Skim

It is clear that the peaks in the chromatogram shown in Fig. 5.8 should be integrated using the exponential skim. The results are shown in Fig. 5.13 for Chromeleon and Figs. 5.14 and 5.15 for Empower, and are actually an improvement. The results are listed in Table 5.2. The integration looks substantially better when using Empower, although, with Chromeleon a clear improvement is also seen. This example is, however, also particularly suited to the exponential skim because the first peak is not changed up to the valley point. An exponential continuation of the peak course is thus appropriate here. In other cases, if the peak being subjected to the exponential skim sits on another peak, the improvement is less, and possibly even a deterioration can occur in comparison with the perpendicular drop.

A two-peak-deconvolution with the integration system HSM (Merck Hitachi), which has not been available for ten years now, brings about a considerable improvement (Fig. 5.16). This two-peak deconvolution uses the baseline found with the perpendicular drop. Highlighting the peak pairs nine times leads to the areas at the top of the peaks. The target areas are reached in practically every case. The deconvolution is described in detail in Chapter 6.

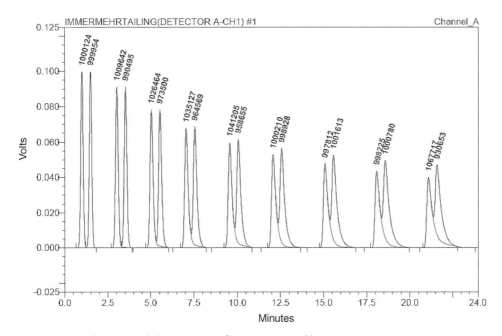

Figure 5.13 The simulated chromatogram of Fig. 5.8 integrated by Chromeleon and exponential skim.

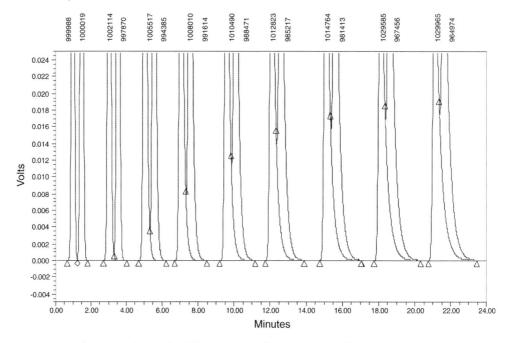

Figure 5.14 The simulated chromatogram of Fig. 5.8 integrated by Empower Traditional Mode and exponential skim.

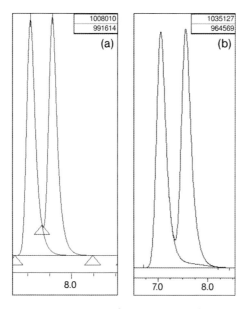

Figure 5.15 Expansion of (a) Fig. 5.14 and (b) Fig. 5.13 to see the fourth peak pair.

Table 5.2 Deviations of the areas and heights with Chromeleon (Fig. 5.13) and Empower (Fig. 5.14).

	Integration of ImmerMehrTailing with exponential skim			
τ/σ	Peak area (%)		Peak height (%)	
	Chromeleon	Empower	Chromeleon	Empower
0	0.01	0.00	−0.27	−0.27
0	0.00	0.00	−0.28	−0.26
0.5	0.96	0.21	−0.24	−0.23
0.5	−0.95	−0.21	−0.27	−0.23
1	2.65	0.55	−0.55	−0.46
1	−2.65	−0.56	−1.66	−0.58
1.5	3.51	0.80	−0.15	−0.11
1.5	−3.54	−0.84	−1.41	−0.42
2	4.12	1.05	−0.13	−0.09
2	−4.13	−1.15	−1.18	−0.67
2.5	0.02	1.28	−0.34	−0.34
2.5	−0.11	−1.48	0.94	−1.16
3	−0.22	1.48	−0.46	−0.42
3	0.16	−1.86	0.67	−1.48
3.5	−0.18	2.96	−0.28	−0.21
3.5	0.08	−3.25	1.14	−2.16
4	6.77	3.00	−0.13	−0.10
4	−6.93	−3.50	−0.51	−2.19

With peaks melted together (see Fig. 5.9) the perpendicular drop cuts the tailing off for the front peak. The exponential skim shall compensate for it. The integration with Empower yields to an increase of the deviations at increasing tailing (Table 5.2). A little part of the area is carried forward from the back peak to the front peak, i.e. an overcompensation. The height of the front peaks is correct, the height of the back peaks marginally too small.

The integration with Chromeleon shows a similar course. The jump in the deviations is astonishing, though. One can just still accept deviations in the height of 1 to 2% at many analyses. The height evaluation with the exponential skim is the best available integration method at tailing peaks not merged too much.

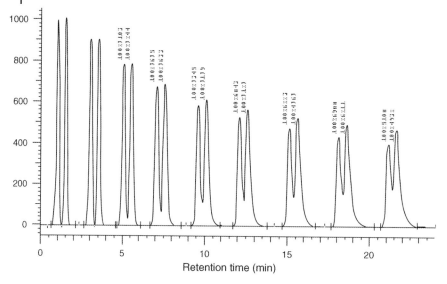

Figure 5.16 The simulated chromatogram of Fig. 5.8 after 2-peak deconvolution with the integration system HSM.

5.6
Integration of Merged Peaks

Integration errors of less than 1% usually disappear within the other errors and are often acceptable, even for the analysis of drug products. Of course they have to be taken into account in measurements where a complete precision of less than 1% must be achieved. If the integration errors are less than 5%, they can usually be tolerated in the analysis of drugs in biological samples. However, attempts must be made to control and reduce them. Errors above 5% usually occur when the integration errors become dominant, and this is not acceptable.

The integration is not a problem for two equal sized Gaussian peaks up to an R of 0.85. There is full compensation for the error – the total area is divided into two equal parts. This only applies exactly to peaks that are also of equal width. In addition, this is valid where there is greater merging, but only when the peak pairs are exactly equal in size. In reality the usual limiting value is sensible for the quantification: $R = 1$.

If the concentration quotient between the peaks is 10, R must be 1.1, to give the same visual separation of a 50% valley. At a concentration quotient of 100, $R = 1$ means there is a shoulder for the smaller peak. In this case R must be 1.275 at once. Namely 1.425 must still be the resolution, at a concentration quotient of 1000.

With tailing peaks it becomes a little more difficult because of the extent to which the tailing plays a role. An A_f of 1.2 ($T_f = 1.1$) does not characterize any visually recognizable tailing. The integration is similarly not a problem, as in the case of Gauss peaks and the above-mentioned limits can be used.

The tailing is clearly recognizable but moderate up to an A_f of 1.7 ($T_f = 1.35$). The higher the valley, the worse the integration result with the perpendicular drop method. A valley of 10% should not be exceeded for an A_f of 1.7. The exponential skim can help at a higher valley. At the latest at a valley of 50% this method is also exhausted.

A tailing of $A_f > 3$ ($T_f > 2$) is like a chromatographic declaration of bankruptcy. When a quantitative evaluation is required, the tailing must not be too significant. If no better chromatographic solution can be found, only free standing peaks that are integrated correctly should be used.

Assessing of the resolution also becomes more difficult with tailing peaks. In every case the concentration quotients have to be taken into account as described above. With regard to tailing peaks there is a difference depending on whether the smaller peak stands in front or comes second. The resolution yields the same value for the same retention time difference.

A much simpler measure for the quality of a separation is the valley height between two peaks, as suggested by Christophe [4]. The European Pharmacopoeia (cp. Chapter 15) allows the use of a reciprocal quotient as a characteristic quantity. This says the same thing but is less descriptive.

All (four) data systems can integrate with both the tangent method and the perpendicular drop. Only in a few special cases does the tangent method offer any advantages. Therefore it should be turned off for the integration of each chromatogram (at time 0). If it is known to be quite safe to integrate a peak with the tangent method, it can be switched on again. Thus this leaves only the perpendicular drop method, which determines the minima in addition to the peak maxima between the peaks.

Of the four data systems used here Chromeleon and Empower offer the exponential skim, which is sometimes advantageous. The exponential continuation of the available peak course is not a universal remedy and is difficult if this function must be assigned to every peak one by one. However, it is an option one should not do without when making a new purchase or changing a data system.

5.7
Integration in Daily Practice

In the report besides t_R, area and height one should have $w_{50\%}$ and $T_{f10\%}$ ($A_{f10\%}$) printed out. If these values are missing, there should be evidence that the valley exceeds the height threshold of 10% or 50% on one of the two sides of the peak. The tailing factor yields the degree of the tailing and with that the threshold demanded. Some recommendations are given in Table 5.3, not to exceed an error of 5% for the area.

It is generally advantageous to use the peak height instead of the peak area. There are only a few cases where the area would be preferred. Each precision test should be done by area and height simultaneously. Then a decision between using the area or height calibration can be made.

Table 5.3 Recommended limits of the valley heights for acceptable integration (error not exceeding 5%).

Tailing	$A_{f10\%}$	$T_{f10\%}$	τ/σ	Maximum valley for drop method (%)
Minimal	1 to 1.2	1 to 1.1	0 to 0.7	50
Considerable	1.2 to 1.7	1.1 to 1.35	0.7 to 1.5	10
Strong	1.7 to 3.5	1.35 to 2.25	1.5 to 4	3
Too great	> 3.5	> 2.25	> 4	0

At minimal tailing ($1 < A_f < 1.2$) the same restrictions are valid, as in the case of merged Gaussian peaks up to a valley of 50%: The integration with the perpendicular drop is sufficiently. Of course one should, if possible, try a follow-up chromatogram simulation to find the weak points and to optimize the integration parameters. The integration error is presumably dominated by the uncertainty of the baseline in complex chromatograms and not by the tailing.

The valley should only be 10% with considerable tailing so as not to produce too big an integration error with the perpendicular drop. The integration can, if possible, be checked in a chromatogram simulated individually with the exponential skim for valleys of between 10 and 50%.

The valley may still be only few per cent with strong tailing to use the perpendicular drop. For a valley up to 10% the exponential skim can bring an advantage. However, the possibilities with today's CDS are exhausted when using this.

For higher fusion of tailing peaks (or considerably tailing peaks with a valley over 50%) a sensible integration can only be carried out with the deconvolution described in the next chapter.

References

1   J. P. Foley, J. G. Dorsey, Equations for Calculation of Chromatographic Figures of Merit for Ideal and Skewed Peaks, *Anal. Chem. 55* (1983) 730–737.

2   S. L. Tamisier-Karolak, M. Tod, P. Bonnardel, M. Czok, P. Cardot, Daily validation procedure of chromatographic assay using gaussoexponential modelling, *J. Pharm. Biomed. Anal. 13* (1995) 959–970.

3   N. Dyson, *Chromatographic Integration Methods*, 2nd ed., The Royal Society of Chemistry, London, 1998.

4   A. B. Christophe, Valley to Peak Ratio as a measure for the separation of two chromatographic peaks, *Chromatographia 4* (1971) 455.

6
Deconvolution

Mike Hillebrand

6.1
Introduction

During the development of an optimal chromatographic method, the aim is for
the various components in a sample to all elute from the baseline separately.
However, in the fields of biotechnology and natural product chemistry this is
frequently not possible. If substances produced by micro-organisms are, under
certain circumstances, modified by enzymes, the by-products are usually very
similar to the primary product and therefore demonstrate similar characteristics
during the chromatographic process. This can lead to the peaks not being completely separated but being merged.

There are only a few methods vailable for the separation of merged peaks as
part of the integration process. In practice, the most frequently used are the
perpendicular drop and the tangential skim methods. Chapter 3 pointed out the
enormous errors that can arise with these two methods, especially in cases of
asymmetric peaks, as Chapter 5 demonstrated. Some integration systems allow
the use of the exponential skim and Gaussian skim methods. These two methods
also lead to large errors because they work through a graphical extrapolation of
the peak, without considering any shift of the sum curve. Westerberg pointed out
the inadequacies of the perpendicular skim method 40 years ago and suggested
a curve approximation [1]. However, today the perpendicular skim method is still
the most commonly used method for the separation of merged peaks. There are
not only errors in the areas produced by the skim methods but also errors in the
amplitudes and retention times resulting from the merge of the peaks. Figure 6.1
shows five merged peaks. The bold curve is the sum curve which you will see in
a chromatogram and is generated from the five single compound curves beneath
it. At peak number 2 the shifts of the retention time and the amplitude are clearly
visible.

Because the chromatogram with merged peaks (the sum curve), represents
the sum of the single peaks, the application of a curve approximation is obvious.
Deconvolution, as a mathematical method, signifies a curve approximation by

Quantification in LC and GC: A Practical Guide to Good Chromatographic Data
Edited by Hans-Joachim Kuss and Stavros Kromidas
Copyright © 2009 WILEY-VCH Verlag GmbH & Co. KGaA, Weinheim
ISBN: 978-3-527-32301-2

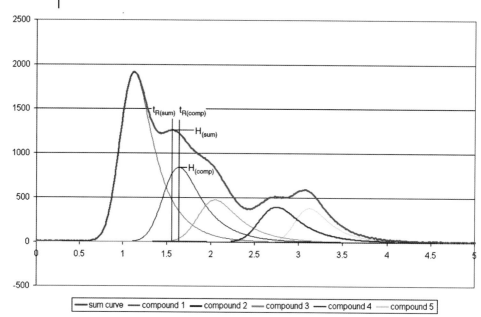

Figure 6.1 Example of a chromatogram with merged peaks
(bold curve: sum curve; thin curves: curves of the single compounds).

means of theoretical models. In other words, the sum curve of a chromatogram is divided with the help of theoretical peak models back into the individual addends, i.e. the single peaks of the components (Fig. 6.1).

The shape of the peaks eluted separately from the baseline is known. This can be described by theoretical models, depending on the symmetry. The most common is the Gaussian peak model Eq. (6.1) [2], for symmetrical peaks.

$$y = \frac{A}{\sqrt{2\pi}\,\sigma} \exp\left[-\frac{1}{2}\left(\frac{t - t_R}{\sigma}\right)^2\right] \tag{6.1}$$

Where y is the value of the signal at time t, A is the peak area, t_R the retention time and σ represents the standard deviation of the peak. This equation is based on a normal distribution and represents the ideal peak of an injected solution that partitions linearly between the stationary and mobile phases with no instrumental distortion.

Often the peaks do not have this ideal symmetry. Since 1959, there has been much effort devoted to finding a better peak model [2]. The exponentially modified Gaussian model (EMG) is the most commonly used model for asymmetrical peaks. The EMG equation is the product of a constant amplitude term, an exponential term, and the integral of a Gaussian function, Eq. (6.2), where t is the time, t_R the maximum (retention time) of the Gaussian component, σ is the standard deviation of the Gaussian component and τ is the time constant of the exponential function [3]. The ratio τ/σ gives the value for the degree of tailing.

$$f^{EMG}(t, t_R, \sigma, \tau) = A \exp(B) \, C(t, t_R, \sigma, \tau) \tag{6.2}$$

If the normalized form is taken, the terms A, B and C are given by Eqs. (6.3) to (6.5) [3].

$$A = \frac{1}{2\tau} \tag{6.3}$$

$$B = \left[\frac{1}{2} \left(\frac{\sigma}{\tau} \right)^2 - \frac{t - t_R}{\tau} \right] \tag{6.4}$$

$$C(t, t_R, \sigma, \tau) = \mathrm{erf} \left[\frac{1}{\sqrt{2}} \left(\frac{t_R}{\sigma} + \frac{\sigma}{\tau} \right) \right] + \mathrm{erf} \left[\frac{1}{\sqrt{2}} \left(\frac{t - t_R}{\sigma} - \frac{\sigma}{\tau} \right) \right] \tag{6.5}$$

The convolution integral, which results from the fundamental mathematical definition of the EMG function as a convolution of the Gaussian function with an exponential decay, is described by Eq. (6.6). Many equations are available for the expression of the EMG function, Di Marco et al. [4] showed 18 different ones. Equation (6.6) is the most common form in the literature.

$$f^{EMG}(t) = \frac{1}{\tau \, \sigma \, \sqrt{2\pi}} \int_0^t \exp \left[-\frac{(t - t_R - t')^2}{2\sigma^2} \right] \exp \left(-\frac{t'}{\tau} \right) dt' \tag{6.6}$$

The statistical moments, M_n, for Eq. (6.6) can be obtain by the Fourier transformation [5].

The zeroth moment M_0, the area, is given directly by the equation if present in the non-normalized form. The first moment M_1, the retention time measured at the center of gravity, is described by Eq. (6.7).

$$M_1 = t_R + \tau \tag{6.7}$$

The peak variance, the second moment M_2, is given by Eq. (6.8); Eq. (6.9) describes the vertical asymmetry or skew, the third moment M_3; and the fourth moment, the excess, is obtained from Eq. (6.10).

$$M_2 = \sigma^2 + \tau^2 \tag{6.8}$$

$$M_3 = 2\tau^3 \tag{6.9}$$

$$M_4 = 3\sigma^4 + 6\sigma^2 \tau^2 + 9\tau^4 \tag{6.10}$$

If n components are merged in a chromatogram, the data can be described by Eq. (6.11) [6].

$$D(t) = c_1 \, f_1^{EMG}(t) + c_2 \, f_2^{EMG}(t) + \ldots + c_n \, f_n^{EMG}(t) \tag{6.11}$$

Many studies have dealt with the EMG function and its applicability to the chromatographic process. It has been shown that various dead volume contributions alter the peak exponentially, producing varying degrees of tailing [6, 7]. Each particular extra-column effect has a certain time constant τ associated with it. The external dispersion is mainly responsible for the tailing. Naish *et al.* [8] showed that the exponential component is not independent of the retention time and assumed an addition of the chemical and physical tailing in the exponential term.

Many studies have shown that the EMG model gives good results [8–10]. Foley [9, 11] found that over 90% of the real data analyzed were valid. Papas [12] noted that, nevertheless, 10% of the peaks are badly characterized by the EMG model. Dyson [2] added that the EMG model is best applied to LC peaks, especially isocratic ones. Capillary GC peaks, but not all packed column GC peaks, are well served by the EMG model.

The EMG model is today the most accurate and applicable peak model, especially when there are asymmetrical peaks present. This model is based on a greater theoretical background and definitely produces better accuracy than the skim methods. Therefore it is used in this chapter to deconvolute merged peaks. Nevertheless, attend that there exist much more models, that can also be used for deconvolution.

6.1.1
Software Applications for Deconvolution

Today's chromatographic data systems (CDS), including the most well known and most expensive ones, do not possess an application for deconvolution. The company Systat Software Inc. has an analysis software (no CDS) called PeakFit, which offers the possibility of detection and separation of peaks, as well as curve approximation. Chromatograms are read in as AIA, ASCII or excel files and can be worked on relatively simply. In the first step the baseline is defined, either manually or automatically. If desired, the data can be smoothed and afterwards the theoretical peak model to be deconvoluted is selected.

Different theoretical models can be selected for curve approximation from spectroscopy and electrophoresis as well as chromatography. For chromatography, many theoretical models are available to illustrate the various peak shapes including exponentially modified Gaussian, half-Gaussian modified Gaussian, Haarhoff–Van der Linde, Giddings, non-linear chromatography, EMG–GMG hybrids, log normal 4-parameter or the extreme value 4-parameter model. The Gaussian is the basis model. It is possible to approximate single peaks in a chromatogram by means of different models. Thus an approximation of symmetrical peaks can be made using the Gaussian model and that of asymmetrical peaks using, for example, the exponentially modified Gaussian model.

The procedure for the detection of hidden peaks depends on the selected mode. One can choose from the residual method, the second derivative method and the deconvolution method. The residual method determines the residuals that result from the difference between the single peaks and the sum curve. The peaks that the residuals were compensating for can then be found. In the second derivative method the smoothed second derivative of the chromatogram is determined. The local minima reveal the hidden peaks. This is the same way as the apex track in the CDS Empower works [13]. The deconvolution method is a complex mathematical procedure to detect hidden peaks [14]. Peaks can be manually added, worked on and deleted. Their utilization in a GMP regulated area, such as the pharmaceutical industry, is at present impossible because, for example, you can change every datapoint from the imported chromatogram and a full audit trail does not exist.

In the report, PeakFit affords, in addition to statistical moments and adapted determinants, chromatographic parameters, such as the number of theoretical plates (certainly by means of the moment theory as well as determination with the Gaussian peak model), the width at the base, asymmetry at 10% height and the resolution.

Which adapted parameters are reported depends on the peak model selected for the approximation. Usually these are the area or amplitude, the retention time, σ and τ.

This program is, however, only an analysis software not a CDS. For the user it would be handy and extremely meaningful to have a software which includes all application possibilities for data capture, processing by curve approximation and reporting.

The first application of this type was developed in the 1990s with the D-7000 HPLC system manager (in the following abbreviated to HSM) by Merck KGaA and Hitachi Ltd. This CDS has a function in version 4.1 that allows the deconvolution of two merged peaks. The curve approximation is performed with HSM using the exponentially modified Gaussian model. The application of the deconvolution is limited to pairs of peaks. The Marquardt non-linear least-squares method is used to deconvolute the two merged peaks [15].

The application is very simple, you just have to mark the pair of peaks in the mode "2d-peak deconvolution" with the mouse and the results are promptly supplied.

As with CDS, HSM reports the chromatographic parameters to the user in the desired way. HSM supplies, along with the application of the approximation using the exponentially modified Gaussian model, the parameters area, retention time, the width (σ) and the distortion (τ) as well as the statistical moments measured from them.

6.2
Influences on Deconvolution

The influences on chromatography and on chromatographic peaks, as the Ishikawa diagram [16] indicates, are very complex and comprehensive. Some of these also affect the quantification. If there exist merged peaks the quantification will be incorrect, except you have the luck that the errors compensate each other. Regardless of which method is used to separate them, whether the skim methods or deconvolution, an error will always exist, but this error should be as small as possible. In order to be able to use and to evaluate the deconvolution and the skim methods, the operator must know where to find the potential sources of error and their influence on the quantification.

6.2.1
The Baseline Path

Knowledge of the path of the baseline is the basis for the quantification of chromatographic peaks. The method parameters give a guide to its path. The almost exact path can be determined from a blank, i.e. a sample containing the matrix being used, but without the analyte. One must be sure that there is no carryover from the system, e.g. from the injection system, hence the blank should not be run directly after a sample. The chromatogram of the blank is nearly identical with the baseline.

Whether an isocratic or gradient run is performed should also be given in the method parameters. If only the chromatogram is given then it is difficult to determine whether the run was isocratic or gradient if the gradient is overlaid by peaks. In such a case there is the possibility of determining this from the resolution. As shown in Chapter 3, there is baseline separation between two peaks if the resolution is equal to or larger than 2. If the peaks are on a gradient and are therefore not merged, the resolution is almost 2. In this case the operator will know where the baseline is.

Knowledge of the presence of a drift of the baseline is also important. Baseline drift means that in, or in parts of, the analysis, the path of the baseline rises or falls linearly. If you are not able to recognize this, because of peak merge of the baseline, the blank will also help here. Baseline irregularities are relatively common.

The influences, mentioned above, are to be recognized and evaluate only if the baseline is stable. That means that the baseline runs over an analysis sequence constantly. In order to ensure this, it is important to adequately equilibrate the system.

The path of the baseline must be known!

6.2.2
Peak Homogeneity – Number of Peaks

Peak homogeneity in the case of merged peaks refers to peaks that are so well hidden that they cannot be recognized as single peaks. They are "swallowed" in the sum curve.

The presence of hidden peaks does not lead to greater problems but their existence must be known for a correct deconvolution to be achieved. If it is unclear whether or not hidden peaks are present under the visible peaks there is a possibility, as described in Chapter 11, of determining them by MS-coupling. If it is known that there are hidden peaks then they must be considered in the deconvolution.

The number of peaks as well as the peak homogeneity must be known!

6.2.3
Noise

There are different types of noise [2]: short-time or high-frequency noise and long-term or low-frequency noise. The short-time noise is easy to determine and, due to its origins, is identical over the whole chromatogram. It leads to an error in the quantification that is larger if the amplitude of the peak is smaller i.e. if the signal to noise ratio is smaller. The limit of quantification (LOQ), depending on the regulatory literature, is at a signal to noise ratio of 10:1 [17] to 50:1 [18]. This is the limit for an acceptable error in the quantification that results from the noise.

The long-term noise leads to larger errors with the use of deconvolution. It is difficult to separate it from the baseline progress.

The signal to noise ratio must be known and should not go below the LOQ!

6.2.4
Peak Symmetry

For the accurate application of deconvolution it is important to know the symmetry of the peaks. It is possible to determine this with more or less precision for peaks that have a maximum and that have a minimum at 50% of the visible peak. The measure of the symmetry is more accurate when the width is determined as close as possible to the baseline. The nearer the measurement is made to the maximum of the peak, the more inaccurate it is. If no peak maximum is present, measuring the symmetry without further experiment is difficult. It is possible to analyse the individual component with the relevant chromatographic method at an appropriate concentration to determine the symmetry of the peak.

The peak symmetry should be known!

6.3
Deconvolution of Gaussian-Shaped Peaks

As an introductory example a simulation 1, Fig. 6.2, was developed with MS-Excel (details are given in Section 3.3), which consists of three Gaussian-shaped merged peaks. The area ratios of the peaks are 2:1 for peaks A and B, and 2:1 for peaks B and C. The simulation is performed without noise, baseline drift or other influences. The target areas of the simulation are, for peak A 60 000 $\mu V\ s^{-1}$, for peak B 30 000 $\mu V\ s^{-1}$ and for peak C 15 000 $\mu V\ s^{-1}$.

The three peaks are separated with the perpendicular, the tangential and the exponential skim methods. The deconvolution with application of the Gaussian peak model was accomplished by PeakFit. Table 6.1 shows a comparison of the area results in $\mu V\ s^{-1}$ and the deviation from the target area. The area can be used to check the fit [9], which will be shown in this chapter. This is possible because

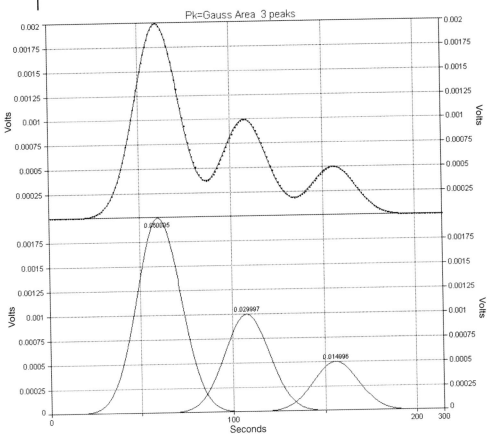

Figure 6.2 Simulation 1 with Gaussian-shaped peaks (PeakFit). The upper trace is the sum curve and the lower trace the single peaks of the approximation.

the target areas are known exactly. The other parameters σ, τ and t_R could also be used, but to minimize the data the area is used to show the goodness of the fit.

The resolution of the peaks is relatively high so they are not hidden too much.

The perpendicular skim method results in a deviation from the target area for peak A of 0.8% because it gains area from peak B. Peak B loses more to peak A than it gains from Peak C so a negative deviation of –0.8% results. Peak C loses 1.6% of its area to peak B.

The purpose of certain separation methods must of course be examined. The application of the tangential skim method does not make sense in a situation where peaks possess this peak pattern, and results in disastrous deviations. Peak A is assigned an area that is 32.0% too large and which it gains from Peak B and Peak C. This effect weakens from peak to peak. Thus the deviation of Peak B is –45.4% and of Peak C is –37.1%.

Table 6.1 Results from simulation 1 with Gaussian-shaped peaks. Comparison of the deconvolution, the perpendicular, tangential and Gaussian skim methods.

	Target area	Decon-volution (PeakFit)	Devia-tion (%)	Perpen-dicular skim	Devia-tion (%)	Tangen-tial skim	Devia-tion (%)	Gauss-ian skim	Devia-tion (%)
Peak A	60000	60005	0.01	60467	0.78	79184	31.97	62075	3.46
Peak B	30000	29997	−0.01	29762	−0.79	16367	−45.44	28977	−3.41
Peak C	15000	14996	−0.02	14761	−1.59	9438	−37.08	13937	−7.09

Despite the use of symmetrical peaks, the peaks that are separated by the Gaussian skim method also show large deviations. This is because this algorithm extends the existing peak, at the valley of the merge, Gaussian-shaped. That the merge rises causing an increase in the peaks (see Fig. 6.1) is thus ignored. The deviation for Peak A with this method is 3.5%.

In addition, this skim method is only considered if the focus is just on one peak and not its neighbors. High errors are obtained because of the senseless progress of the skim. This is shown for Peak B in Fig. 6.3, which presents the results with the Gaussian skim. The deviation here is 3.4%. In this case the third peak was

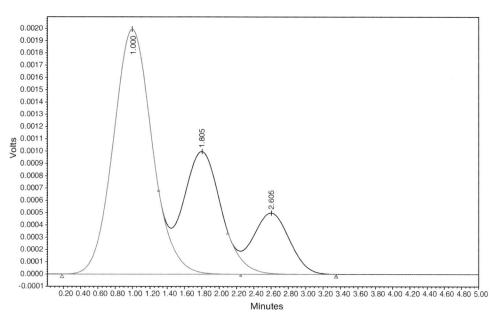

Figure 6.3 Simulation 1 with Gaussian-shaped peaks processed with the Gaussian skim method (Empower).

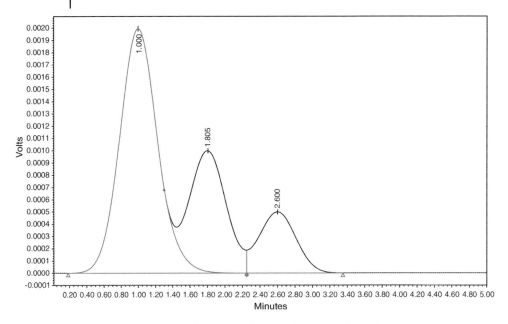

Figure 6.4 Simulation 1 with Gaussian-shaped peaks processed with the Gaussian and perpendicular skim methods (Empower).

also subjected to this senseless type of skim. If you have this pattern it is better to skim the third peak with the perpendicular skim (Fig. 6. 4). It will still look bad, but the deviation of 1.6% is better than 7.1%.

Note, that the resolution is reciprocally proportional to the errors resulting from the skim methods. If the example was chosen with lower resolutions, the results were much more dramatic for the skim methods.

The approximation with the Gaussian peak model leads to deviations to the target area from 0%. The target areas of all three peaks were found nearly exactly. The results from the deconvolution are, with distance, the best.

6.3.1
Westerberg Report Simulation

In the work of Westerberg 1969 [1], mentioned in the introduction, he reported a real isocratic chromatographic run. By using the parameters retention time, area and σ, a simulation using the Gaussian model was carried out. The retention time and the area are given in the report. σ must be determined by Eq. (6.12) [19] using the area and the amplitude.

$$\sigma = \frac{A}{H \cdot \sqrt{2\,\pi}} \tag{6.12}$$

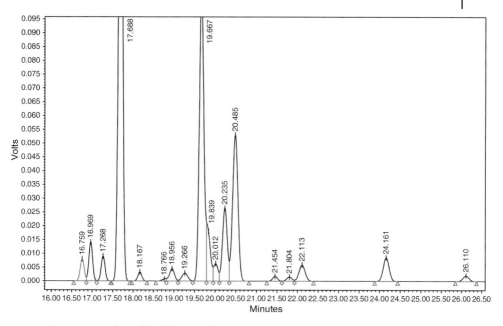

Figure 6.5 Simulation from the Westerberg report processed with the perpendicular skim method (Empower).

The added noise is 10 µV. The chromatogram (Fig. 6.5) contains 18 peaks that are partially heavily merged. Peaks 6, 10 and 11 are shoulder peaks with very little resolution.

Figure 6.5 shows this chromatogram subjected to the perpendicular skim method. The results are bad for the peaks that have little resolution. Thus peak 6 lost 15.0% of its area to peak 7, which gained 3.4% in area. Peak 9 gains 6.8% when using this skim method. This gain is the result of a loss to peak 10 of 26.4%. The deviation of peak 11 from the target area amounts to –11.3%. In cases where the peaks are baseline separated and therefore the skim method does not come into play the deviation is close to 0%. The individual results are summarized in Table 6.2. The results of the approximation with the Gaussian method are also included. Using this latter method, in all instances, whether the resolution of the peaks is small or whether they are baseline separated, there is a deviation of 0%.

The application of the deconvolution supplies the smallest deviation from the target area. This is almost 0% for all peaks. This result is definitely better than those from the separation methods, but the peaks in the simulation are ideal symmetrical ones. How does it look if the peaks are asymmetrical, for example with tailing?

Table 6.2 Simulation from the Westerberg report subjected to with the perpendicular skim method (Empower) and deconvolution (PeakFit).

Peak	Target area	Perpendicular skim	Deviation (%)	Deconvolution (PeakFit)	Deviation (%)
1	947.35	935.92	−1.21	947.41	0.01
2	1544.05	1555.42	0.74	1544.09	0.00
3	1015.70	1015.32	−0.04	1015.74	0.00
4	22941.67	22941.50	0.00	22941.59	0.00
5	395.37	395.15	−0.05	395.17	−0.05
6	132.73	112.78	−15.03	132.78	0.04
7	621.22	642.32	3.40	621.15	−0.01
8	497.73	495.87	−0.38	497.57	−0.03
9	12795.50	13661.27	6.77	12795.46	0.00
10	3019.67	2223.55	−26.36	3019.82	0.01
11	881.40	781.50	−11.33	881.27	−0.01
12	3583.83	3555.65	−0.79	3583.79	0.00
13	8130.83	8188.82	0.71	8130.90	0.00
14	245.20	244.82	−0.16	245.29	0.04
15	236.33	237.45	0.47	236.38	0.02
16	1040.27	1038.87	−0.13	1040.23	0.00
17	1467.47	1467.15	−0.02	1467.42	0.00
18	367.75	367.42	−0.09	367.74	0.00

6.4
Deconvolution of Peaks with Tailing

To arrange the peaks more realistically and also to show the influence of the noise, simulation 2 is provided, Fig. 6.6, which possesses three hidden peaks. These peaks are developed by means of the Exponentially Modified Gaussian model. The area ratios of the peaks here are also 2:1 for peaks A and B, and 2:1 for peaks B and C. The target areas of the simulation are, for peak A 60 000 $\mu V\ s^{-1}$, for peak B 30 000 $\mu V\ s^{-1}$ and 15 000 $\mu V\ s^{-1}$ for peak C.

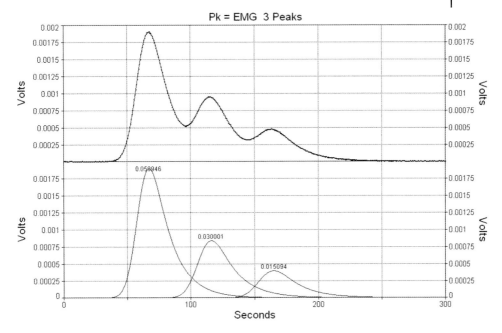

Figure 6.6 Simulation 2 – peaks with tailing (PeakFit). The upper trace is
the sum curve and the lower trace the single peaks of the approximation.

Table 6.3 Results from simulation 2 – peaks with tailing – comparison of the
deconvolution and the perpendicular, tangential and exponential skim methods.

	Target area	Decon-volution (PeakFit)	Devia-tion (%)	Perpen-dicular skim	Devia-tion (%)	Tangen-tial skim	Devia-tion (%)	Expo-nential skim	Devia-tion (%)
Peak A	60000	59946	−0.09	54820	−8.63	87318	45.53	64040	6.73
Peak B	30000	30001	0.00	31487	4.96	11516	−61.61	23097	−23.01
Peak C	15000	15094	0.62	18711	24.74	6186	−58.76	17882	19.21

The three peaks were separated with the perpendicular, tangential and exponential skim methods. The approximation with the application of the exponentially modified Gaussian model by PeakFit was accomplished here. Table 6.3 shows a comparison of the area results in $\mu V\ s^{-1}$ and the deviation from the target area.

Application of the perpendicular skim method to peaks with tailing results in much bigger deviations than if Gaussian-shaped peaks were present. The reason is that the drop cuts the tail of the peak off. So the area of peak A is about 8.6% too small, and although it gains some area from peak B, the loss of the tail is much larger. This tail from peak A is added to peak B, which simultaneously loses its tail to peak C. Because the area of peak A is twice that of peak B, it results in a deviation of 5.0%. Because of the small area ratio of peak C the gain from the tail of peak B leads to a deviation of 24.7%.

In this example, the tangential skim method is also the worst, even though we have exponentially modified Gaussian peaks, the area ratios are not close to the ratios needed to make the tangential skim method of interest. The skim method is only of interest in rare special cases. Here the deviation for peak A is 45.5%, for peak B –61.6% and for peak C –58.8%.

The Gaussian skim method is not a suitable choice for peaks with tailing. However, there is also a skim method that works through graphical extrapolation of the peaks – the exponential skim method (Fig. 6 7). The reason for the error in the case of the Gaussian skim method also applies here. The algorithm extends the still existing peak, at the valley of the merge, exponentially. That the merge rises and causes increase in the peaks (see Fig. 6.1) is thereby ignored. The deviation for peak A based on this is 6.7%. As for the Gaussian skim method, the use of the exponential skim method only makes sense if the focus is on one single peak and not its neighbors. Peaks B and C were separated using the perpendicular skim method. Their deviations are –23.0% for peak B and 19.2% for peak C.

It can be seen here, that the deviations of the deconvolution using the exponentially modified Gaussian model are clearly smaller than those of the skim methods. The deviation for the deconvolution is almost near to 0% for all peaks.

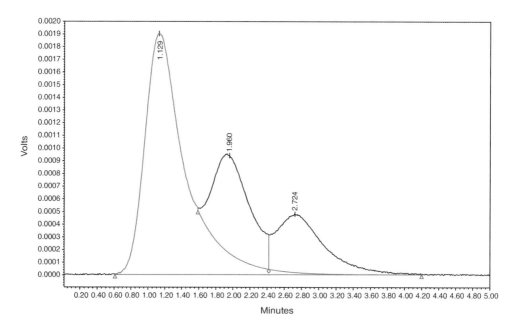

Figure 6.7 Simulation 2 – peaks with tailing, processed with the exponential and the perpendicular skim methods (Empower).

6.4.1

Peaks with Different Tailing

This example demonstrates that the application of deconvolution can be very favorable for cases of peaks with tailing. The realistic situation is that peaks do not tail with the same intensity every time; the exponential decay can differ. This results from the different behavior of the various components in the chromatographic process. The question is whether or not deconvolution will be able to accomplish this approximation successfully as well?

To examine this influence on the results of the deconvolution, a further example (Fig. 6.8) with seven pairs of peaks all with different tailing was provided (according to the possibility of simulating exponentially modified Gaussian peaks described in Section 3.3). The area ratio of the first five pairs is 1 to 1 and of the last two pairs 1 to 0.1. The target area is given as 100 000 $\mu V \, min^{-1}$ and for peaks 13 and 14 as 10 000 $\mu V \, min^{-1}$. The deviation from the target area of the peaks can be read off directly from the chromatogram.

The first pair of peaks have Gaussian shape. Pairs two and three have more tailing (exponential decay of 0.9 and 0.95). The peaks of pairs 4 and 5 have different exponential decay. The first peak of pair 4 has an exponential decay of 0.9 and the second a decay of 0.95. In pair 5 the exponential decay is reversed. The

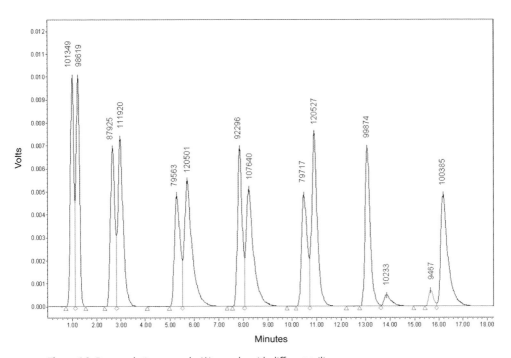

Figure 6.8 Deconvolution example (A) – peaks with different tailing, processed with the perpendicular skim method.

Table 6.4 Response and exponential decay values from the deconvolution example (A) – peaks with different tailing.

Peak pair	1		2		3		4		5		6		7	
Exp. decay	0.0	0.0	0.9	0.9	0.95	0.95	0.9	0.95	0.95	0.9	0.95	0.9	0.95	0.9
Response	1	1	1	1	1	1	1	1	1	1	1	0.1	0.1	1

exponential decay of the peaks of pairs 6 and 7 is identical with that of pair 5. These results are clearly represented in Table 6.4.

Figure 6.8 shows the effect on the example of processing with the perpendicular skim method. The error increases when the tailing of the peaks increases. It goes from a deviation of 1.3% in peak pair 1, which are Gaussian-shaped peaks, to 20.5% in peak pair 3 which has a decay of 0.95.

If the peak with the larger tailing is the pre-peak, there is a larger deviation, because the drop cuts the tail off, as mentioned above. So the deviation in peak pair 5 is 20.5%, much larger than the deviation from the target area of 7.6% for peak pair 4, where the first peak has the lesser tailing.

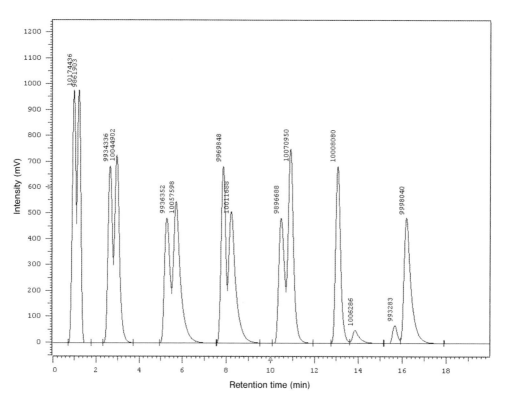

Figure 6.9 Deconvolution example (A) – peaks with different tailing, processed by HSM.

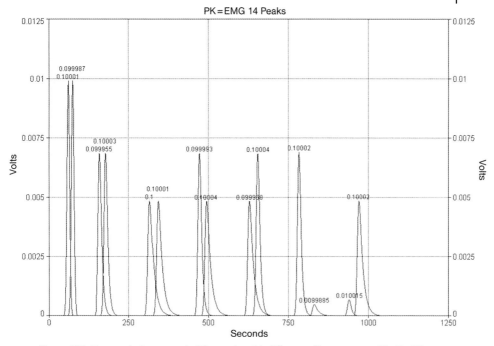

Figure 6.10 Deconvolution example (A) – peaks with different tailing, processed by PeakFit.

When the area ratio is 1 to 0.1 and the first peak in the peak pair has the larger tailing, with a decay of 0.95 to 0.9 for the post-peak, the first peak loses area. This is always the same on application of the perpendicular skim method, when peaks are separate and the first one has tailing. Independent of the size ratio, the first peak loses area.

The deconvolution (here initially with HSM in Fig. 6.9) clearly shows smaller deviations than the perpendicular skim method. The maximum deviation is 1.7% for peak pair 1, which are the Gaussian-shaped peaks, because the curve approximation is performed with HSM by the exponentially modified Gaussian model. All other peaks with tailing, in addition to the small peaks of peak pair 6 and 7, have deviations less than 1%.

The curve approximation was also done with Peakfit, see Fig. 6.10, in order to compare both programs. The area results are indicated on the peaks in V min^{-1}. For the deconvolution the exponentially modified Gaussian model was used. From the algorithm used by PeakFit results a deviation of 0% was found in all cases. A connection between the various degrees of tailing and the size of the deviation is not seen. The deconvolution is significantly favored over the perpendicular skim method.

In Fig. 6.11 the peaks for example (A) are pushed much closer together to provide examples (B) and (C) that much larger overlays. By use of the perpendicular skim method the deviations for the peaks with tailing are doubled from example (A) to example (C), reaching for example almost 40% for peak pair 5 in (C).

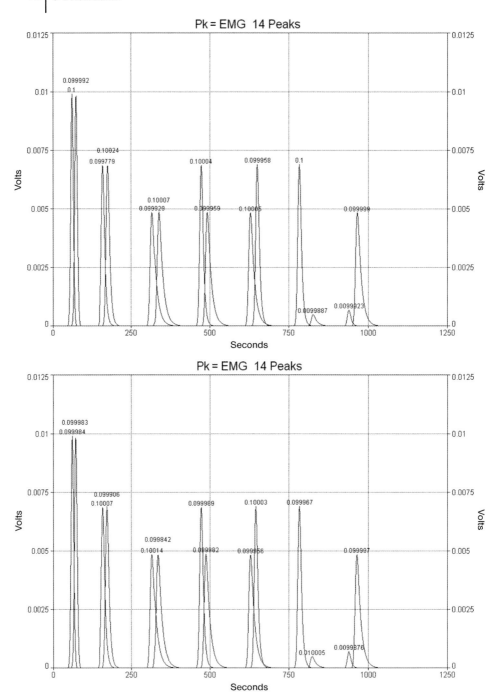

Figure 6.11 Deconvolution examples (B) and (C) – peaks with different tailing, processed by PeakFit.

Also here, as in example (A), the approximation with both programs leads to very small deviations, in all cases 0%. A connection between the extend of the merge and the deviation is not seen.

6.5
Deconvolution of Gradient Runs

As described in Section 6.2.1, knowledge of the path of the baseline is important. If an isocratic method is not used the parameter of the gradient determines the baseline. Information about the effect of the gradient could be obtained either from the method parameters or the blank.

If a gradient, such as that in Fig. 6.12a is applied to the deconvolution example (A) the chromatogram in Fig. 6.12b is obtained.

The exposure of this baseline can be done in various ways. The common way is to extrapolate the baseline linearly from the start to the end of the peak pair. Because of the path of the gradient, as in the example, the linear extrapolation of a curve leads to an error. PeakFit initially offers the possibility of approximating the baseline with mathematical functions. The range of functions contains all common gradient curves.

The second option is to subtract the blank before application of the deconvolution. From such a subtraction the corrected chromatogram as shown in Fig. 6.13 is obtained. The following deconvolution leads to the same deviations as given in Section 6.4.1 for the example (A), that is 0%.

6.5.1
Complex Baseline

In Chapter 3 a chromatogram that contains three main peaks and a large number of small peaks with concentrations at 0.1–1% of those of the main peaks was discussed, in order to show the use of the resolution to find the baseline. To demonstrate how to handle a complex baseline, an almost identical simulation was developed from this chromatogram, but with fewer parts of the baseline shifts being used, because the handling is still the same. It is a chromatogram with three different areas of baseline drift, but it could also be present if a gradient is being run. There is a total of 26 peaks which were partly strongly merged. The baseline runs negatively.

The first step is always to find the baseline. In this case it is very difficult, because there are clusters of peaks that overlay the baseline. If the operator does not have any additional information, it is possible that they will integrate the chromatogram in the way shown in Fig. 6.14.

If someone treats this chromatogram in this way, the results obtained will be drastically wrong, especially for the peaks appearing in the ninth minute. It is necessary to obtain more information to achieve good results! As in the preceding simulation, the blank could be subtracted. If it is not available as a data file,

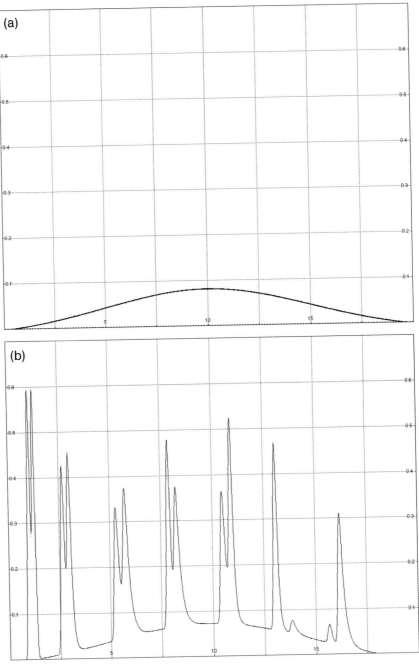

Figure 6.12 (a) Blank of deconvolution example (A) with gradient.
(b) Deconvolution example (A) with gradient.

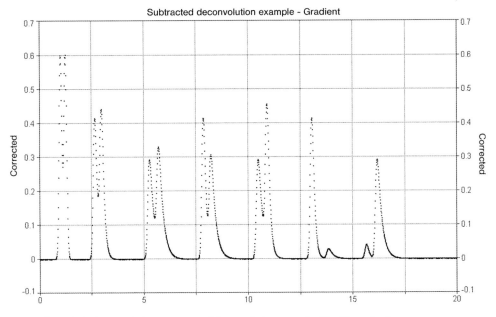

Figure 6.13 Corrected chromatogram of deconvolution example (A) with gradient.

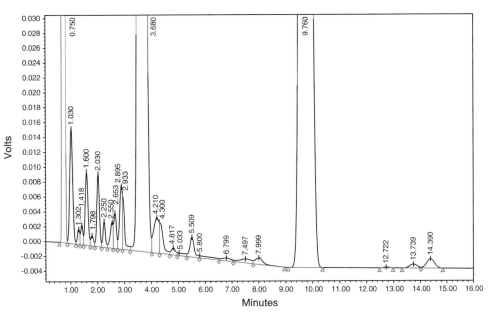

Figure 6.14 The simulation with a complex baseline, without any informations, processed by with the perpendicular skim method (Empower).

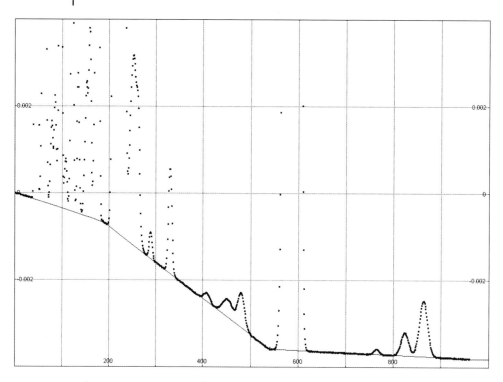

Figure 6.15 Determination of the baseline of the simulation with a complex baseline (PeakFit).

the path, the chromatographic conditions or parameters such as the resolution should provide the missing information, as described in Section 6.2.1. Here the drift runs up to the third minute with a slope of -200 μV min^{-1}, up to the ninth minute with a slope of -500 μV min^{-1} and after that with a slope of -30 μV min^{-1}. Using this information, the baseline can be determined (Fig. 6.15).

After the determination of the baseline, the approximation with the exponentially modified Gaussian model was carried out. The automatic detection and deconvolution algorithm discovered 24 peaks, but the results for the overlay of the sum curve of the approximation and the sum curve of the simulation are not equal. This is because the peaks in the clusters are very well hidden. To compensate for these differences, the approximation must be modified. The two missing peaks are hidden so much that the detection around the second deviation is not successful. Based on this, knowledge of the number of peaks in a chromatogram is essential. The first hidden peak is the smallest in the chromatogram, which is directly under peak 9. The approximation shows no residual, because the peak is so small. A manual addition can only be guessed at. If the focus is on this peak, these chromatographic conditions are wrong anyway. Therefore, it was inevitably that it would be missed out, but it should be kept in mind that peak 9 consists of two peaks. The second hidden peak is peak 22, which is a high, hidden shoulder,

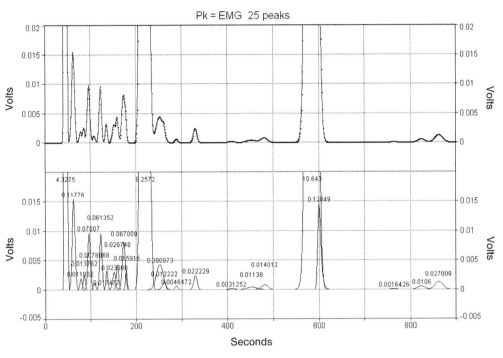

Figure 6.16 Simulation with a complex baseline (PeakFit). The upper trace is the sum curve, the lower trace shows the single peaks of the approximation.

post-peak of peak 21. The residual of the approximation shows that there could be one more peak and if this is correct, (that is the information on the peak number and location!) as in this case, then the missing amount is added.

The complete treatment with PeakFit using the exponentially modified Gaussian model (Fig. 6.16) results in 25 single compounds, with the knowledge that peak 9 contains two peaks. The results are summarized and compared with the deviation from the perpendicular skim.

The deviations from the target area by the deconvolution are under 5% in all instances. The biggest error, 4.3%, is with peak 22, which is the high, hidden shoulder, post-peak of peak 21. Also, in intervals where many peaks are clustered and highly merged, as in the fifth minute, the approximation obtains small deviations, which is shown in the right-hand column of Table 6.5. The deconvolution leads to larger errors when the peaks are hidden so much that they have no maximum. These errors are considerably smaller than those obtained using the skim methods. In this case the deviations of the skim methods are disastrously large, as can be seen in Table 6.5. As mentioned the integration can be improved, however, the skim methods often lead to unacceptable deviations in cases of high merge.

Table 6.5 Simulation of a complex baseline (PeakFit) deviations in comparison with the integration.

Target area	Deconvolution	Deviation with deconvolution (PeakFit) (%)	Deviation with integration (%)
4 327 431	4 327 526	0.00	0.00
117 469	117 762	0.25	−1.30
11 720	11 862	1.21	−10.98
13 608	13 762	1.13	−1.58
69 813	70 070	0.37	−2.93
7 758	7 807	0.63	−30.61
60 974	61 352	0.62	−3.80
17 364	17 477	0.65	−5.82
898			
22 789	23 368	(Sum 9 & 10) −1.35	
25 910	26 748	3.23	91.73
66 558	67 009	0.68	24.55
15 981	15 915	−0.41	
8 255 658	8 257 217	0.02	0.06
80 821	80 873	0.06	8.23
12 211	12 222	0.09	
4 702	4 647	−1.17	−5.55
22 264	22 229	−0.16	−0.79
3 200	3 125	−2.34	−35.51
11 789	11 380	−3.47	−65.41
13 740	14 012	1.98	−33.36
10 644 186	10 642 565	−0.02	1.16
123 204	128 492	4.29	
1 626	1 643	1.02	12.24
10 582	10 600	0.17	2.55
27 034	27 009	−0.09	−9.36

6.6
Real Chromatogram

The simulations were all developed from the peak models used for the approximation. The deconvolution calculates back to give the retention time, the width (σ) and the distortion (τ) of the single peaks. In a real chromatogram these ideal peaks are not present.

Therefore the next example (Fig. 6.17), not based on a simulation, should demonstrate that the application of deconvolution for real cases is just as interesting and affords good results. Using a mixture of four different esters a chromatogram was obtained that contains four merged peaks. To arrive at the target area for each material a calibration curve was generated. The areas of the individual components within the four-component mixture were read off from the calibration curves. These target areas have errors unlike those in the simulation.

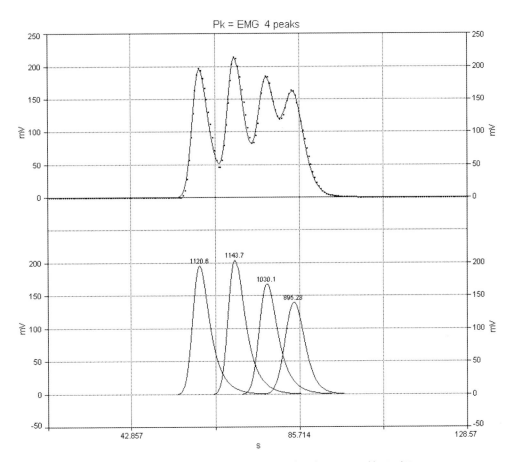

Figure 6.17 Example of a real chromatogram with merged peaks processed by PeakFit. Upper trace the sum curve, lower trace the single peaks of the approximation.

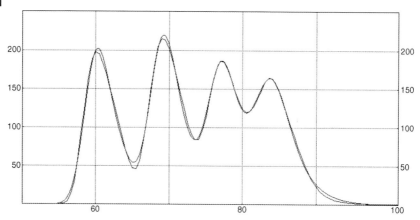

Figure 6.18 Overlay of the sum curve of the approximation and the sum curve of the chromatogram (PeakFit).

With the available peak pattern, the perpendicular skim method for the separation of the merged peaks would appear to be preferable. Therefore a comparison was made between the perpendicular skim method and deconvolution with the exponentially modified Gaussian model by means of PeakFit.

The use of the perpendicular skim method is favorable when the areas of the compounds are nearly equal as here. The tailing of the peaks is, however, unfavorable for the perpendicular skim. The deviation for peak A is –11.8%, because it lost its tail to peak B. Peak B gains an equivalent amount to that which it loses, which results in a deviation of 0.2%. Peak C has a deviation of –5.5% and peak D a very large one of 21.5%, because it gains the area from peak C.

The automatic approximation leads to deviations for peak A of –3.9%, for peak B of 5.8%, for peak C of –3.3% and for peak D of 3.8%. However, the overlay in Fig. 6.18 of the sum curve of the approximation and the sum curve of the chromatogram shows that there are still some differences between them.

Because of this the approximation was manually modified, which means that the parameters of the single compound peaks were adjusted so that both the sum curves become equal. The result is shown in Fig. 6.17.

Table 6.6 shows the deviations from the target area for the perpendicular skim method in comparison with the approximation.

The approximation gives smaller deviations than the perpendicular skim method for all peaks. With the perpendicular skim, peaks A, C and D have deviations larger than 5%. With the approximation only peak A has a deviation larger than 5%. The deviations with the approximation are: peak A –6.2%, peak B 0.1%, peak C –1.7% and peak D 1.4%. Note that the manual approximation is operator dependent, because it is a subjective adjustment. However, even with an automatic approximation the results are better than with the perpendicular skim method even if the deviation for peak B is smaller, because it is based only on the area equality. Please note, that the errors from the perpendicular skim rise if the ratio of the peak high differs more. Deconvolution here is also favorable.

Table 6.6 Example of a real chromatogram with merged peaks treated with the perpendicular skim method (Empower) vs. deconvolution (PeakFit).

Peak	Target area	Perpendicular skim	Deviation (%)	Deconvolution (PeakFit)	Deviation (%)
A	1 194 812	1 054 147	−11.77	1 120 570	−6.21
B	1 145 209	1 147 845	0.23	1 143 687	−0.13
C	1 048 159	990 802	−5.47	1 030 112	−1.72
D	907 931	1 103 317	21.52	895 283	−1.39

6.7
Outlook

It has been shown that the application of deconvolution is an excellent method for the separation of merged peaks in comparison with the skim methods. Gaussian-shaped peaks or peaks with different tailing can all be determined with very small deviations from the target areas. It can also be used for chromatograms where a complex baseline and peak pattern are present. This shows that deconvolution should be included in the CDS. It is unfortunate that it is not part of the CDS today, even though the work of Westerberg [1] over 40 years ago and the inclusion of HSM showed that it is possible and the leading results are very good. But what possibilities are offer by the CDS today? There is the perpendicular skim with its inadequacies and poor results, particularly in the presence of peaks with tailing, and the tangential skim, which is not a real possibility, because it is only useable in specific rare cases. For some CDS the possibilities end here. "Better" CDS offer, in addition, graphical extrapolation skim methods – the Gaussian and exponential skim methods. At first this sounds promising but, as mentioned above, these two skim methods are also preprogrammed with an error and so offer no real possibility of good results.

PeakFit shows how the method of deconvolution advanced after HSM so that it is applicable to more than one peak pair and there are more possibilities for its application. The results for both programs, however, are almost the same. The possibilities and algorithms, as present in PeakFit, must be include in the CDS. It is the time to boost the CDS with the deconvolution!

In addition, the results show that the use of the approximation leads to better accuracy than the skim methods. These methods are supported, in contrast to deconvolution, by established guidelines [20]. The approximation should also be supported by the guidelines to give operators the possibility of generating more accurate analytical results.

The often quoted sentence: *"Clever algorithms are not an acceptable substitute for good chromatography"* still holds true today and for the future.

Deconvolution, the same as separation methods, should not be used alone if an optimization of the chromatographic method is to be achieved. If this is the case, an exact understanding of the chromatogram and its statements is absolutely necessary for good results. Even using deconvolution or separation methods, understanding of the chromatogram and the chromatographic conditions is enormously important for the generation of accurate results. The operator must analyze his chromatographic run and decide, in the cases of merged peaks, how to separate them. This chapter should show that the application of deconvolution is very interesting and is preferable in most cases. The possibility of simulating ones own chromatogram with MS-Excel and the knowledge gained from it about the target areas, put the operator in a position to test all methods and then choose which method to use to get the most accurate results. The company Systar offers the possibility of downloading a full version of PeakFit from www.systat.com/downloads/, valid for 30 days.

References

1 A. W. Westerberg, Detection and resolution of overlapped peaks for an on-line computer system for gas chromatographs, *Anal. Chem.*, **1969**, *41*, 1770.

2 N. Dyson, *Chromatographic Integration Methods*, 2nd ed.,The Royal Society of Chemistry, Cambridge, **1998**.

3 D. Hanggi, P. W. Carr, Errors in Exponentially Modified Gaussian Equations in the Literature, *Anal. Chem.*, **1985**, *57*, 2394–2395.

4 V. B. Di Marco, G. G. Bombi, Mathematical functions for the representation of chromatographic peaks, *J. Chromatogr. A*, **2001**, *931*, 1–30.

5 E. Grushka, Characterization of Exponentially Modified Gaussian Peaks in Chromatography, *Anal. Chem.*, **1972**, *44*, 1733–1738.

6 J. C. Sternberg, *Advances in Chromatography*, J. C. Giddings and R. A. Keller, Eds., Marcel Dekker, New York, **1966**, p. 205.

7 S. L. Tamisier-Karolak, M. Tod, P. Bonnardel, M. Czok, P. Cardot, Daily validation procedure of chromatographic assay using gaussoexponential modelling, *J. Pharm. Biomed. Anal.*, **1995**, *13*, 959–970.

8 P. J. Naish, S. Hartwell, Exponentially Modified Gaussian Functions – a Good Model for Chromatographic Peaks in Isocratic HPLC?, *Chromatographia*, **1988**, *26*, 285–296.

9 J. P. Foley, J. G. Dorsey, A Review of the Exponentially Modified Gaussian (EMG) Function: Evaluation and Subsequent Calculation of Universal Data, *J. Chromatogr. Sci.*, **1984**, *22*, 40–46.

10 M. S. Jeansonne, J. P. Foley, Review of the Exponentially Modified Gaussian Chromatographic Peak Model Since 1983, *J. Chromatogr. Sci.*, **1991**, *29*, 258–266.

11 J. P. Foley, Equations for Chromatographic Peak Modeling and Calculation of Peak Area, *Anal. Chem.*, **1987**, *59*, 1984–1987.

12 A. N. Papas, Accuracy of Peak Deconvolution Algorithms within Chromatographic Integrators, *Anal. Chem.*, **1990**, *62*, 234–239.

13 Empower 2 – Einführung in die Software sowie Erfassung und Verarbeitung von Daten", *Waters Corporation*, **2005**.

14 PeakFit – Peak separation and analysis software – User's Manual, *SeaSolve Software Inc.*, **2003**.

15 L.-H. Zang, A. Ou, *New 2D Peak Decon-volution Tool for Processing Chromatogra-phic Peaks*, Hitachi Instruments, **1998**.

16 K. Ishikawa, *Introduction to Quality Control*, Kluwer, Dordrecht, **1991**

17 V. R. Meyer, Fallstricke und Fehlerquellen in der HPLC in Bildern, Wiley-VCH, Weinheim, **2006**.

18 M. W. Dong, *Views, Perkin-Elmer Newsletter*, **1994**, S. 7.

19 S. Kromidas, H.-J. Kuss, *Chromato-gramme richtig integrieren und bewerten*, Wiley-VCH, Weinheim, **2008**.

20 *Chromatographische Trennmethoden*, European Pharmacopoeia 5.0, **2005**, *2.2.46, 87*.

7
Interpretation of Chromatograms

Hans-Joachim Kuss

The most difficult integration problem is of course an ill-defined baseline. The baseline, for example, cannot be identified under a "forest" of peaks in some chromatograms. One then has to choose the position of the baseline where you feel it should be. The integration systems can only use the measured data points as support points for the baseline. However, small changes in the baseline lead, for small peaks, to large differences in the peak area and to differences approximately half as large as these for the peak height.

Peaks merged together are a great problem particularly with respect to tailing. It is difficult to quantify a peak shoulder. However, when a shoulder appears, there can be no doubt that at least two analytes are present. A worse scenario is when peaks lie directly on top of one another, and there is no clue as to whether there is more than one analyte under the peak. This possibility should always be anticipated. No integration system can help here, a more specific detector, injection onto an orthogonal column or a multi-detection system is required.

Noise is not usually a big problem for the integration. If an S/N of less than 100 is avoided, then the integration error caused by noise does not need to be taken into account. Only when measurements are taken close to the limit of quantification is the integration error considerable and then it can be dominant. In a very sensitive analysis this could, of course, be a major problem.

A general problem is that in recent decades, integration systems have hardly improved with respect to integration of the peaks. The exponential skim is the only function that has actually been added. Only when there is sufficient demand for many users will the situation change.

7.1
Using the Peak Height

One can expect the best quantitative results in chromatography with tall free-standing peaks. The integration problems will begin if the peaks have merged together to a greater or lesser extent. The problems become greater as the difference in height between the merged peaks increases.

Quantification in LC and GC: A Practical Guide to Good Chromatographic Data
Edited by Hans-Joachim Kuss and Stavros Kromidas
Copyright © 2009 WILEY-VCH Verlag GmbH & Co. KGaA, Weinheim
ISBN: 978-3-527-32301-2

By far the greater number of analyses are carried using with the integrated areas, as the height evaluation is regarded as inferior. In the early days of chromatography, height was used, when peaks had to be measured with a ruler. It is possible that the peak height is less precise than the area if the flow is not very stable. This indicates a chromatographic problem. However, if the peak height is measured by an integration system it is usually very exact. There are heads of laboratories who generally prefer the height evaluation.

The peak height often shows substantially lower deviation, even shoulder peaks match the target values of the height quite well. However, this practical evidence is difficult to reconcile with the theoretical considerations.

The influence on the height of a "wrong" base line is also usually lower by approximately a factor of 2, generally because more area than height is located in the broad foot of the peak. The area integration but not the height always needs a separating line between the peaks. The height remains stable provided the neighboring peak is not larger than the peak of interest. Indeed, height is at least a simpler measurand than the area. It is simply the signal height at the maximum of a peak after the baseline has been subtracted.

Many methods of analysis can be used to judge whether a limiting value is exceeded. Therefore, the decisive difference could be that the peak height can never get too small – it is an original measure. On the other hand, the area of small Gaussian peaks that have merged with larger Gaussian peaks could be significantly smaller than they should be. Also, the area of the first peak in a series of tailing peaks merged together can be substantially smaller after integration with the perpendicular drop. The exponential skim improves the area and height similarly, i.e. the height is similarly advantaged.

Let us assume that a chromatographic analysis must be presented in front of judge in a court. The judge asks: "Is it completely impossible that your concentration is too low or must we include a higher value amongst the conceivable possibilities?" If the answer is "no", one has to be 100% certain of this.

The peak height is the safer measurand.

7.2
Evaluation

In chromatographic analysis, in 99% of the cases a linear relationship is expected between the concentration x and the signal y. Presupposing linearity, one calibration point at sufficiently high concentration is sufficient, because the second point for the straight line $y = b\,x$ would be the origin.

From several calibrating points, a straight line through zero can be extrapolated. The general way to find a linear relationship is linear regression with an intercept. For validation, generally, a calibration (straight) line has to be shown. In a daily routine, on the other hand, one often confines oneself to a one point calibration.

7.2.1

Evaluation Methods

The minimum form of a calibration is the one point calibration. One calculates the response factor RF as the quotient of the signal value y and the concentration x and assumes that this relationship is valid for the calibration concentration x_K and also for all analyte concentrations x_A.

$$RF = \frac{y_K}{x_K} \tag{7.1a}$$

$$RF = \frac{y_A}{x_A} \tag{7.1b}$$

Equations (7.1a) and (7.1b) combine to give x_A:

$$x_A = \frac{y_A}{RF} = \frac{y_A \, x_K}{y_K} \tag{7.2}$$

The measurement y can be the area or the height of the peak and the quotient of the area (height) divided by the area (height) of the internal standard IS in the same chromatogram.

$$y_K = A_K \tag{7.3a}$$

$$y_K = H_K \tag{7.3b}$$

$$y_K = \frac{A_K}{F_{IS_K}} \tag{7.3c}$$

$$y_K = \frac{H_K}{H_{IS_K}} \tag{7.3d}$$

Government regulations for chromatographic calibrations require straight lines with an intercept, which really is only necessary for measurements with a background signal. In chromatography, at the retention time of the substance of interest, there must be a zero signal. Otherwise the method is inadequate. Therefore, it is a requirement that the intercept is not significantly different from zero. However, if this is the case, there is not actually any statistical argument against the straight line through zero either.

One point calibration $\quad x_A = \dfrac{y_A}{RF} \tag{7.4a}$

Straight through zero $\quad x_A = \dfrac{y_A}{b} \tag{7.4b}$

$$\text{Linear regression} \qquad x_A = \frac{y_A - a}{b} \qquad (7.4c)$$

Using these very similar equations the results of the analysis can be calculated. In the linear regression with an intercept, there is no difference except for the usually low correction of the signal value y_A by a. Small signal values are corrected much more relative to the large values.

The additions method is also a linear regression of course, where the intercept is necessary to calculate the concentration of the sample without any addition. Therefore at least two points are needed to extrapolate the straight line. The calculation of the concentration using the additions method is described in DIN 32633:

$$x_A = \frac{a}{b} \qquad (7.5)$$

An internal standard can of course also be used in the additions method. As a minimum, at least two analyses must be carried out: once without any addition, and once with addition Z.

$$a = y_A \qquad (7.6a)$$

$$b = \frac{y_{A+Z} - y_A}{x_Z} \qquad (7.6b)$$

A linear regression must be undertaken at more than two different concentrations. The two characteristic values a and b arise directly from the straight line equation. A more detailed description, including examples calculated with Excel, can be found in Kuss [1, 2].

7.3
Calibration

For the calibration, the slope b (sensitivity) and the intercept a (background) of the straight line equation with the linear regression are required.

$$y = b\,x + a \qquad (7.7)$$

One needs the area or height of the corresponding peaks for the calibration in ascending concentrations in single chromatograms. Then one can determine the calibration range AB_x limited by the lowest concentration x_U and the highest concentration x_O.

$$AB_x = \frac{x_O}{x_U} \qquad (7.8)$$

7.3.1
Linear Regression

The following example from Johnsson [3] covers, with triple measurement of every concentration, a working range of 600, which can itself cause particular problems. Table 7.1 shows only part of Johnsson's table, namely the lower 7 calibration points (and the respective average) shown initially to cover a working range of 100. The (corrected) signal values are quotients from the area of the analyte and the appropriate area of the internal standard. The values were multiplied by 1000 to make the table more readable. As only relative calculations are carried out, this corrects itself in the evaluation.

In Excel it is easy to determine the two values ($a = 8.4614$; $b = 5.3486$) with <Insert> <Function> <Slope> and <Intercept>. They are used to calculate the $y(x)$ values on the straight line.

The differences between the measured y values and the calculated $y(x)$ values are called the residuals. The linear regression calculates the straight line minimizing the sum of the residuals squared, for which the smallest possible value (here: 64.462) results. Each residual draws a straight line towards the measured point, because the squares of the deviation are summed. If it is assumed that in the lower part of the straight line only small deviations appear and in the upper part there are only large deviations (inhomogeneous variances), then the lower points are ignored because they do not have any weight and the straight line would be too far away. The deviations at the smaller concentrations have earned "the higher weighting" because they can be measured (in absolute units) substantially more precisely.

With the chart wizard one can represent the measurements and the straight line (Fig. 7.1) that yields the best approach towards the measurements after marking the first three columns of Table 7.1.

Table 7.1 The first seven calibration points averaged for each concentration (from Johnsson [3]).

x	y	y(x)	y – y(x)	[y – y(x)]²
5	35	35.20	−0.20	0.04
10	59	61.95	−2.95	8.69
25	143	142.18	0.82	0.67
50	274	275.90	−1.90	3.59
100	550	543.33	6.67	44.48
250	1343	1345.63	−2.63	6.94
500	2683	2682.81	0.19	0.04
134	727	Average	SAQ =	64.462

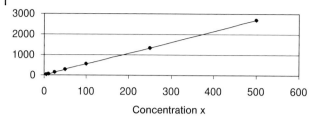

Figure 7.1 Linear regression for the values of Table 7.1.

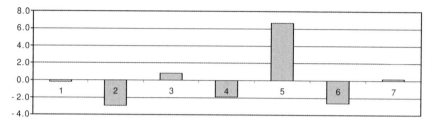

Figure 7.2 Residuals from Table 7.1.

One should always apply the residuals (Fig. 7.2) graphically, to see whether they are distributed randomly. From Fig. 7.2 it can be assessed that the average residual is about 3. Relative to the mean average value of $y = 727$ this is less than 0.5%.

The RSD as a y deviation (the difference between measured and calculated y values) is converted just like other y values with the linear equation in x deviations. The correction with the intercept does not need to be taken into account here, because in the calculation, the difference is eliminated.

$$RSD = \sqrt{\frac{[y - y(x)]^2}{n - 2}} = 3.591 \tag{7.9}$$

Therefore one only divides by the slope b and yields the method standard deviation. This is related to the mean average value of the concentration and yields the method coefficient of variation (relative standard error (%)).

$$Method \ CV \ (\%) = \frac{RSD}{b \, \bar{x}} = 0.5\% \tag{7.10}$$

Now one has a clear measure for the spread of the signal values around the straight line, analogous to the coefficient of variation for multiple measurements at one concentration. Half a per cent is a very good value, which corresponds to the visual impression of Fig. 7.1 in which virtually no residual deviations can be seen.

The calculation of analysis values from the signal y_A is carried out with the linear equation, in which the slope (sensitivity) is the essential value and a constant correction factor $-a/b$ (background in concentration units) is added, which in this instance is −1.58, i.e. below the smallest concentration.

$$x_A = \frac{y_A}{b} - \frac{a}{b} \tag{7.11}$$

Taking all (3 times 10) calibration points from the Johnsson table – the three values were each represented in Fig. 7.3 – the slope increases ($b = 5.7362$) and the straight line cuts the y-axis at a = −72.389. Therefore a method CV of 3.47% and a correction factor −a/b of 12.62 are calculated, i.e. the values become seven times as high by including the additional three calibration points.

The straight line with the ten calibration points measured in triplicate (Fig. 7.3) looks plausible at first. However, if one looks at the lower four calibration points in Fig. 7.3a one can see that the straight line misses them, although the three values coincide at each concentration. The residuals show exclusively positive deviations up to the fifth calibration point (15th value) and are the greatest for the highest concentration (Fig. 7.3b).

These large deviations cause a significant leverage of the upper three calibration points and the straight line is drawn higher. As the lower calibration points with their smaller deviations have less influence (lower weight), a considerable nega-tive axis section results. It can be recognized from Fig. 7.3a, that the straight line does not represent the lower calibration points and very large deviations arise for the small concentrations. The correction factor of 12.62 is added to the calibration sample values so that the lower concentrations are far too high, which correlates with what Johnsson has pointed out. It does not agree with the CV of the signal

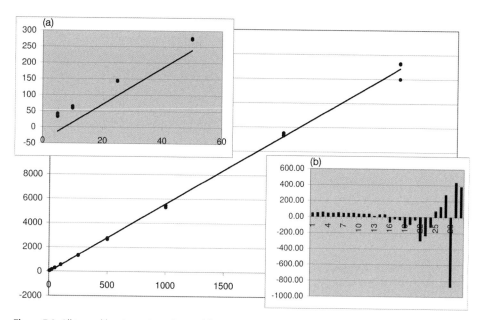

Figure 7.3 All ten calibration points, from Table 7.2, which were each measured in triplicate. Inset (a), magnification of main figure. Inset (b) residuals for the calibration points in the main figure.

Table 7.2 Data from Johnsson [3] – the signal was multiplied by 100.

x	y	M_y	S_y	VK (%)	x(y)	Dev. (%)
5	32				13.1	162.62
5	35	36	5.13	14.1	13.7	173.28
5	42				14.9	198.13
10	60				18.1	81.03
10	59	61	3.21	5.2	17.9	79.25
10	65				19.0	89.91
25	143				32.8	31.36
25	143	144	1.15	0.8	32.8	31.36
25	145				33.2	32.78
50	277				56.6	13.26
50	274	276	1.53	0.6	56.1	12.20
50	276				56.5	12.91
100	530				102	1.55
100	550	544	12.2	2.2	105	5.11
100	552				105	5.46
250	1302				239	−4.55
250	1343	1325	21.1	1.6	246	−1.64
250	1331				244	−2.49
500	2642				477	−4.69
500	2683	2688	49.2	1.8	484	−3.23
500	2740				494	−1.21
1000	5295				948	−5.24
1000	5360	5374	87.4	1.6	959	−4.08
1000	5468				978	−2.17
2000	11297				2013	0.67
2000	11352	11384	106	0.9	2023	1.15
2000	11502				2050	2.49
3000	15981				2845	−5.17
3000	17286	16834	739	4.4	3077	2.56
3000	17235				3068	2.25

values of 14.1% at 5 ng ml^{-1}, which is a high deviation but, for the lowest concentration, acceptable. It should be noted that the linear regression with intercept produces a value 170% too high with a calculated uncertainty of 14%.

One can also calculate the significance of the axis intersection. This at least answers the question: "Can one for certain distinguish the axis intersection from zero with the given spread?" Both with the seven and with the ten calibration points the axis intersection is not significant because the ten times bigger axis section is connected to a much broader spread, which results from the residuals of the three topmost calibration points.

A significant intercept is a clear warning sign, which should lead to a check of the chromatographic procedure. For the assessment of the calibration, however, the above-mentioned correction factor is an important measure. Using a large working range the intercept can be insignificant and lead, in spite of this, to a non-tolerable correction factor. This is the case if a/b is larger than the lowest concentration because this is then distorted by more than 100%.

The standard deviation of the y values S_y is shown at all concentrations in the fourth column of Table 7.2, i.e. the absolute values in units of the signal values. Dividing S_{yk} by $\bar{y}_K$, the relative values of the standard deviation (CV%) arise. The absolute values cover a much larger range (1.15 to 739) than the relative values with a range of 0.55% to 14.12%. The F value (quotient of the absolute values at the upper and lower ends of the field of study) in the above example amounts to 20 748.

Linear regression is an arithmetic process in which the squared deviations (variances) are minimized. This is only permitted if it can be assumed that the absolute deviations do not depend on the concentration. However, this is clearly the case for Johnsson's data as the fourth column of Table 7.2 shows.

For calibration in the context of a validation one has to measure the individual calibration points repeatedly only at the highest and lowest calibration points, where 6 or 10 samples are measured. Through an F test one can then find out whether the squared standard deviations are significantly different. With $F = 20\,748$ the limiting value of 5.35 is exceeded by orders of magnitude.

This so-called variance inhomogeneity is the reason why, in Fig. 7.3a, the straight line missed the lower four calibration points because these are really not taken into account, i.e. have very little weight. It seems reasonable to compensate for this by multiplying the variances by a higher (weight-) factor at the low concentrations and a lower factor at the high concentrations.

7.3.2
Weighted Linear Regression

In the generally valid form of the linear regression the difference between the measurement y_K and $y(x_K)$ calculated from the linear equation must be standardized on the corresponding variances. For homogeneous variances this is not necessary. Normally the standard deviation S_y is not known at every calibration point. It would be advisable from a statistical point of view to measure ten calibration points ten times. From the point of view of the analytical chemist this would be an unacceptable effort.

$$\sum_{K=1}^{n} \frac{[y_K - y(x_K)]^2}{S_{yk}^2} = \text{Minimum} \tag{7.12}$$

A significant F test is already frequently found in chromatography for a working range above ten. The relative error is fairly constant and the absolute error increases considerably in chromatography with sample preparation.

$$VK = \text{const.} = \frac{S_y}{y} \tag{7.13}$$

At a constant relationship between standard deviation of the signal and the signal itself, the measured signal can be used as an estimate of the standard deviation. The above equation becomes:

$$\sum \frac{[y_K - y(x_K)]^2}{y_K^2} = \text{Minimum} \tag{7.14}$$

This is simply the known $1/y^2$ weighting of the variances which can be used in the integration programs. Typically, further weightings are $1/y$, and also $1/x$ and $1/x^2$ and $1/y^{\frac{1}{2}}$. As an estimated value is needed for SD_y, a y weighting makes more sense than an x weighting.

The case of the homogeneous variances can be explained as "instrument variance" by carrying out a simple physical measurement. The y^2 weighting due to a constant CV can be termed the "volume dependent variance". This describes the situation for injection volume differences or deviations in the sample preparation. Volume errors must be concentration dependent, because the concentration is volume related.

All situations between these two extremes of a weighting w with 1 for all concentrations and a y^2 weighting are imaginable, which can be described generally by $1/y^{WE}$ (the weighting exponent WE can vary between 0 and 2). All y weightings mentioned above are included but in addition uneven weightings are permitted. The unweighted (ordinary linear) regression is given as a special case by the weighting exponent $WE = 0$.

$$\sum w \, [y_K - y(x_K)]^2 \, \text{min} \tag{7.15a}$$

$$\text{with } g = \frac{1}{y^{WE}} \tag{7.15b}$$

$$\text{and } w = \frac{n \, g}{\sum g} \tag{7.15c}$$

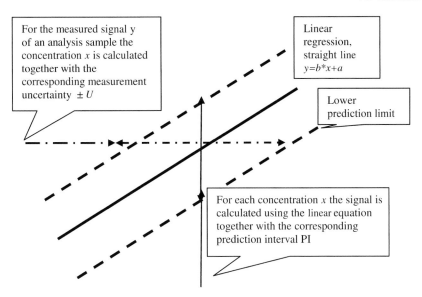

Figure 7.4 Graphical explanation of the prediction interval and the measurement uncertainty.

Table 7.3 Values calculated from the data of Table 7.2.

20 748	F-value
4.3170	$\log(F)$
463	AB_y
2.6659	$\log(AB_y)$
1.62	WE

The problem consists in assessing WE as appropriately as possible from the existing data. WE can be calculated from the logarithm of the variance quotient (F value) divided by the logarithm of the working range in y dimensions using:

$$WE = \frac{\log F}{\log AB y} = \frac{\log Sy^2_{K_n} - \log Sy^2_{K_1}}{\log y_{K_n} - \log y_{K_1}} \tag{7.16}$$

This method can be called the variance quotient weighting.

If one puts the values from Table 7.3 into Eq. (7.16), one gets $WE = 1.62$, i.e. as expected a better approximation of the "volume dependent variance" than the "instrument variance", because an intensive sample preparation was carried out by Johnsson. Using the integration systems one must decide between $1/y$ or $1/y^2$ weighting, both of which are inadequate.

With the weighting given by $WE = 1.62$ the lower four calibration points lie almost exactly on the straight line in Fig. 7.5. The resulting linear equation

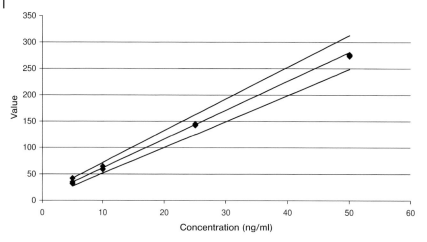

Figure 7.5 Johnsson's data calculated with weighted regression (WE = 1.62).

$y = 5.48\,x + 7.25$ has a positive axis intersection, which corresponds to 1.3 ng ml^{-1}. The back calculated results for the calibration points are given in the column $x(y)$ weight in Table 7.2. The comparison of $x(y)$ with and without weighting shows that inadequate results are possible with the normal linear regression. With the straight line through the origin the recalculated values have only slightly higher deviations than the accurate values. The chromatographically useless intercept generates a big problem for low concentrations.

Another essential reason to carry out a weighting lies in the adequate mathematical reproduction of realistic measuring uncertainties U. On this topic Doerffel wrote: "Deshalb muß die gewichtete Regression immer dann angewandt werden, wenn aus den Messungen Aussagen zur Präzision der Ergebnisse gefordert sind", which means: "Therefore the weighted regression must always be used if, from the analysis, statements concerning the precision of the results are demanded."

The known equations for the prediction interval PI (y deviations) and the measuring uncertainty U (x deviations) are only slightly complicated by the weighting:

$$PI = tRSD \sqrt{\frac{1}{w_k} + \frac{1}{\sum w} + \frac{(x_k - \bar{x}_w)^2}{\sum(x_k - \bar{x}_w)^2}} \tag{7.17}$$

$$U = \frac{tRSD}{b} \sqrt{\frac{1}{w_a} + \frac{1}{\sum w} + \frac{(y_a - \bar{y}_w)^2}{b^2 \sum(x_a - \bar{y}_w)^2}} \tag{7.18}$$

The ordinary linear regression is a special case of the weighted regression using the weighting factors $w = 1$ and $\sum w = n$.

Both equations are similar because U is only another projection of PI. The uncertainty of the calculated concentrations results directly from the uncertainty of the measured signal. The same border line is graphically read once in the *y* and once in the *x* direction.

With weighting, the lower concentrations also are in the permitted range of the measuring inaccuracy as Fig. 7.5 demonstrates.

One carries out a laborious optimized analysis using normal linear regression that in the lower concentration range is more erroneous than necessary (even with a positive *F*-test). This breaks the chromatographer's heart.

References

1 Kuss, H. J., Weighted Least-Squares Regression in Practice: Selection of the Weighting Exponent; *LCGC Europe*, December 2003a.

2 Kuss, H. J., Quantifizierung in der Chromatographie, in: *HPLC-Tipps II* (Ed. S. Kromidas), Hoppenstedt Bonnier Zeitschriften, Darmstadt, 2003b.

3 Johnson, E. L., Reynolds, D. L., Wright, D. S., Pachla, L. A., *J. Chromatogr. Sci.* **26**, 372 (1988).

4 Doerffel, K., *Statistik in der Analytischen Chemie*, Deutscher Verlag für Grundstoffindustrie, Leipzig 1990.

5 Miller, J. N., Miller, J. C., *Statistics and Chemometrics for Analytical Chemistry*, Pearson Education Limited, Essex, 2000.

8
General Interpretation of Analytical Data

Joachim Ermer

Analytical data are often used as the basis on which decisions are made, for example controlling a production process or the usage of materials (release/rejection), or for evaluating the suitability of an analytical procedure itself. These evaluations are often linked to defined rules such as acceptance criteria.

Analytical results always display (at least) random variability [1]. This must be taken into consideration when interpreting the data and especially during the process of establishing the decision rules (see Section 16.2). Therefore, general properties need to be concluded from individual analytical results, i.e. the "normal" distribution of the data.

8.1
(Normal) Distribution of Analytical Data

Analytical measurements can be regarded as random drawings of single data from a pool made up of all possible (indefinite) results (see Section 8.2). The more data that are available, the more obvious the picture of the overall distribution will be. This is illustrated in Fig. 8.1 as histograms. For such a presentation, the whole range of the results is divided into constant intervals (bins) and all data included in the respective bin are counted (frequency). The number of bins should correspond to roughly the square root of the number of data. Such histograms can be generated by means of statistical software, or with Microsoft Ecxel® (Tools/Data Analysis/Histogram; the "Analysis ToolPak" must be installed via "Tools/Add-Ins"). Starting with Fig. 8.1 (a), followed by Fig. 8.1 (b) and (c), it becomes obvious that a large number of data are needed to get a clear picture of the (true) distribution. In addition to the histogram, the theoretical curve of a Gaussian normal distribution, which is the basis of the presented data, is drawn in Fig. 8.1 (c).

Simulation of normally distributed data (Excel®):

$$x = \text{STANDNORMINV(RANDOM())} {*} \sigma + \mu \tag{8.1}$$

Quantification in LC and GC: A Practical Guide to Good Chromatographic Data
Edited by Hans-Joachim Kuss and Stavros Kromidas
Copyright © 2009 WILEY-VCH Verlag GmbH & Co. KGaA, Weinheim
ISBN: 978-3-527-32301-2

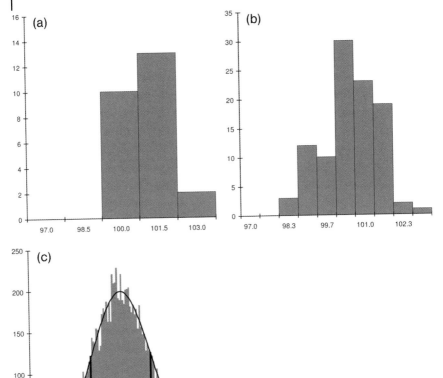

Figure 8.1 Histograms for 25 (a), 100 (b) and 10 000 (c) normally distributed data. Additionally, the theoretical Gaussian curve is displayed in C. Histogram bins representing the limits of the 2 and 3 σ intervals around the true mean μ are highlighted in black. The normally distributed data were calculated using Microsoft EXCEL® for $\mu = 100$ and $\sigma = 1$ (Eq. 8.1).

So why is the normal distribution – which is usually assumed for physico-chemical measurements – so popular?

Because the whole distribution can be completely described in a (relatively) simple mathematical equation with only two parameters, the location (mean μ) and the dispersion (variability, standard deviation σ). (Note: The parameters of the overall distribution, also called true parameters, are usually denoted by Greek letters, whereas Latin letters describe the parameters calculated from a (more or less) limited number of data (sample parameters)). The true mean μ of the overall distribution corresponds to the apex of the curve. With increasing distance from μ, the probability of obtaining results decreases symmetrically on both sides. The dispersion parameter σ describes the width of the distribution and corresponds to the horizontal distance between μ and the inflection point of the Gaussian curve. Small and large standard deviations result in tall, sharp and broad, flat curves, respectively.

It is very important to take into consideration that within a range of $\pm 1\,\sigma$ (only) 68% of all data can be expected. The amount is increased to 95% for $\pm 2\,\sigma$ and to 99.7% for $\pm 3\,\sigma$. (Note: this relates to the true parameters μ and σ, in contrast to the experimental or sample parameters, see Section 8.2.)

These properties are utilized to check the suitability of a method, called the method capability index: The (nearest) acceptance limit must be at least three standard deviations away from the mean.

Confidence interval limits:

of the mean
$$CL(P)_{\bar{x}} = \bar{x} \pm s \cdot \frac{t(P, df)}{\sqrt{n}} \tag{8.2}$$

of the standard deviation
$$CL(P)_{s,\,lower} = s \cdot \sqrt{\frac{df}{\chi^2(1 - P, df)}} \tag{8.3a}$$

$$CL(P)_{s,\,upper} = s \cdot \sqrt{\frac{df}{\chi^2(P, df)}} \tag{8.3b}$$

Where $t(P, df)$ = Student-t-factor for a defined statistical confidence P (often 95%) and degrees of freedom df. Excel®: $t = \mathrm{TINV}(\alpha, df)$; $\alpha = 1 - P$. $\chi^2(P, df)$ = Chi-square factor for a defined statistical confidence P (often 95%) and degrees of freedom df. Excel®: $\chi^2 = \mathrm{CHIINV}(\alpha, df)$; $\alpha = 1 - P$.

The parameters calculated (theoretically) from all data of the normal distribution are identical with the true parameters. With a limited number of data (sample size), the uncertainty is increased: it is now only possible to give a range (with a defined statistical confidence), the true value can be expected within the so-called confidence interval (Eqs. 8.2 and 8.3). Confidence intervals form the basis of many statistical tests (see Section 16.3.4).

The Student-t-factor can be regarded as a correction factor for the experimental (sample) standard deviation s. As the tails of the Gaussian curve are indefinite, i.e. data can also occur at a large distance from μ (although with a very small probability), it is necessary to focus the statements to a practically relevant range. For this purpose, the tails of the distribution are "cut", for example 2.5% of the lowest and highest data are ignored to give a statistical confidence P of 95% (or 0.95). Often, the complementary error probability $\alpha = 1 - P$ is used. The error probability and the power of the decision are linked: In the case of a very small error probability (high confidence level), the decision power is very weak and therefore of limited value. Assuming a mean of 100, a standard deviation of 1 and six results, the true mean can be expected to be between 99.0 and 101.0 with a confidence level of 95%. However, with 99.99% confidence, the expected range is widened from 95.4

to 104.6%. Or, using an example from everyday life, a forecast that tomorrow's temperature will be between 1 °C and 40 °C has a very small error probability but is not very helpful in deciding what clothing to wear.

The degrees of freedom (df) represent the number of data that are necessary to perform the respective calculation, in addition to the absolute minimum required. In order to obtain a mean, for example, at least one result is needed, thus $df = n - 1$. For a straight line, at least two data points are required, i.e. $df = n - 2$. In the case of several series (or runs, k) of data (n), the overall degrees of freedom are calculated as $df = k (n - 1) = k n - k$, i.e. the overall number of data less the number of series.

From Fig. 8.2 it is obvious that the confidence limits become tighter with an increasing number of data, i.e. the reliability increases. Of course, it is the same in the opposite direction: the uncertainty becomes large with a small number of data, especially for standard deviations. The upper limit of the 95% confidence interval for three results ($df = 2$) corresponds to 4.4 times the calculated standard deviation! It certainly does not make much sense to base an evaluation or decision on such a weak result. Unfortunately, such calculations are not uncommon in the literature. However, the overall degrees of freedom are relevant, so that repeating the series of three measurements will also result in an appropriate reliability: $df = k (n - 1) = 2 k$.

Note: Appropriate tests can also be designed with only a small number of data. However, they must take the large uncertainty into account, i.e. the limits must be fairly tight. For example, assuming that a standard deviation of 1, obtained from

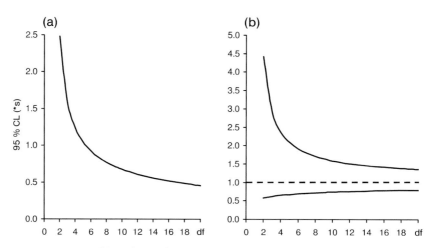

Figure 8.2 95% Confidence limits of the mean (a, one-sided) and of the standard deviation (b) as a function of the degrees of freedom (df). The confidence limits (CL) are displayed as units of the standard deviation. For example, for a series of 6 data ($df = 5$), the 95% confidence interval of the mean ranges ±1.05 standard deviations around the mean (symmetrical). That of the standard deviation ranges from 0.67 to 2.09 times the calculated standard deviation. The latter is indicated in B as a dashed line.

six results, is acceptable, it can be calculated using Eq. (8.3b) that 0.47 would be an equivalent standard deviation from $n = 3$ (resulting in the same upper confidence limit).

8.2
Problems of Prediction

Confidence intervals (only) describe the range where the true parameter can be expected, based on the actual data (sampling). However, predictions of future results are often of interest, for example, how many results can be expected beyond a defined limit, etc.

The future behavior of data can be investigated by means of statistical simulations (see Eq. 8.1). Of course, only random effects are taken into consideration. Such simulations have the advantage of knowing the true parameters, which is not the case with experimental investigations, as well as the possible large number of data sets.

In Table 8.1, five simulated data sets comprised of 3, 6 and 10 data each are listed, together with the calculated sample means, standard deviations and ranges. Even with so few data sets, the dispersion of the parameters is obvious, especially for the means and standard deviations calculated from a small number of data. In contrast, the ranges become wider with increasing number of data. The range will eventually approximate the ±3 σ limit known from the normal distribution.

For a more complete picture, in Figs. 8.3 to 8.5 the results of 50 000 simulations are shown for a number of data up to 20. The upper and lower limits of the ranges which include 90, 95 and 99% of all simulated results are indicated. According to the required assurance, the respective limit(s) can be used as a prediction for the individual parameters, i.e. as acceptance limit(s). The limits are displayed two-sided, i.e. the upper and lower 5, 2.5 and 0.5% of the results are ignored. If only upper limits are of interest (as is usually the case for acceptable standard deviations and ranges), the acceptance limits must be defined one-sided. Therefore, the upper 90% limit in Figs. 8.4 or 8.5 corresponds to a 95% assurance, because only the upper 5% of the results are ignored.

The prediction intervals are related to a true mean and standard deviation of 100 and 1, respectively, and can easily be recalculated using the parameters of a specific application. However, two conditions must be taken into consideration:

1. The simulated data are related to the true standard deviation σ. In order to make a reliable estimation of acceptable dispersion ranges, a very reliable experimental standard deviation, a maximum acceptable true standard deviation, or a target standard deviation for the respective application needs to be known (see Section 8.3).

2. All effects are of a random nature and all relevant variability contributions are included in the respective standard deviation (precision levels, see Section 8.3.1).

Table 8.1 Simulated series of normally distributed data for sample sizes of 3, 6 and 10. The maximum results for the calculated parameters (or the maximum deviation in the case of means) are highlighted in bold. The data were simulated using Microsoft EXCEL® for $\mu = 100$ and $\sigma = 1$ (Eq. 8.1).

Simulated data										Mean	SD	R
100.23	100.10	99.12								99.81	0.61	1.11
100.58	100.26	100.10								100.31	0.24	0.47
101.65	99.94	100.45								100.68	0.88	1.71
98.63	101.53	102.42								**100.86**	**1.98**	**3.79**
100.45	99.90	99.45								99.93	0.50	1.00
100.69	99.13	100.12	101.22	101.13	100.31					100.44	0.77	2.09
98.32	101.89	100.15	97.15	101.57	100.24					99.89	**1.84**	**4.74**
102.13	99.82	100.15	101.13	99.90	99.74					**100.48**	0.95	2.39
98.79	98.75	99.52	99.89	99.38	99.54					99.31	0.45	1.13
100.01	99.09	99.98	99.15	99.20	100.71					99.69	0.65	1.62
100.11	98.97	100.41	100.79	101.16	101.26	99.97	100.24	100.08	102.04	100.50	0.85	3.07
101.32	99.38	99.84	100.54	99.96	100.93	100.29	100.58	99.38	99.46	100.17	0.68	1.94
100.62	99.04	100.71	99.15	99.59	97.76	99.81	98.96	100.6	98.02	**99.44**	1.07	3.01
98.51	99.68	103.66	100.85	101.01	100.49	98.49	101.69	99.81	98.16	100.24	**1.69**	5.51
99.62	99.40	99.80	101.08	99.14	100.31	100.54	98.67	98.73	98.83	99.61	0.82	2.41

SD = standard deviation; R = range (minimum – maximum)

The range, i.e. the difference between the smallest and the largest value, is a suitable evaluation parameter, if only few data are available or for the evaluation of the dispersion of residuals (residual plot, see Chapter 7). The expected range will increase with increasing number of data up to six times the standard deviation, i.e. the $\pm 3\,\sigma$ limits of the normal distribution (see Fig. 8.5 and Section 8.1). An acceptable difference between two single determinations can also be obtained numerically (see Eq. (8.4) [2]). In the case of reliable standard deviations, the expected range for a 95% confidence level corresponds to about three times the standard deviation. For a very large degree of freedom, the range limit approximates to 2.8, corresponding to the solid line in Fig. 8.5 for $n = 2$. In practical applications, it is essential to use the correct precision level (see Section 8.3.1). For example, an injection precision must be used to calculate a limit for duplicate injections,

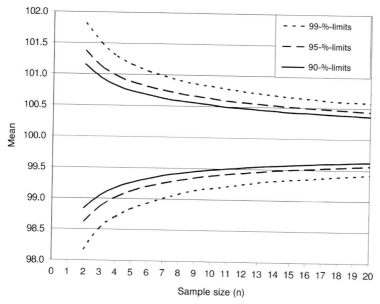

Figure 8.3 Upper and lower limits of expectation intervals for means of simulated data sets in dependence on the sample size. The limits which include 90%, 95% and 99% of all simulated means are shown (two-sided). 50 000 data sets were calculated using Microsoft EXCEL® for $\mu = 100$ and $\sigma = 1$ (see Eq. 8.1).

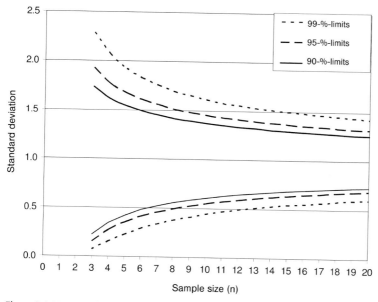

Figure 8.4 Upper and lower limits of expectation intervals for standard deviations of simulated data sets as a dependence on the sample size. The limits which include 90%, 95% and 99% of all simulated standard deviations are shown (two-sided). 50 000 data sets were calculated using Microsoft EXCEL® for $\mu = 100$ and $\sigma = 1$ (see Eq. 8.1).

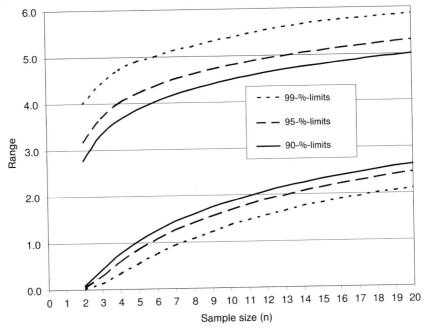

Figure 8.5 Upper and lower limits of expectation intervals for ranges (minimum to maximum) of simulated data sets as a dependence on the sample size. The limits which include 90%, 95% and 99% of all simulated ranges are shown (two-sided). 50 000 data sets were calculated using Microsoft EXCEL® for $\mu = 100$ and $\sigma = 1$ (see Eq. 8.1).

the standard deviation of the repeatability will result in a limit for independent sample preparations (repeatability limit), etc.

$$\text{Range limit } (n = 2): \quad R = s \cdot t(P, df) \cdot \sqrt{2} \tag{8.4}$$

8.3
Analytical Variability

8.3.1
Variability Contributions and Precision Levels

Each individual step of an analytical procedure contributes to the overall precision according to its variability. If all (important) contributions are known, the overall uncertainty can be calculated by applying the rules of error propagation, the so-called "bottom-up" approach [3, 4] (see also Chapter 9). These contributions and their links can be visualized as "fishbone" or Ishikawa diagrams (Fig. 8.6). For the reliability of such a calculation, it is of course crucial to identify all relevant contributions, which may turn out to be rather complex.

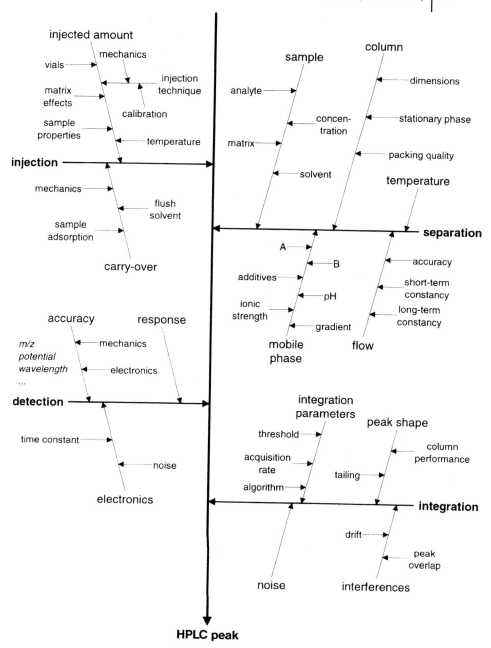

Figure 8.6 Ishikawa diagram for LC system precision (reproduced from [6], p. 39).

In pharmaceutical analysis, usually a "top-down" approach is used. Here, the various contributions are experimentally summarized as precision levels [5]. For an (overall) evaluation of the precision, it is not essential to know the individual contributions, although it is quite helpful in the case of trouble shooting (at least to know the major contributions, corresponding to the horizontal branches in Fig. 8.6).

For the precision levels, basically short- and long-term effects are distinguished. The former includes system (injection) precision and repeatability, the latter, intermediate precision and reproducibility.

Variance and standard deviation:

$$s^2 = \frac{\sum(x_i - \overline{x})^2}{(n-1)} \qquad s = \sqrt{s^2} \tag{8.5}$$

$$s^2 = \frac{\sum(x_{i,1} - x_{i,2})^2}{2\,k} \tag{8.6}$$

where x_i, $\overline{x}$ = single data and mean of a series/run, $x_{i,\,1/2}$ = duplicate results of a sample, batch, etc., k = number of samples, batches, etc. analyzed in duplicate.

Usually, standard deviations (or variances) are calculated from the squared differences between individual data and the corresponding mean (Eq. 8.5). It is essential to report clearly the experimental conditions the standard deviation is based on, such as from repeated injections, from means of injections, etc., otherwise, no meaningful interpretation is possible. Preferably, standard deviations should always be related to individual results (injections), because only then can all further variability parameters be calculated. For example, the standard deviation of a mean from six injections can be calculated by dividing the standard deviation of individual injections by the square root of 6 (the so-called standard error $s/\sqrt{n}$).

In order to facilitate the evaluation, the standard deviation is usually divided by the respective mean. The relative standard deviation as a normalized (percentage) parameter is easier to compare with other applications.

A standard deviation can also be calculated from the ranges of duplicate results (Eq. 8.6). This is an excellent possibility to extract precisions from routine analyses, without additional effort (see Section 8.3.3). In order to avoid weighting effects, the means must not vary too much, up to about 10% should be acceptable.

8.3.2
System or Injection Precision

When considering system precision, the same sample is analyzed many times, for example in chromatography the same analyte solution is injected repeatedly. The often used term "injection repeatability" should be avoided, because it runs the risk of confusion with "repeatability", i.e. the next precision level. Unfortunately, this can be observed in numerous publications and may lead to the wrong conclusions, because the thus-obtained precision does not include all contributions relevant to the routine application.

In Fig. 8.6, the various individual contributions to the system precision become quite obvious. The major contributions are separated in Eq. (8.7) applying the rules of error propagation (see Chapter 9). The relative amount of the contributions may change with the concentration of the analyte (see Section 8.3.6). For example, at higher concentration, the variances of detection and integration are negligible (see Fig. 8.10). The remaining variance contributions of injection and separation can be separated by means of repeated injections of a mixture of substances, originally described as the "Maldener Test" to qualify LC-systems [7]. The variance contributions of the separation include mainly the short-term flow constancy of the LC-pump. Rearranging Eq. (8.7) and subtracting the variance of the relative peak areas from those of the absolute peak areas (corresponding to the overall system precision) results in the variance of the injection (Eq. 8.8), here somewhat simplified as the squared relative standard deviation. Performing 10 injections of a mixture of methyl, ethyl, propyl and butyl esters of 4-hydroxybenzoic acid, precisions of the relative peak areas of between 0.04 and 0.12% were obtained, corresponding to relative variance contributions between 5 and 22%. The relative peak areas are mainly influenced by short-term fluctations of the flow. The smaller the overall precision, the larger this contribution becomes. However, the major part of the variability is caused by the injection system, at least in the case of simple chromatographic separations.

Variance contributions to system precision:

$$s_i^2 = s_{Injection}^2 + s_{Separation}^2 + s_{Detection}^2 + s_{Integration}^2 \tag{8.7}$$

Injection precision:

$$rsd_{Injection} = \sqrt{rsd_{abs}^2 - rsd_{rel}^2} \tag{8.8}$$

The system precision corresponds to the lowest precision level. As variability can only increase in the other levels, it makes sense to establish acceptance limits here. Therefore, injection precision is part of the system suitability test requirements of the compendia (see Section 16.1). In the literature, the variability of auto samplers is reported between 0.3 and 0.7%, for system suitability tests between 0.5 and 1.2% [8] or between 0.7 and 1.0% [9]. Reviewing about 20 publications ([10], p. 41) relative standard deviations of between 0.06 and 1.90% were found, with an average of 0.92%. About three-quarters of all results were below 1.0%. However, it has to be taken into consideration that interpretation was sometimes difficult because the information provided was insufficient, for example with respect to the difference between test concentration and quantitation limit.

In some cases, the objective was simultaneous determination of several drugs. During an investigation on the concentration dependence of the precision [11], an average system precision of 0.31% was obtained for the larger concentrations. In stability studies of a lyophilisate, from 132 duplicate injections, a pooled system precision of 0.28% was calculated, with values for the six storage intervals (9 to 48 months) of between 0.14 and 0.49%. A compilation of 53 system precisions from the annual performance qualification of 28 LC systems resulted in values between 0.10 and 0.87%, with an average of 0.34%.

8.3.3
Repeatability

Samples are independently prepared and analyzed within a short timeframe, i.e. within the same series (run). In addition to the variance contributions of the system precision, the variability of the respective sample preparation is included. As the precision of the routine application represents the relevant information, the experimental design of the precision study should reflect these routine conditions as far as possible. Therefore, authentic samples should be used (for exceptions see Section 8.3.6).

The ICH guideline [5] prescribes at least six determinations, in order to achieve an acceptable reliability (see Fig. 8.2b).

Usually, during validation Eq. (8.5) is used to calculate an (individual) repeatability. Using Eq. (8.6), repeatability can also be calculated from analyses of duplicate sample preparations in routine testing. The differences between the two sample preparations for each batch are squared and summed over several batches. This makes it possible to include a large number of data over time and thus to increase the reliability of the repeatability estimate considerably [12]. Such a repeatability corresponds (if obtained from more than about 20 batches) to the pooled repeatability of the variance analysis (Eq. 8.9). In order to justify the summation of the differences, the mean batch results must not differ too much, up to 10% should be acceptable.

If authentic samples are not available, spiked samples may be used. In the case of impurity determinations this can be combined with linearity, i.e. spiking over the working range with at least nine determinations [5]. Performing an unweighted linear regression, the residual standard deviation can be used, normalized with the mean concentration (see Chapter 7). This corresponds to the relative standard deviation of repeatability, provided that the range between maximum and minimum concentrations does not exceed a ratio of about five, in order to avoid weighting effects. Such an approach allows one to combine accuracy, precision, linearity and quantitation limit, provided that the concentration ranges are not too large. As a minimum, subsets of data from the same experimental run can be used. The disadvantage of such an approach is that the sample preparation differs somewhat from the routine procedure and possibly inhomogeneous distribution of the impurities in the material cannot be detected.

What is acceptable for repeatability?

When establishing acceptance limits, it is essential to include all results based on a normal variability. Therefore, the upper limit of the respective distribution must be taken (see Fig. 8.4). Including 99% of the whole distribution, the upper limit for standard deviations calculated from six results corresponds to about twice the true standard deviation. If a reliable experimental standard deviation (target standard deviation [13], Eq. 8.9) is used, the 99% level covers also the uncertainty of this determination.

Of course, the acceptance limits of the specification establish minimum requirements for the variability (see Section 16.2). Further, the current analytical state-of-the-art should be considered. Often, in an LC assay a limit for repeatability of 2.0% is used, some sort of "collective intuition" because this limit cannot be traced back to hard facts. In publications, repeatability results mostly range between 0.5 and 1%, but sometimes they are distinctly larger, up to 4% and, in exceptional cases, up to 12% (for review, see [10], p. 43). It is often difficult to inspect such extreme results more closely because of the lack of information, for example, on the difference between the test concentration and quantitation limit (see Section 8.3.6).

A compilation of 458 repeatabilities from validation and transfer studies at Sanofi-Aventis and from stability studies (project of the German Pharmaceutical Society [14]) allowed further differentiation. Repeatability depends more on the respective sample type (drug product) than on the analyte. The upper limits of the distribution of individual repeatabilities and the averages were investigated for various drug product types. The latter can be taken as typical or target for the respective group. The upper distribution limits and averages for drug substances, injection solution and lyophilisates were found to be 1.2–1.4% and 0.5–0.6%, respectively. This is in good agreement with the target standard deviation of 0.6% from a collaborative trial of the European Pharmacopeia for the LC assay of Cloxacillin [15]. The assay of tablets displays higher variability, up to 1.8%, with an average repeatability of 0.9%. The average for semi-solid formulations is slightly larger at 1.0%, the upper limit for a pharmaceutical bath corresponds to 3.4%. For other drug products, such as emulsions and chewing gums, repeatabilities up to 5% and 15%, respectively, are described in the literature (see [10], p. 48).

The dependence of repeatability on the type of drug product is probably linked to the sample/sample preparation. The more complex they are, the larger is the variability. Of course, special analytes or methods may have other precisions, but the investigation supports the often used 2% limit for a lot of applications. Another interesting finding is the agreement of the ratio between upper distribution limit and true standard deviation of about 2 from simulations with that between the upper limit and the average for experimental repeatabilities.

8.3.4
Reproducibility

The next precision level includes factors that are important for the long-term variability, such as different time, operators, equipments, reagents, etc. When varying these factors within the same laboratory, the term "intermediate precision" (or intra-laboratory) is used, involving other laboratories then "reproducibility" (inter-laboratory) is used. The distinction between the two is somewhat arbitrary, especially if larger time frames are taken into consideration, such as covered in stability studies. This leads us to an almost philosophical question: For how long a time does a laboratory remain the same? Over five years, whole companies can change! Therefore, in the following, the term reproducibility will be used.

Collaborative trials (which include several companies or organisations) should be discussed separately from the within-company applications. Here, additional effects can be expected, such as (larger) differences in equipment, know-how and interpretation of documentation ("cultural" effects).

8.3.4.1 Number of Series

As several series/runs are performed in a reproducibility study, for chromatographic assay procedures the variability of the calibration is an important contribution. Consequently, the whole analytical procedure including calibration must be independently repeated for each series.

In the ICH guideline, the required number of series is not specified, only the calculation of precision for each level is requested. Therefore, the minimum approach consists in two series with six determinations each, for example performing the second series with another operator on another LC-system. Here, the varied factors are confounded. Of course, as discussed before, the reliability of the reproducibility will also increase with the number of series. As this precision level is the most relevant for the very routine applications, such as batch release or stability, a pragmatic compromise on the number of series should be made. Increasing the number of series (k), the number of determinations within a series (n) may be reduced, because the overall degrees of freedom are of importance ($df = k(n-1)$).

8.3.4.2 Analysis of Variances

Preferably, the calculation is performed by means of an analysis of variances (ANOVA) [2]. Here, results for several precision levels are obtained, as requested by the ICH guideline [5], also allowing an evaluation of the ruggedness [16] of various steps of the analytical procedure.

Two precision levels can be obtained applying a one-way (or one-factor) ANOVA (Eqs. 8.9 and 8.11 [2, 17]), such as (average) repeatability and reproducibility in the case of a reproducibility study. The former summarizes the variability within (all) series (intra-serial variance), the latter includes additionally the variance between the series (inter-serial variance), caused by the varied factors of the study, such as calibration, operator, equipment, etc. Owing to the larger degrees of freedom, the pooled repeatability (Eq. 8.9) is a better estimate of the true repeatability standard deviation. The example in Table 8.2 resulted in a pooled repeatability of 0.87% and 0.85%, using the first and second injection, respectively. The injections were taken separately, so that the calculated standard deviation is based on single injections.

Table 8.2 Reproducibility study of an LC assay (in percent label claim) including five series with four sample preparations and two injections each. The one-way ANOVA was calculated separately using the first and second injections (see also ANOVA.xls on the Bonus-CD).

Sample preparation	Sample preparation 1		Sample preparation 2		Sample preparation 3		Sample preparation 4	
Injection	1	2	1	2	1	2	1	2
Series 1	100.22	99.80	99.86	99.80	99.44	99.53	100.57	100.47
Series 2	101.83	101.88	101.20	100.81	101.78	101.58	101.20	101.03
Series 3	99.51	99.46	100.57	101.09	102.10	101.38	99.06	99.07
Series 4	100.55	100.60	100.76	100.56	100.06	100.38	100.52	100.27
Series 5	99.57	99.68	97.88	97.57	100.03	100.83	100.79	99.09

One-way ANOVA (1. Injection)	Intra-series variance ($df = 15$)	0.7575	Repeatability	0.87%
	Inter-series variance ($df = 4$)	0.3259	Reproducibility	1.04%
One-way ANOVA (2. Injection)	Intra-series variance ($df = 15$)	0.7200	Repeatability	0.85%
	Inter-series variance ($df = 4$)	0.3789	Reproducibility	1.05%
Two-way ANOVA	Injection variance (level 3) ($df = 20$)	0.1265	System precision	0.35%
	Sample preparation variance (level 2) ($df = 15$)	0.5996	Repeatability	0.85%
	Inter-serias variance (level 1) ($df = 4$)	0.3689	Reproducibility	1.04%

Intra-series variance:

$$s_r^2 = \frac{\sum_{j=1}^{k} \sum_{i=1}^{n_j} (x_{j,i} - \overline{x}_j)^2}{\sum n_j - k} \quad \text{or} \quad s_r^2 = \frac{\sum (s_j^2)}{k} \quad \text{(for same } n\text{)} \tag{8.9}$$

Inter-series variance:

$$s_g^2 = \frac{\sum_{j=1}^{k} (\overline{x}_j - \overline{\overline{x}})^2}{k - 1} - \frac{s_r^2}{n} \quad \text{(for same } n\text{)} \tag{8.10}$$

Overall variance and precision:

$$s_R^2 = s_r^2 + s_g^2 \quad \text{if } s_g^2 < 0: \quad s_R^2 = s_r^2 \quad s_R = \sqrt{s_R^2} \quad (8.11)$$

where $x_{j,i}$ = individual result x in series/run j; $n_j, s_j, \bar{x}_j$ = number of results, standard deviation, and mean for series/run j; k = number of series/runs, respectively.

The calculation just takes series of data into account. Therefore, it is essential that the user is aware of the experimental design, i.e. which precision levels are addressed. For example, the repeated injection of several independently prepared sample solutions (series = sample preparations) would allow one to calculate the pooled system precision (Eq. 8.9) and the repeatability (Eq. 8.11).

The original statistical analysis of variances includes tests for the homogeneity of intra-serial variances and for significant differences between the means. The latter is the very aim of such an analysis, the identification of deviating groups of data. However, our intention is the calculation of precisions. Therefore, the above mentioned statistical tests should not be performed, or their results ignored. Variability that is not acceptable is better controlled by establishing upper limits for individual repeatabilities and for reproducibility. The latter will also include differences between the means, but instead of testing statistical significance, an upper precision limit will take the practical relevance into account.

Applying multiple-factor ANOVA, the variances of multiple steps of an analytical procedure, i.e. multiple precision levels can be separated [18, 19]. For example, if duplicate injections are performed in a reproducibility study, system precision, repeatability, and reproducibility can be calculated (see Table 8.2 and ANOVA.xls on the Bonus-CD). With other designs, it is possible to separate the contribution of individual factors, such as operator, equipment, etc. Such results can be used for method optimization (see Section 8.3.5).

8.3.4.3 Benchmarking for Reproducibility

It is much more challenging to find results for intermediate precision and reproducibility in the literature. Relative standard deviations between 0.5 and 3.5% are reported (for review see [10], p. 46). The aforementioned compilation of parameters from Sanofi-Aventis and the German Pharmaceutical Society includes 224 reproducibilities from validation and transfer studies and from stability investigations [14]. The pooled reproducibilities are 1.4- to 2-fold larger than the respective pooled repeatabilities, reflecting the additional variance contributions. The averages and the upper limits of the reproducibilities were 1.0% and 1.7% for lyophilisates and drug substances, 1.1% and 2.5% for injection solutions, 1.2% and 2.3% for tablets, and 1.6% and 3.0% for creams, respectively (see [10] p. 48). The smaller the repeatability standard deviation (higher precision), the larger the variance contributions of the varied factors, as in the case of injection solutions, and *vice versa*, for tablets. This may easily result in the former case in statistically significant differences between the means, despite quite acceptable overall precision.

8.3.5
Consequences for the Design of an Analytical Procedure

Knowledge of the various variance contributions can be used to identify the main contribution to the overall precision (see Fig. 8.7) and thus the best leverage to optimize the analytical procedure. This may be achieved either experimentally (e.g. by the sample preparation, by better control test descriptions, etc.), or "statistically" by increasing the number of determinations. The equations for standard deviations discussed previously relate to single determinations (injections). If the individual variance contributions are known to be reliable, a "synthetic" standard deviation of the complete analytical procedure can be calculated using Eq. (8.12). Determining the precision exactly according to the control test, for example with duplicate injections, two sample preparations, two external reference standards repeated (altogether) a sufficient number of times (minimum six times) would result in an equivalent standard deviation of the analytical procedure. However, this precision would only be valid for the performed control test design.

Precision of the analytical procedure:

$$s_{AP} = \sqrt{\frac{s_i^2}{mnk} + \frac{s_p^2}{nk} + \frac{s_g^2}{k}} \tag{8.12}$$

where s_i^2, s_p^2, s_g^2 = system variance, variance of the sample preparation, inter-series variance, respectively, m = number of multiple injections, n = number of sample preparations, k = number of series (with own calibration), usually $k = 1$.

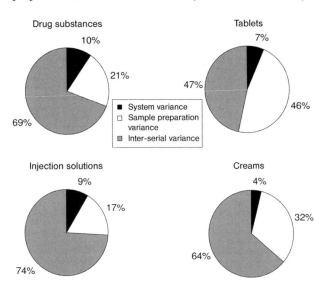

Figure 8.7 Presentation of the relative amount of variances contributing to reproducibility for various drug products or sample types. The average variances of the examples given in [10], Table 2.1-5, were used, complemented by a system precision of 0.31% [11], which is assumed to be the same for all sample types.

With m, n, and $k = 1$, the standard deviation of a single injection is obtained. The variance of the repeatability results from the first two terms in Eq. (8.12). The number of series (k) is given for completeness, but will usually be 1. Usually, a final result will not be composed of several individual series with independent calibration, apart from maybe bioassays.

For the purpose of optimization, the number of multiple injections and of sample preparations can be varied in order to investigate their influence on the overall precision. Of course, the standard deviation will continuously decrease with increasing numbers, leading to an effort–benefit optimization. The optimum has to be defined by the user. In Fig. 8.8, various scenarios are shown for injection solutions and tablets (from [10], Table 2.1-5) (see also Optimisation.xls on the Bonus-CD). It is obvious, in the tablet example, that repeated injections do not make sense, because the system variance has only a small contribution (Fig. 8.8b), in contrast to an increase in the number of sample preparation procedures. In case of the injection solution (Fig. 8.8a), increasing both numbers is effective. However, the starting point is already at a fairly high precision.

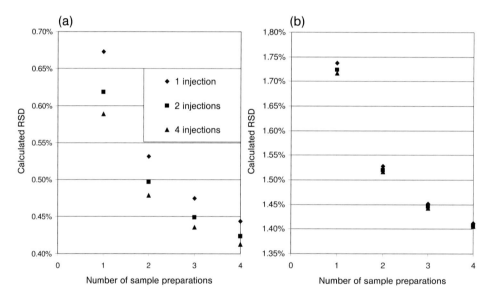

Figure 8.8 Calculated relative standard deviation of the analytical procedure as a dependence on the number of multiple injections and sample preparations (see also Optimisation.xls on the Bonus-CD.
(a) LC assay of an injection solution (from [10]) with $s_i^2 = 0.0961$ (16%), $s_p^2 = 0.0159$ (3%), and $s_g^2 = 0.4960$ (82%).
(b) LC assay of tablets (from [10]) with $s_i^2 = 0.0961$ (3%), $s_p^2 = 1.274$ (42%), $s_g^2 = 1.650$ (55%). The variances were normalized to refer to a mean of 100.

8.3.6
Concentration Dependence of the Precision

With decreasing concentration, the absolute variability will also decrease, but to a lesser extent. Consequently, the relative standard deviation increases. William Horwitz *et al.* reviewed a large number of collaborative trials of various analytes and analytical procedures and found a strikingly simple mathematical function of the concentration dependence of the (collaborative) reproducibility (Eq. 8.13). The relative standard deviation is only dependent on the concentration of the analyte in the respective matrix. The so-called Horwitz function is used in collaborative trials as a benchmark to estimate an acceptable reproducibility. The results should range between half and twice the Horwitz prediction for the respective concentration fraction. The intra-laboratory variability (i.e. the repeatability) can be expected to be between half and two-thirds of the reproducibility [20]. For example, analyzing a drug substance in a dosage of 4 mg in a 400 mg tablet (i.e. a concentration fraction of 0.01), the collaborative reproducibility and the repeatability can be expected to be between 2.0 and 8.0% and between 1.0 and 5.3%, respectively.

Horwitz function:

$$RSD_R(H) = 2\,c^{-0.1505} \tag{8.13}$$

where $RSD_R(H)$ = relative standard deviation of the collaborative reproducibility predicted by the Horwitz function. Experimental results should range between one half and twice this prediction. c = concentration fraction of the analyte in the original material.

The Horwitz relationship is valid for a broad range of concentrations and applications. In addition, the scope includes collaborative trials where additional effects take place, leading to a variability increase [21]. Applications in pharmaceutical analytics, applied within a company, can be expected to be under better control, and, at least, to be more consistent.

What are the consequences for the concentration dependence of precision in these applications?

In a model investigation using Glibenclamide within a matrix of typical tablet excipients [11], repeatabilities based on five to seven determinations were calculated for a concentration range between 0.025 and 125% of the label claim (see Fig. 8.9).

It can be seen in Fig. 8.9a that the repeatabilities are below the Horwitz predictions for higher concentrations. Approaching the quantitation limit, they are found at the upper end. Above about 10% of the label claim, both system precision and repeatability are constant, with a clear difference (see Fig. 8.9b). This difference is due to the variability of the sample preparation. The corresponding variance of $(0.63\%)^2$ can therefore be calculated from the pooled repeatability variance of $(0.70\%)^2$ and the pooled system variance of $(0.31\%)^2$. When considering the variance contributions to the system precision discussed in Section 8.3.1 (see

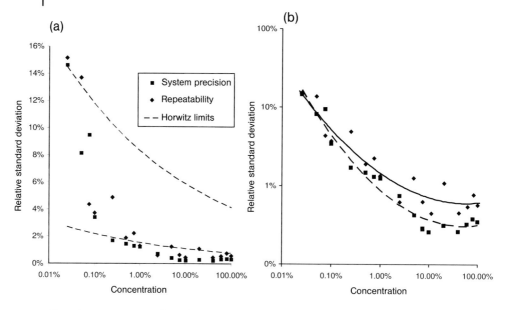

Figure 8.9 Concentration dependence of system precision (squares) and of repeatability (diamonds) of the LC assay of Glibenclamid tablets (data from [11]). (a) The dashed lines represent the limits of the predicted repeatability according to the Horwitz relationship [20]. (b) Double logarithmic presentation. The trends of the system precision and repeatability are displayed by dashed and solid lines, respectively. They were calculated by quadratic regression of the logarithm of the respective standard deviations versus the logarithm of the concentrations.

Eq. 8.7), at this concentration only the injection variance is of importance. With decreasing concentration, the relative standard deviations of system precision and repeatability increase and approach each other. As both sample preparation and injection variability remain constant, the increase in the overall variability must be caused by the detection and integration variability. In Fig. 8.10, the concentration dependence of the major relative variance contributions is shown. Up to about 0.1% (corresponding to two to four times the quantitation limit), the detection and integration variability is the overwhelming variance contribution. The variances of sample preparation and injection can be neglected and, consequently, system precision and repeatability are the same. At concentrations above 10% (corresponding to about 200 times the quantitation limit), the detection and integration variance can be neglected and the system precision corresponds to the injection variability (see Fig. 8.10b). It is essential to consider the trend of the contributions, because the relative contributions for individual concentrations are heavily influenced by the scatter of the individual standard deviations. The increase in the integration variability above 80% concentration is probably due to injection precisions larger than the average by chance (the results at 125% were also used for calculation but are not shown in the figures for scaling reasons).

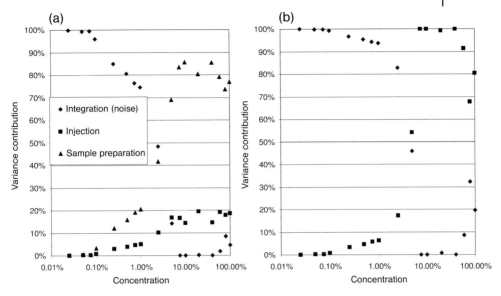

Figure 8.10 Concentration dependence of the relative variance contributions on repeatability (a) and on system precision (b), for explanation see the text.

8.3.7
Conclusions

The use of internal standards will only result in an improvement in the overall precision if the injection variability is the major contribution. Of course, the compensation of matrix or sample preparation effects is another practically relevant justification.

If no authentic samples containing impurities in sufficient concentration are available, the precision of the impurity determination can be obtained by repeated injections. However, authentic samples are still preferable, because only here can inhomogeneities in the material or in the sampling be recognized. If they are of an acceptable size, such inhomogeneities must be regarded as part of the normal material variability.

Precision is always linked to the overall concentration. Therefore, in the case of recovery from analyte-containing matrices, the added and found amounts must always be related to the overall concentration. Otherwise, the – perhaps much larger – variability of the overall concentration is attributed to the small spiked amount of analyte. This may lead to large deviations from the theoretical recovery of 100%. The error can be neglected if the spiked amount exceeds the initially present concentration by more than a factor of about 4 (see [10], p. 71).

8.4
Key Points

- The reliability of parameters is strongly dependent on the number of data the calculation is based on. For instance, the true value may be up to 4.4 times the standard deviation calculated from $n = 3$ (upper limit of the 95% confidence interval)!

- Because of the normal random variability of experimental data, parameters determined in the future also scatter. In order to include all (normal) future results, only the (usually upper) limit of the respective distribution is a meaningful and safe acceptance limit. For example, the upper distribution limit for standard deviations calculated from six results can be estimated to be twice that of a reliable experimentally determined standard deviation (target standard deviation, from reproducibility studies).

- Each step of an analytical procedure contributes to the overall variability. The precision levels system precision, repeatability, and reproducibility contribute various sub-steps. Standard deviations should preferably be related to single determinations in order to allow further calculation and a straightforward interpretation.

- The variance contributions of the various precision levels can be calculated separately by means of an analysis of variances (ANOVA). Based on such information, it is possible to optimize the overall precision of the analytical procedure varying the number of weighings and multiple injections.

- Precision depends on concentration. Within the same method, the relative composition of the variance contribution changes with concentration. Decreasing the concentration, the relative amount of the noise variance increases until it becomes the dominating variability below about 20 times the quantitation limit. In contrast, with increasing concentration, the injection variance becomes more and more important for the system precision until it dominates above about 200 times the quantitation limit. Consequently, this represents the minimum concentration for a meaningful system suitability test.

References

1 United States Pharmacopeia 29, National Formulary 24, Section <1010> Analytical Data – Interpretation and Treatment, United States Pharmacopeial Convention, Rockville, **2006**.

2 DIN ISO 5725-2: Accuracy (trueness and precision) of measurement methods and results; a basic method for the determination of repeatability and reproducibility of a standard measurement method, **1990**.

3 EURACHEM/CITAC Guide: Quantifying uncertainty in analytical measurement, 2nd ed., **2000**. http://www.measurementuncertainty.org

4 T. Anglov, K. Byrialsen, J. K. Carstensen *et al.*, *Accred. Qual. Assur.* **2003**, 8, 225–230.

5 ICH. Q2 (R1): Validation of Analytical Procedures: Text and Methodology, **2005**.

6 V. R. Meyer, *Pitfalls and Errors of HPLC in Pictures*, Wiley-VCH, Weinheim, **2006**.

7 G. Maldener, *Chromatographia* **1989**, *28*, 85–88.

8 S. Küppers, B. Renger, and V. R. Meyer, LC-GC Europe, **2000**, 114–118.

9 B. Renger, *J. Chromatogr. B* **2000**, *745*, 167–176.

10 J. Ermer, Chapter 2: Performance parameters, calculations and tests, in *Method Validation in Pharmaceutical Analysis, A Guide to Best Practice* (J. Ermer and J. H. McB. Miller, Eds.), Wiley-VCH, Weinheim, **2005**.

11 U. Schepers, J. Ermer, L. Preu, and H. Wätzig, *J. Chromatogr. B* **2004**, *810*, 111–118.

12 J. Ermer, H.-J. Ploss, *J. Pharm. Biomed. Anal.* **2005**, *37*, 859–870.

13 A. G. J. Daas, J. H. M. Miller, *Pharmeuropa* **1997**, *9*, 148–156.

14 J. Ermer, C. Arth, P. De Raeve, D. Dill, H. D. Friedel, H. Höwer-Fritzen, G. Kleinschmidt, G. Köller, H. Köppel, M. Kramer, M. Maegerlein, U. Schepers, and H. Wätzig, *J. Pharm. Biomed. Anal.* **2005**, *38*, 653–663.

15 A. G. J. Daas, J. H. M. Miller, *Pharmeuropa* **1998**, *10*, 137–146.

16 United States Pharmacopeia 29, National Formulary 24, Section <1225> Validation of Compendial Methods, United States Pharmacopeial Convention, Rockville, **2006**.

17 MVA – Method Validation in Analytics. Software for calculation of analytical validation data, **2005**. NOVIA GmbH, Frankfurt, Germany. www.novia.de

18 DIN ISO 5725-3: Accuracy (trueness and precision) of measurement methods and results – Part 3: Intermediate measures on the precision of a test method, **1991**.

19 K. Koller, H. Wätzig, *Electrophoresis* **2005**, *26*, 2470–2475.

20 R. Albert, W. Horwitz, *Anal. Chem.* **1997**, *69*, 789–790.

21 W. Horwitz, *J. Assoc. Anal. Chem.* **1977**, *60*, 1355–1363.

9
Metrological Aspects of Chromatographic Data Evaluation

Ulrich Panne

9.1
Introduction

Measuring and testing laboratories typically analyze chromatographic data at several hierarchical levels. The first step is a numerical analysis of the data using mathematical and statistical methods. The objective here is to avoid such well-known flaws as, for instance, non-linear relationships or data transformation in order not to generate "invalid results from valid data". Other multivariate chemometrical methods such as factor analysis and classification algorithms often run the risk that researchers either apply them inappropriately or interpret their results incorrectly. The abundance of information provided by modern methods such as LC-MS/MS thus seems to lose some of its value (e.g. [1, 2] and the literature therein). A detailed discussion of how to achieve more appropriate data analysis can be found in the preceding chapters.

The next level in evaluating the quality of chromatographic measurements is concerned with metrological aspects, especially uncertainty and traceability of the final results. *Metrology* is the science of measurement and is often used synonymously with *measurement engineering* or *measurement technology*. In general, it is referred to as the system of rules and procedures that helps guarantee the correctness and comparability of measuring results. Important elements of this system are the International System of Units (SI), metrological traceability, the calculation and statement of measurement uncertainty, comparisons between laboratories and the principles of quality management. All parties involved contribute their specific expertise: national metrology institutes assure traceability to SI units by providing reference measurements of the highest quality, calibration laboratories verify the quality of measuring instruments through comparison measurements, and testing laboratories guarantee the quality of testing results by means of quality assurance in compliance with the principles laid down in the ISO/IEC 17025 standards. *Metrology in chemistry* thus ensures that all analytical results are accurate, reliable, and comparable.

Quantification in LC and GC: A Practical Guide to Good Chromatographic Data
Edited by Hans-Joachim Kuss and Stavros Kromidas
Copyright © 2009 WILEY-VCH Verlag GmbH & Co. KGaA, Weinheim
ISBN: 978-3-527-32301-2

At the international level, however, mutual recognition of analytical results cannot always be taken for granted, especially when it comes to such sensitive issues as consumer protection, health care, or environmental protection. This, of course, can have serious consequences for international trade, which is why national metrology institutes worldwide signed the CIPM Mutual Recognition Agreement (MRA) in 1999. The aim of the agreement is to establish the mutual recognition of national standards as well as of calibration and measurement certificates issued by national metrology institutes. This also affects analytical chemistry (see section in this agreement on "Amount of Substance"). The mutual recognition of national measurement standards is meant to be a foundation for wider agreements related to international trade, commerce, and regulatory affairs. However, recognition in this sense is far more than just a formal act. It is also a liability, because the respective national institutes are required to make use of an adequate quality system and to give proof of their competence and reliability by submitting their results to international key comparisons (KC). In the interest of the greatest possible transparency, the Bureau International des Poids et Mésures (BIPM) in Paris makes the results of these interlaboratory comparisons publicly available on its website (www.bipm.org).

While national measurement standards for physical quantities have been in place for decades, the establishment of such standards for chemical analysis is only just beginning. In the classic fields of metrology, it is the calibration laboratories which ensure traceability in accordance with national standards. In analytical chemistry, however, having such an intermediate level is the exception rather than the rule. Here, it is common practice that the national metrology institutes offer metrological services directly to the users, i.e. calibration, measurement or testing laboratories.

Thanks to the establishment of these requirements in ISO/IEC 17025 [3], the traceability of measurements and their corresponding uncertainty have become the two major paradigms for the work of measuring and testing laboratories. As traceability is not a mathematical property, it has only to be demonstrated, whereas uncertainties can be calculated numerically. Not surprisingly, metrological procedures in chemical analysis are often said to be somewhat artificial and unrealistic [4, 5].

Understandably enough, this is especially true when compared with metrology in physics, where the focus primarily lies on reducing uncertainties. Neither the definition of analytes and quantities nor their traceability seem to pose a major problem. It is considered self-evident that the length of an object, for instance, has to be measured independently of the object itself and that analytical balances are a totally inadequate instrument for this type of measurement. In analytical chemistry, however, measurement tasks are often much more complex and generally lack the sense of self-evidence that has just been described as characteristic of physical metrology. Usually, the exact composition of the sample as a whole is not known and measurements therefore focus on a single analyte or a certain class of analytes. The measurement is often seriously influenced by the matrix, i.e. the determination of an analyte varies with the object of the measurement.

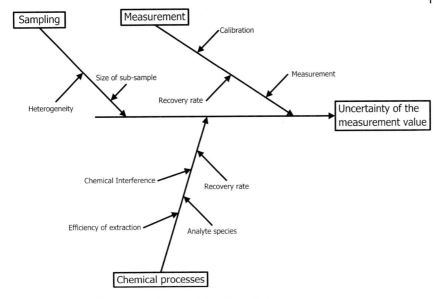

Figure 9.1 Cause/effect diagram for traceability of chemical measurements.

On the other hand, the identity of an analyte cannot be determined free of doubt due to the underlying chemistry in the matrix, which is not sufficiently known or understood, so that the methods used cannot bring about valid results. The sample itself, or *what analyte* is to be measured in *what kind of material*, thus becomes the decisive element in chemical metrology. This in turn increases the importance of reference materials and their availability for ensuring traceability.

Measurement results are typically determined by three analytical elements (see Fig. 9.1): the measurement instrument, i.e. the repeatability of a controlled measurement set-up, the analytical method, i.e. the sample treatment and the selectivity resulting from it, and the sampling. Interestingly, in evaluating uncertainty, analysts focus almost exclusively on the first two processes mentioned above, in spite of the significance of sampling with regard to the overall uncertainty of the analysis. This of course entails a significant restriction as to what type of samples can be analyzed with the methods established by a given testing laboratory. In other words, the available validated methods only cover a certain range of sample heterogeneity.

9.2
Measurement Uncertainty

Uncertainty is a fundamental characteristic of measurement results and is intrinsically tied to them [6–9]. In analytical chemistry, these results usually indicate the concentration of a specific analyte. However, they can be noticeably more complex in cases where the dependent variable is method-dependent, e.g. in measuring adsorbable organic halogens (AOX). Indicating uncertainty is neither an optional feature of analytic measurements, nor does it have any sort of implications regarding the validity of the measurement. On the contrary, the knowledge of uncertainty implies increased confidence in the validity of measurement results. Even if all sources of systematic error have been eliminated, random variation of an isolated measurement with regard to the expected value may already result in creating uncertainty. The concept of "error" or "error of measurement" must be clearly differentiated from that of "uncertainty". Error of measurement can be defined as the difference between an individual result and the true value of the quantity in question, i.e. a single value. Uncertainty on the other hand is an interval. Once it has been determined for a specific analytical method and a specific type of sample, it can be applied to all measurements under the same specifications. A measurement result can randomly approach the (true) value of a quantity and thus have a negligible error. Yet, the uncertainty of the measurement can still be very high because the true value is an idealized value and thus remains unknown.

Consequently, the GUM (*Guide to the Expression of Uncertainty in Measurement* [10]) defines measurement uncertainty as a *"parameter, associated with the result of a measurement that characterizes the dispersion of the values that could reasonably be attributed to the measurand"*. This definition applies to values obtained under repeatability, intermediate, or reproducibility conditions (see below). Other factors that may contribute to this dispersion are the corrections of known systematic errors and the theoretical approximations to systematic errors. Thus, all experimental or theoretical values which have not been identified as false are attributed to the quantity. *Measurement uncertainty expresses the range of possible quantity values and characterizes, together with the measurement result itself, our knowledge of the measurand.*

Measurement uncertainty is therefore intrinsically tied to the *reliability* and *comparability* of analytical results and backs up the confidence placed in subsequent decision processes. Uncertainties have to be quantified independently of the procedure and of the parameter in question. Moreover, the uncertainties of a measurement result can be used to determine the uncertainty of other derived quantities. For a general overview, references to the corresponding ISO materials, and exemplary quantifications, see the EURACHEM/CITAC guide cited in [11] or the respective BAM document in [12].

Numerical treatment of uncertainty begins with the definition of the following concepts: The *standard uncertainty, u,* is the uncertainty of a measurement result expressed as the standard deviation. The *combined standard uncertainty, u,* is the standard uncertainty of a measurement result, when that result is obtained from the values of a number of other quantities. It is equal to the positive square root of a sum of terms, the terms being the variances or covariances of these other quantities weighted according to how the measurement result varies with changes in these quantities. The *expanded measurement uncertainty, U,* is a quantity defining an interval about the result of a measurement that may be expected to encompass a large fraction of the distribution of values that could reasonably be attributed to the measurand. The *coverage factor, k,* is a factor used as a multiplier of the combined standard uncertainty in order to obtain an expanded uncertainty. The *accuracy* of a measurement is defined as the closeness of agreement between a quantity value obtained by measurement and the true value of the measurand. Subordinate to this generic term are two other terms: *precision,* i.e. the closeness of agreement between a quantity value obtained by independent measurements and the true value of the measurand, and *trueness,* i.e. the closeness of agreement between the average of a large number of independent measurements and the true value of the measurand. According to the way in which it is determined, *precision* can be differentiated into repeatability conditions, intermediate conditions, and reproducibility conditions. Repeatability conditions refer to the same measurement procedure, the same laboratory, the same operator, the same measuring system and replicated measurements over a short period of time. Reproducibility conditions on the other hand refer to the same measurement procedure, but to different laboratories, operators and measuring systems. Repeatability conditions and reproducibility conditions are therefore instances of minimal and maximal variability. Trueness, i.e. the idealized true value, is much more difficult or even impossible to determine in chemical analysis. Suitable reference objects (standards, artefacts, reference materials) or parallel use of reference procedures are alternatives.

The GUM classifies components of measurement uncertainty as either *Type A* or *Type B,* based on the evaluation methods employed. Type A is evaluated through a statistical analysis of a measurement series [13], Type B from means other than a statistical analysis. The conventional Type A evaluation assumes a probability distribution, in many cases the normal distribution, to account for the random dispersion of quantity values. The alternative Type B evaluation is often used to estimate uncertainties that result from unknown systematic deviances and is based on experience. A typical example of a Type A evaluation is to calculate an estimated value for the standard deviation σ of an assumed normal distribution. Based on the values $x_1, x_2, ..., x_n$ as estimates of the quantity X, the standard deviation σ of this normal distribution can be used to estimate the standard deviation s

$$s = \sqrt{\frac{\sum_{i=1}^{n}(x_i - \bar{x})^2}{n-1}} \tag{9.1a}$$

$$\text{with} \quad \bar{x} = \frac{\sum_{i=1}^{n} x_i}{n} \tag{9.1b}$$

The standard uncertainty $u(\bar{x})$ of the quantity is calculated as

$$u(\bar{x}) = \frac{s}{\sqrt{n}} \tag{9.2}$$

If the measurement procedure is guaranteed to be free of systematic deviation and to operate under constant dispersion within the measurement range, the empirical standard deviation of the quantity value $\{x_1, x_2, ..., x_n\}$ can be used as an estimate for the standard measurement uncertainty of other measurements within the same measurement range. In this case it must be taken into consideration whether the quantity value is a single value or an average obtained from multiple independently measured single values. For single values, standard uncertainty corresponds to s, for (arithmetic) means obtained from m single values to $s/\sqrt{m}$. An example of a Type B evaluation is the transformation of a maximum/minimum value into a standard uncertainty. If, for a reference value, only a minimum value x_{min} and a maximum value x_{max} are known and if all values within the interval can be assumed to be equally probable, the reference value x and its standard uncertainty $u(x)$ can be expressed by the standard deviation of the continuous uniform distribution with x_{min} and x_{max} as boundaries.

$$x = \frac{(x_{max} + x_{min})}{2} \tag{9.3a}$$

$$\text{and} \quad u(x) = \frac{(x_{max} - x_{min})}{\sqrt{12}} \tag{9.3b}$$

In general, measurement uncertainties (see Fig. 9.2) consist of several components, some of which are evaluated using Type A evaluation, others using Type B. A classification into Types A and B therefore only makes sense with regard to individual uncertainty components.

The GUM does not differentiate between undetected systematic uncertainty components on the one hand and random ones on the other. Known systematic

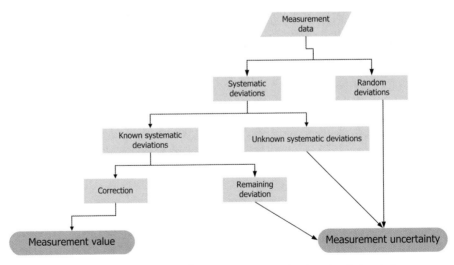

Figure 9.2 Different components of measurement uncertainty in a chemical analysis.

errors of measurement are compensated by technical measures or mathematical calculations. The remaining component in the uncertainty balance thus encompasses the uncertainties of both technical adjustments and corrections of the measurement. All uncertainty components are required to be treated equally since the decision as to whether effects are systematic or random in nature depends on the specific measurement. A random effect becomes a systematic one when the result of an input variable becomes the input of a second measurement. Conversely, systematic effects produced by a specific measurement procedure in a specific laboratory become random effects when, for instance, the results of interlaboratory comparisons display a great number of laboratories which are subject to different systematic effects.

In calculating the combined standard uncertainty, all components of uncertainty must be treated equally. In the past, this practice was sometimes ignored in a "conservative" estimate, which was obtained by combining the random components in quadrature (see below) and the systematic components linearly. Today, it is recommended to choose an appropriate coverage factor k. In addition, a maximum value estimate for the uncertainty of the results is also suggested, for example when comparing with threshold values. The expanded uncertainty $U(y)$ for the measurement result y:

$$U(y) = k \, u(y) \tag{9.4}$$

obtained as the product of the standard uncertainty $u(y)$ and a coverage factor k, defines a confidence interval

$$y - U(y) \leq Y \leq y + U(y) \tag{9.5}$$

within which the true value Y of the measurand is believed to lie with a specified probability p (e.g. $p = 95\%$). This interval thus contains the fraction p of all values which can reasonably be attributed to the measurand. The calculation for the confidence interval assumes the distribution function of the measurement values is known. Given that, generally speaking, this is rarely the case, the GUM recommends choosing a coverage factor k which lies between 2 and 3 ($k = 2$ corresponds to $p = 95.45\%$, $p = 95\%$ to $k = 1.96$). In either case, the coverage factor k should always be stated so that the standard uncertainty u can be recovered.

The calculation of measurement uncertainty is based on the assumption that the result of a measurement is corrected for all recognized systematic effects. The rest of the process is divided into different steps (see Fig. 9.3).

1. *Specify the measurement result y and the procedure used to quantify it.* In analytical chemistry, the measurement result can depend on the type of empirical method in use. These procedures primarily serve to achieve comparability between laboratories and thus to define the procedure as a standard method on a local, national, or international level. For this reason, no corrections are made for systematic effects in the method or for matrix effects.

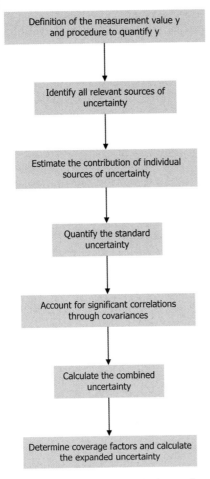

Figure 9.3 Flow chart for the calculation of uncertainty.

2. *Identify all relevant sources of uncertainty.* Here, all relevant sources of uncertainty are initially acknowledged without quantification. Cause and effect diagrams are typically used to illustrate the hierarchical relationships between the influencing variables. The factors used include the relevant measurands of the measurement result, indirect influencing variables (e.g. pressure and temperature), reference quantities (values taken from reference materials) and literature data and parameters for the individual sub-steps of the whole measurement procedure (e.g. recovery in preparing the sample). The aim is to formulate an equation $y = F(x_1, x_2, ..., x_N)$, which describes the measurement result in relation to all relevant influencing variables.

3. *Estimate the contribution of individual sources of uncertainty.* In some cases, particularly when only a few data are available to characterize the method, the

most suitable procedure may be to evaluate each uncertainty component separately. For empirical methods, an interlaboratory comparison is normally used to obtain an estimate. A collaborative study to validate a reference procedure (carried out according to AOAC/IUPAC or ISO 5752 standards) is a valuable source of data (e.g. for reproducibility or systematic effects based on CRM studies) to support an uncertainty estimate. By using the appropriate Certified Reference Material (CRM), all systematic errors can be estimated as the whole measurement procedure is carried out. If the systematic errors are not significant, then the systematic uncertainty represents the standard uncertainty of the corresponding CRM value. As a general rule, if two contributions from influence factors differ by a factor of 1/5, then the smaller contribution can be ignored in favor of the larger one. As a result of the quadratic addition (see below) a standard uncertainty which is smaller by a factor of $1/p$ only contributes a fraction of approximately $1/(2\,p^2)$ of the larger standard uncertainty to the uncertainty ($p = 5$ corresponds to approximately 2%). Small uncertainty contributions that appear in greater numbers or correlated contributions which result in a linear addition of uncertainty contributions instead of a quadratic one must be taken into account.

4. *Quantify the standard uncertainty.* The standard uncertainty $u(x_i)$ is identified for all input variables x_i ($i = 1, 2, ..., N$) as far as experimental data is available. Although the Type A evaluations are reputed to achieve greater objectivity, the empirical standard deviations can result in an imprecise estimate value for the standard uncertainty when very few measurement values are available. In this case, a Type B evaluation, which is based on previous experience and scientific judgment, may be preferred. Furthermore, differential or difference quotients are used to determine the sensitivity coefficients c_i for the dependence of the measurement result $y = F(x_1, x_2, ..., x_N)$ on the input variables x_i.

$$c_i = \frac{\partial F}{\partial x_i} \tag{9.6a}$$

$$\text{or} \quad c_i = \frac{\Delta y}{\Delta x_i} \tag{9.6b}$$

The uncertainty contribution of an input variable x_i to the combined standard uncertainty of y is thus obtained through $u_i = c_i\, u(x_i)$.

5. *Account for significant correlations through covariances.* A correlation between two input variables x_i and x_k can be expected when the relevant input variables are interdependent, or are both dependent on a third variable (e.g. when the same stock solution is used to calibrate two different measurements). Correlations, as products of $u_{ik} = c_i\, c_k\, u(x_i, x_k)$ of the covariance $u(x_i, x_k)$ with the relevant sensitivity coefficients c_i and c_k, contribute to the combined standard uncertainty of the measurement result. The calculation can, on the one hand,

be carried out through a propagation of uncertainty relative to the common variables, that is, the variables x_i and x_j are dependent on the uncorrelated variables $z_1, z_2, ..., z_K$ and hence the following applies:

$$u(x_i, x_j) = \sum_{k=1}^{K} \left(\frac{\partial x_i}{\partial z_k}\right) \cdot \left(\frac{\partial x_j}{\partial z_k}\right) \cdot u(z_k)^2 \tag{9.7}$$

If the variables $z_1, z_2, ..., z_K$ are correlated for their part, then their covariances must also be accounted for with

$$u(x_i, x_j) = \sum_{k=1}^{K} \left(\frac{\partial x_i}{\partial z_k}\right) \cdot \left(\frac{\partial x_j}{\partial z_k}\right) \cdot u(z_k)^2$$
$$+ 2 \sum_{k=1}^{K-1} \sum_{l=k+1}^{K} \left(\frac{\partial x_i}{\partial z_k}\right) \cdot \left(\frac{\partial x_j}{\partial z_l}\right) \cdot u(z_k, z_l) \tag{9.8}$$

Alternatively, parallel measurements under repeatability conditions provide an opportunity to estimate empirical covariances of the mean value. To determine whether covariances have to be accounted for at all, note that $u(x_i, x_j)$ is then between $u(x_i) u(x_j)$ and $-u(x_i) u(x_j)$.

6. *Calculate the combined uncertainty.* The contributions to the standard uncertainty of the measurement result are summarized as follows:

$$u(y)^2 = \sum_{i=1}^{N} u_i^2 + 2 \sum_{i=1}^{N-1} \sum_{k=i+1}^{N} u_{ik} \tag{9.9}$$

Equation (9.9) can also be written as

$$u(y)^2 = \sum_{i=1}^{N} \left(\frac{\partial F}{\partial x_i}\right)^2 u(x_i)^2 + 2 \sum_{i=1}^{N-1} \sum_{k=i+1}^{N} \left(\frac{\partial F}{\partial x_i}\right)\left(\frac{\partial F}{\partial x_k}\right) \cdot u(x_i, x_k) \tag{9.10}$$

If no correlations are observed or if the contributions of the correlations are neglected, Eq. (9.10) is simplified to

$$u(y)^2 = \sum_{i=1}^{N} \left(\frac{\partial F}{\partial x_i}\right)^2 u(x_i)^2 \tag{9.11}$$

This equation corresponds to the well known Gaussian law of propagation of uncertainty. Both equations are series expansions which are terminated after the linear term.

In the case of nonlinear correlations, either further terms of the series expansion, i.e. higher degrees of deviants, have to be analyzed or other analysis methods, such as Monte Carlo simulations, have to be applied. If a simple mathematical relationship exists between the measurement result y and the input variables x_i then the sensitivity factors may be precisely determined through differentiations, otherwise calculus of differences are to be aimed for. The covariances $u(x_i, x_k)$ are linked with the standard uncertainties of the appropriate input variables by

$$u(x_i, x_k) = r(x_i, x_k) \cdot u(x_i) \cdot u(x_k) \tag{9.12}$$

where $r(x_i, x_k)$ is the correlation coefficient. If all input variables were equally correlated, the result would be the combined standard uncertainty as the linear sum $u(y) = \Sigma\, u_i$ of the uncertainty contributions. However, in the case of completely uncorrelated input variables the uncertainty contributions are quadratically added as per $u(y)^2 = \Sigma\, u_i^2$ (cf. Eq. 9.9). Summing in quadrature tends to result in considerably smaller values for the combined standard uncertainty $u(y)$ than does the linear addition. This is why the linear addition can be applied in estimating the highest values for the combined standard uncertainty, without the need for the correlations to be individually checked.

7. *Determination of the coverage factor and calculating the expanded uncertainty.* The uncertainty of the measurement result can alternatively be stated either as the standard uncertainty $u(y)$ or as the expanded standard uncertainty $U(y) = k\, u(y)$. The expanded uncertainty is chosen in order to set the boundaries for a range which can be expected to contain the true value of the measurement result with a greater degree of certainty. Typically (i.e. normal distribution) a k is selected between 2 and 3, with $k = 2$ being recommended. If there is sufficient knowledge of the frequency distribution of the measurement result, then k can be calculated as a "confidence factor" and fixed as a confidence level. For this the confidence level 0.95 (95%) is recommended. The necessary number of effective degrees of freedom can be calculated from the standard uncertainties and the degrees of freedom of the distribution for the values of the input variables. Calculated or estimated values of the measurement uncertainty are stated together with the value of the measurement result y, either as standard uncertainty $u(y)$ or as expanded uncertainty $U(y) = k\, u(y)$. If the expanded standard uncertainty is given, then the coverage factor and the corresponding confidence level must also be indicated. Both the standard uncertainty and the expanded standard uncertainty can be stated either directly or as relative values (e.g. in percent).

The complexity of ascertaining the measurement uncertainty based on a detailed uncertainty budget makes it particularly well suited to measurement procedures with a broad range of application, i.e. those with a considerable range of variation for measurement objects and the measurement conditions. In practice, the suitability of analytical methods is proven by a method validation [14], i.e. there are already data to hand on the relevant influence factors. For measurement procedures with a narrow range of application, the more sensible alternative therefore is to estimate the measurement uncertainty using the laboratory's own validation data or data from round robin tests (see for instance [15, 16]). This option is particularly appealing as, in practice, the comparability of methods is more often guaranteed through global parameters such as linearity, limit of determination, recovery rates, etc., without requiring a formal, mathematical model. Because the largest sources of uncertainty arise mostly from discrete results (not differentiable ones cf. Eq. (9.6)) and can only be deduced through experimental studies, they are often a part of a global method validation approach. A separate breakdown of individual sources of variation from such data is therefore not possible [5, 17, 18].

Example

A calibration standard is prepared from a high purity substance X with a stock solution of 1000 mg L^{-1}. The measurand is the concentration of the solution c_X and is formed from

$$c_X = \frac{m\ P}{V}$$

(B1)

with m the weighed mass of X, P the purity as mass portion, and V the volume of the solution. To calculate the combined standard uncertainty, the following uncertainty components have to be taken into consideration:

The purity of substance X is stated in the certificate as $99.99 \pm 0.02\%$, so that $P = 0.9999 \pm 0.0002$. As there is no information available on the uncertainty, a rectangular distribution is assumed. The standard uncertainty $u(P)$ then emerges with Eq. (9.3b) as

$$u(P) = \frac{0.00002}{\sqrt{3}} = 0.000115$$

(B2)

A further step involves the weighing of the substance. The volumetric flask is weighed both with and without the sample, yielding $m = 100.42$ mg, whereas the uncertainty of 0.06 mg is taken from the manufacturer's data.

The uncertainty of the volume of the solution arises from the certified inner volume of the flask and variations from the filling of the flask. The difference between the temperature of the flask and the solution and the temperature at which the volume of the flask was calibrated is neglected. The manufacturer states the volume of the flask as being 100 ± 0.1 mL. As there is no confidence level for the uncertainty, the standard uncertainty is calculated by assuming a triangular distribution. The last appears to make the most sense, because in quality-assured mass production of such items as a flask the nominal value is more probable than the extreme values. The uncertainty is then

$$\frac{0.01}{\sqrt{6}} = 0.04 \text{ mL}$$

(B3)

The possible variance in the filling of the flask can be estimated by using a repeatability experiment. A series of 10 filling and weighing tests resulted in a standard deviation of 0.025 mL, which is then used directly as the standard uncertainty. $u(V)$ is then given by:

$$u(V) = \sqrt{0.04^2 + 0.025^2} = 0.05 \text{ mL} \tag{B4}$$

with the above values c_X is calculated as 1004.1 mg L^{-1} and the uncertainty as per Eq. (9.11) is calculated by the differentiation and conversion of the multiplicative expression in Eq. (B1) as

$$u(c_X) = c_X \sqrt{\left(\frac{u(P)}{P}\right)^2 + \left(\frac{u(m)}{m}\right)^2 + \left(\frac{u(V)}{V}\right)^2}$$

$$= 10004.1 \text{ mg L}^{-1} \sqrt{\left(\frac{0.000115}{0.9999}\right)^2 + \left(\frac{0.06}{100.42}\right)^2 + \left(\frac{0.05}{100}\right)^2}$$

$$= 0.8 \text{ mg L}^{-1}$$

Further complex calculation examples can be found in [11].

This approach, also known as the "top-down approach" in the literature when compared with the "bottom-up approach" of the GUM (see Fig. 9.3), was successfully demonstrated for several chromatographic applications. Studies on robustness, which are for instance quite typical in pharmaceutical analysis, can simulate the variations between laboratories in an appropriate experimental design. This could be demonstrated in numerous studies [19–21] for a range of applications, for example in the area of pharmaceutical and food analysis [22–28]. The various aspects of the method validation allow insights into the problems involved in calculating uncertainty in chromatographic procedures. The significance of the sampling for the whole uncertainty was made clear in [29], while in [20] the recognition of the second order terms in Eqs. (9.9) and (9.10) was referred to. Barwick [30] and Ambrus [31] both offer a good overview of possible sources of error in liquid and gas chromatography. Significantly simpler are calculations for primary methods or reference procedures based on an isotope dilution analysis using GC-MS [32–37].

A direct method to ascertain measurement uncertainty consists of taking the measurement procedure in question for reference objects (standards, material measures, reference materials) and comparing the results determined under the comparability conditions of one's own laboratory with the published reference values available. The simplest instance of this is the use of a single reference object. A variant, that broadly follows the same principle, consists in applying the measurement procedure parallel to a reference procedure on suitable measurement objects and comparing the results of the procedure to be analyzed with those of the reference procedure. In both variants, the measurement uncertainty is calculated from known values of the trueness and known values of precision. The sequence of individual steps then consists of: (1) checking the precision; (2) checking for systematic deviations and making the appropriate corrections; (3) establishing the measurement uncertainty for the measurand. Tests on the precision and trueness of a measurement procedure are normally repeated regularly. If the data of a current examination is comparable with the data from previous examinations, they can be consolidated in order to improve the statistical basis of the estimated values in question.

In procedures concerning regulatory compliance of standards, trueness and precision tend to be established in interlaboratory comparisons (cf. ISO 5725-2). The known data arising from such procedures lend themselves for use in determining what is known as the "reproducibility standard deviation" (s_R) as an estimator for the measurement uncertainty (see details in ISO/TS 21748). As it already contains systematic effects through the various working methods of the laboratories involved, an additional inclusion of systematic uncertainty contributions is generally not required. If the laboratory has successfully participated in such comparisons with a test method as part of accreditation procedures, then it can use the results to estimate its own measurement uncertainty. Alternatively, interlaboratory comparisons for accreditation provide an opportunity to check a measurement uncertainty which has otherwise been determined.

If only a limiting value is of interest for the analysis, e.g. in maintaining preset specifications and requirements, then the following simplifications apply: The uncertainty contributions $u_i = c_i\, u(x_i)$ of the input variables are added linearly; the correlation contributions $u_{ik} = c_i\, c_k\, u(x_i, x_k)$ are dropped. In the uncertainty contributions $u_i = c_i\, u(x_i)$ of the input variables only the highest values of the possible measurement deviations $|\Delta x_i|_{\max}$ are taken instead of $u(x_i)$. These approximations result in the following two equations, each of which may be used to calculate the highest values for the measurement uncertainty as required.

$$|\Delta y|_{\max} = \sum_{i=1}^{N} \left|\frac{\partial F}{\partial x_i}\right| \cdot u(x_i) \tag{9.13a}$$

$$|\Delta y|_{\max} = \sum_{i=1}^{N} \left|\frac{\partial F}{\partial x_i}\right| \cdot |\Delta x_i|_{\max} \tag{9.13b}$$

As discussed above, the linear approximation in Eq. (9.9) loses its validity in nonlinear correlations between the measurement result y and an input variable x_i. Furthermore, the probability distribution of the measurement result y can deviate significantly from a normal distribution, with the result that a k of 2 underestimates the coverage factor for 95%. Using the Monte Carlo methods, a suitable distribution (e.g. normal distribution, rectangular distribution, or triangular distribution) is estimated for each input variable. A pseudo-random value is simulated from each of these distributions and from this set of generated input data a value for the measurement result is calculated. This process is repeated as many times as

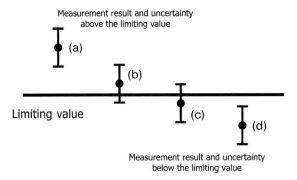

Figure 9.4 Measurement of uncertainty and limiting values.

necessary until a set of data is formed which presents a sample of the possible values of the measurement result, independently of variations in the input variables as corresponding to their distribution. The mean and the standard deviation of this sample are estimated values for the value of the measurement result and its standard uncertainty. In the case of larger deviations from the normal distribution a more realistic confidence interval can be determined from this distribution.

Making a comparison with limiting values often requires evidence that the measurand lies within certain boundaries. Further to this, the uncertainty in the analytical result has to be taken into account, while in the definition process of the limiting values the measurement uncertainties are often neglected. Figure 9.4 illustrates the correlations for the instance of a limiting value being set without considering the uncertainties. (a) The result exceeds the limiting value by more than the determined uncertainty or (b) the result exceeds the limiting value by less than the determined uncertainty. (c) The result falls below the limiting value by less than the determined uncertainty or (d) the result falls below the limiting value by more than the determined uncertainty. The instances (a) and (d) are typically looked on as not being concordant with the specifications. In practice, the instances (b) and (c) need to be individually taken into consideration in view of possible agreements with the user of the data. An exception is made in the monitoring of a limiting value by means of a fixed method. In this case, it is explicitly assumed that the uncertainty, or at least the reproducibility, of the fixed method is small enough to be ignored in practice. In a suitable quality control process, compliance can thus only be indicated through the value of the special measurement result.

9.3
Traceability of Analytical Measurements

For comparability of results between various different laboratories or of results from the same laboratory at different points in time, the use of the same scale of measurements or the same reference points has to be guaranteed. In many cases this is achieved through a chain of calibrations which ultimately is traced back to a national or international primary standard and, ideally, an SI unit. While this traceability is comparatively easy to realize for physical quantities (for example analytical balance and the kilogram), trying to establish an unbroken chain of comparisons in analytical chemistry to form a known reference value is, in many instances, very difficult indeed. Traceability is no new concept, the significance of preparing stock solutions for the purposes of titrimetry and gravimetry has been well known ever since the beginning of analytical chemistry. For modern instrumental methods, an adequate traceability to an SI unit is significantly more difficult to demonstrate.

Traceability is formally defined [38] as the *"property of the result of a measurement or the value of a standard whereby it can be related to stated references, usually national or international standards, through an unbroken chain of comparisons all having stated uncertainties"*. From this naturally arises a reference to uncertainty, as the agreement between laboratories is limited. Traceability is thus intimately linked to uncertainty, which characterizes the performance of the links in the chain and the agreement to be expected between laboratories making similar measurements. A sensible estimation of a measurement result y with the input variable $x_1 \dots x_m$ is possible through the functional correlation of f:

$$y = f(x_1, x_2, \dots x_m) \,\big|\, x_{m+1}, x_{m+2}, \dots x_n \tag{9.14}$$

when the conditions $x_{m+1} \dots x_n$ are specified and confirmed through a validation of the method. Then y is traceable to $x_1 \dots x_n$ provided the input variables are also traceable or clearly defined. In practice, this means that the input variables have to be clearly defined through the method and that critical quantities can be traced through calibration to reference values. The measurement result can be a defined quantity, such as analyte mass or concentration, or determined through the analytical method itself. The results can only then be compared, when they can be expressed in the same unit. This chain of comparisons then leads to the corresponding primary unit or to a corresponding realization of a unit at the end of the chain.

Calibration is a fundamental step in establishing a traceability chain and allows comparison between the results and the values of a reference standard. Validation of an analytical procedure not only plays an important role in calculating the uncertainty where necessary, but is also decisive for the traceability. Validation studies are also necessary for reference procedures, i.e. in checking whether the method has been appropriately applied and thereby whether the necessary boundary conditions $x_{m+1} \dots x_n$ have been specified. Typically, this occurs by way of a certified reference material or an appropriate set of interlaboratory comparisons.

The traceability of the results of an analysis procedure can thus be guaranteed through the combination of the following calibration steps:

1. *Use of a traceable standard to calibrate the measuring equipment.* From a metrological perspective, reference materials (RM) are measures of the material

composition and other material known values. Certified reference materials (CRM) are reference materials that carry a certificate, in which one or more property values have been certified with an uncertainty and the corresponding confidence level as a result of a defined analytical method. Traceability of the values is achieved to an exact realization of the unit. CRMs serve for establishment of reference values of these quantities as well as the realization of pure substances. Pure substance CRMs are employed for the analyte calibration of measurement equipment and the manufacture and trueness control of calibration solutions. Calibration solutions are used for the analyte calibration of measurement equipment and the trueness control of analysis results through spiking solutions (recovery experiments). Matrix reference materials are used in the trueness control of analyses, as well as in the validation of analysis procedures and quality control processes. In certain areas, matrix RM are used in the analyte calibration of laboratory equipment for the purposes of observing matrix effects (for an overview of the reference materials available cf. www.comar.bam.de)

2. *Use of a primary method or a comparison to the results thereof.* A primary method of measurement is a method with a high metrological quality, so that all steps in the procedure are detailed in full and are expressed in SI units. The results can thus be recognized without the need for a reference to a standard of the same quantity.

3. *Use of a pure substance RM or an appropriate matrix CRM.* Traceability can be demonstrated by a comparison of the measurement results with a CRM. If the value of the CRM is traceable to SI, then the traceability to SI is also proved for the measurement. The concept of a pure substance is therefore an ideal concept, so that the use of "pure" substances requires meticulous checking. Since these substances are often also the end of the traceability of CRMs, the entire end of the chain can be endangered by imperfect or missing pure substances by a circular argument.

4. *Through the use of an accepted, well defined procedure.* Reference procedures are thoroughly characterized, validated and provenly controlled quantitative analytical measurement procedures used in assessing other procedures for similar tasks, in the certification of reference materials or in determining reference values with a measurement uncertainty appropriate to the proposed application. Reference procedures are also used in the validation of routine procedures, in the certification of reference materials, in the determination of reference values (target values), in accreditation, analyses with far-reaching consequences (e.g. in referee analyses in legal disputes), in determining reference values of material constants or properties (e.g. in determining extractable heavy metal content in soil with aqua regia exposure). The statement of traceability is achieved through referencing to the applied standards for calibration and the statement of reference standards used in controlling the measurement conditions.

References

1 M. Daszykowski and B. Walczak, Use and abuse of chemometrics in chromatography, *Trends Anal. Chem.* (2006) **25**, 1081–1096.

2 M. Badertscher and E. Pretsch, Bad results from good data, *Trends Anal. Chem.* (2006) **25**, 1131–1138.

3 GUM, *General Requirements for the Competence of Calibration and Testing Laboratories*, 1999, ISO, Genf.

4 J. L. Love, Chemical metrology, chemistry and the uncertainty of chemical measurements, *Accred. Qual. Assur.* (2002) **7**, 95–100.

5 A. Rios and M. Valcarcel, A view of uncertainty at the bench analytical level, *Accred. Qual. Assur.* (1998) **3**, 14–19.

6 P. De Bievre and H. Günzler, *Measurement Uncertainty in Chemical Analysis.* Springer Verlag, Berlin, 2003.

7 F. Adunka, *Messunsicherheiten.* 2000, Essen: Vulkan.

8 L. Kirkup and B. Frenkel, *An Introduction to Uncertainty in Measurement.* Cambridge University Press, Cambridge, 2006.

9 M. Drosg, *Dealing with Uncertainties.* Springer Verlag, Berlin, 2006.

10 *Guide to the Expression of Uncertainty in Measurement*, 1995, ISO, Genf.

11 EURACHEM/CITAC GUIDE, *Quantifying Uncertainty in Analytical Measurement*, 2000: www.eurolab-d.bam.de.

12 W. Hässelbarth, *BAM-Leitfaden zur Ermittlung von Messunsicherheiten bei quantitativen Prüfergebnissen.* N. W. Verlag, Bremerhaven, 2004.

13 M. Rosslein, Evaluation of uncertainty utilising the component by component approach, *Accred. Qual. Assur.* (2000) **5**, 88–94.

14 S. Kromidas, *Handbuch Validierung in der Analytik.* Wiley-VCH, Weinheim, 2000.<

15 M. Patriarca, F. Chiodo, M. Castelli, and A. Menditto, Estimates of uncertainty of measurement from proficiency testing data: a case study, *Accred. Qual. Assur.* (2006) **11**, 474–480.

16 P. Dehouck, Y. V. Heyden, J. Smeyers-Verbeke, D. L. Massart, R. D. Marini, P. Chiap, P. Hubert, J. Crommen, W. Van de Wauw, J. De Beer, R. Cox, G. Mathieu, J. C. Reepmeyer, B. Voigt, O. Estevenon, A. Nicolas, A. Van Schepdael, E. Adams, and J. Hoogmartens, Interlaboratory study of a liquid chromatography method for erythromycin: determination of uncertainty, *J. Chromatogr. A* (2003) **1010**, 63–74.

17 A. Maroto, R. Boque, J. Riu, and F. X. Rius, Evaluating uncertainty in routine analysis, *Trends Anal. Chem.* (1999) **18**, 577–584.

18 S. L. R. Ellison and V. J. Barwick, Estimating measurement uncertainty: reconciliation using a cause and effect approach, *Accred. Qual. Assur.* (1998) **3**, 101–105.

19 E. Hund, D. L. Massart, and J. Smeyers-Verbeke, Comparison of different approaches to estimate the uncertainty of a liquid chromatographic assay, *Anal. Chim. Acta* (2003) **480**, 39–52.

20 V. J. Barwick, S. L. R. Ellison, C. L. Lucking, and M. J. Burn, Experimental studies of uncertainties associated with chromatographic techniques, *J. Chromatogr. A* (2001) **918**, 267–276.

21 E. Hund, D. L. Massart, and J. Smeyers-Verbeke, Operational definitions of uncertainty, *Trends Anal. Chem.* (2001) **20**, 394–406.

22 E. C. Gimenez and S. Populaire, Use of validation data for fast and simple estimation of measurement uncertainty in liquid chromatography methods, *J. Liq. Chromatogr. Relat. Technol.* (2005) **28**, 3005–3013.

23 V. J. Barwick, S. L. R. Ellison, M. J. Q. Rafferty, and R. S. Gill, The evaluation of measurement uncertainty from method validation studies – Part 2: The practical application of a laboratory protocol, *Accred. Qual. Assur.* (2000) **5**, 104–113.

24 J. O. De Beer, P. Baten, C. Nsengyumva, and J. Smeyers-Verbeke, Measurement uncertainty from validation and duplicate analysis results in HPLC analysis of multivitamin preparations and nutrients with different galenic forms, *J. Pharm. Biomed. Anal.* (2003) **32**, 767–811.

25 T. Anglov, K. Byrialsen, J. K. Carstensen, F. Christensen, S. Christensen, B. S. Madsen, E. Sorensen, J. N. Sorensen, K. Toftegard, H. Winther, and K. Heydorn, Uncertainty budget for final assay of a pharmaceutical product based on RP-HPLC, *Accred. Qual. Assur.* (2003) **8**, 225–230.

26 D. W. M. Sin, Y. C. Wong, C. Y. Mak, S. T. Sze, and W. Y. Yao, Determination of five phenolic antioxidants in edible oils: Method validation and estimation of measurement uncertainty, *J. Food Compos. Anal.* (2006) **19**, 784–791.

27 N. Ratola, L. Santos, P. Herbert, and A. Alves, Uncertainty associated to the analysis of organochlorine pesticides in water by solid-phase microextraction/gas chromatography-electron capture detection – Evaluation using two different approaches, *Anal. Chim. Acta* (2006) **573**, 202–208.

28 F. Cordeiro, G. Bordin, A. R. Rodriguez, and J. P. Hart, Uncertainty estimation on the quantification of major milk proteins by liquid chromatography, *Analyst* (2001) **126**, 2178–2185.

29 S. Leito, K. Molder, A. Kunnapas, K. Herodes, and I. Leito, Uncertainty in liquid chromatographic analysis of pharmaceutical product: Influence of various uncertainty sources, *J. Chromatogr. A* (2006) **1121**, 55–63.

30 V. J. Barwick, Sources of uncertainty in gas chromatography and high-performance liquid chromatography, *J. Chromatogr. A* (1999) **849**, 13–33.

31 A. Ambrus, Reliability of measurements of pesticide residues in food, *Accred. Qual. Assur.* (2004) **9**, 288–304.

32 M. J. T. Milton and R. I. Wielgosz, Uncertainty in SI-traceable measurements of amount of substance by isotope dilution mass spectrometry, *Metrologia* (2000) **37**, 199–206.

33 B. King and S. Westwood, GC-FID as a primary method for establishing the purity of organic CRMs used for drugs in sport analysis, *Fresenius J. Anal. Chem.* (2001) **370**, 194–199.

34 M. J. T. Milton and R. I. Wielgosz, Use of the international system of units (SI) in isotope ratio mass spectrometry, *Rapid Commun. Mass Spectrom.* (2002) **16**, 2201–2204.

35 A. S. Brown, M. J. T. Milton, C. J. Cowper, G. D. Squire, W. Bremser, and R. W. Branch, Analysis of natural gas by gas chromatography – Reduction of correlated uncertainties by normalisation, *J. Chromatogr. A* (2004) **1040**, 215–225.

36 H. Hasegawa, Y. Shinohara, T. Hashimoto, R. Matsuda, and Y. Hayashi, Prediction of measurement uncertainty in isotope dilution gas chromatography/mass spectrometry, *J. Chromatogr. A* (2006) **1136**, 226–230.

37 C. Planas, A. Puig, J. Rivera, and J. Caixach, Analysis of pesticides and metabolites in Spanish surface waters by isotope dilution gas chromatography/mass spectrometry with previous automated solid-phase extraction – Estimation of the uncertainty of the analytical results, *J. Chromatogr.* (2006) **1131**, 242–252.

38 International Vocabulary of Basic and General Terms in Metrology, 1993, ISO, Genf.

Part 2
Characterization of the Evaluation of Different Chromatographic Modes

Quantification in LC and GC: A Practical Guide to Good Chromatographic Data
Edited by Hans-Joachim Kuss and Stavros Kromidas
Copyright © 2009 WILEY-VCH Verlag GmbH & Co. KGaA, Weinheim
ISBN: 978-3-527-32301-2

10
Evaluation and Estimation of Chromatographic Data in GC

Werner Engewald
(Translated from the German by Mike Hillebrand)

10.1
Introduction

Why does the evaluation of GC-data require a chapter to itself? In gas chromatography the same hard- and software are used for the signal collection and processing as for HPLC. The calibration procedures do not differ and the A/D transducer does not know with which version of chromatography the arriving signal was produced. Exactly! However, chromatography in the gas phase usually requires higher temperatures for injector, column and detector. This leads to some substantially different aspects and sources of error for qualitative and quantitative analysis, which the operator should know about and have already accounted for during the development of the method, in order to produce correct and reproducible results. During the evaluation of gas chromatograms errors in the preceding steps of the analysis (sample preparation, injection, separation, detection, signal processing) cannot be cancelled. At best some can be identified. The scenario can extend from no or insufficient separation, peak forms that are difficult to evaluate, up to fewer findings for "critical" sample components. Also, additional peaks can occur, which do not originate from the sample. The evaluation of a chromatogram should therefore not take place in isolation from the previous analysis steps and the conditions used. A critical evaluation and validation of the received results should also be performed in order to avoid wrong statements, which can sometimes lead to serious consequences. Therefore this chapter points out what is different in gas chromatography, or more precisely the popular capillary gas chromatography CGC, compared with HPLC and what consequences result from these differences. Some new developments which are important for qualitative and quantitative analysis are presented. For a detailed discussion of possible errors, their identification and avoidance, which are affected by the instrumental techniques available, the reader is referred to the literature [1–13].

Quantification in LC and GC: A Practical Guide to Good Chromatographic Data
Edited by Hans-Joachim Kuss and Stavros Kromidas
Copyright © 2009 WILEY-VCH Verlag GmbH & Co. KGaA, Weinheim
ISBN: 978-3-527-32301-2

10.2
How Does GC Differ from HPLC?

In gas chromatography the mobile phase is a gas, usually hydrogen, helium or nitrogen, which is the so-called carrier gas. Gases are characterized, in contrast with liquids, by a smaller density and viscosity and by a considerably higher diffusion coefficient. The stationary phase is either a high-boiling liquid (gas-liquid chromatography, GLC) or a solid (gas-solid chromatography, GSC). In contrast to HPLC, packed columns play only a subordinate role in GC. Capillary columns with an internal diameter of less than 1 mm are mainly used, in which the stationary phase is either a thin film or a thin layer of a fine-grained adsorbent on the inner wall of the column. Because of the high permeability of capillary columns, longer columns can be used, giving significantly higher separation power. The use of gases as the mobile phase, which transports the material through the column, means that the sample must be transferred in a vapor or gas phase form, usually at column temperatures between room temperature and 350 °C. Hence the range of application of GC is limited to volatile and thermally stable compounds that can be volatilized quickly without decomposition or reaction with the other components in the sample.

In CGC it is more difficult and more problematic to introduce the sample into the column than in HPLC, where the sample solution only needs to be transferred into the flow of the liquid eluent. In place of a universally applicable injection technology, various injectors as well as a range of injection techniques are available to the GC operator, from which a particular technology must be selected, depending on the specific analytical problem and the characteristics of the sample. The operating parameters must be suitable for the particular sample.

In liquid chromatography, phase equilibria, chemical equilibrium (e.g. ion exchange) or steric effects (exclusion) form the basis of the separation. With phase equilibria, different intermolecular interactions of the analyte with both the stationary and the mobile phases are responsible for the separation. This competition between the sample components and the solvent molecules for positions on the surface of the stationary phase, which is called appropriately the "interactions effect triangle", means that retention and separation of the components depend not only on the structure of the stationary phase but also on the composition of the mobile phase. In contrast, in GC the intermolecular interactions of the analyte and of the stationary phase with the small gas molecules, which serve as the mobile phase, are insignificantly small. Hence, in GC, only the interactions between the stationary phase and analyte are responsible for retention and separation of the sample components. With liquid stationary phases (gas-liquid chromatography, GLC) the distribution of the components between a high-boiling solvent and the gas phase, or with solid stationary phases (gas solid chromatography, GSC), the adsorption of the components from the gas phase at the surface of the solids is responsible. In both instances, the appropriate equilibrium depends strongly on the temperature, hence the column temperature plays a decisive role in GC. Thus, this is, along with the stationary phase, the most important variable with respect to analysis time and an accessible separation.

In the volatilized condition the molecules of the analyte are not solvated and possess, depending upon the polarity, a more or less distinctive affinity for adsorption on cold as well as on active surfaces. On the other hand, on hot surfaces pyrolysis can occur, even if the carrier gas provides an inert atmosphere. Therefore the temperature and the inertness of all surfaces with which the hot sample can come into contact (injector, column, detector) have a large influence on the analytical result (particularly in the trace determination of polar compounds). The retention time (contact time) of the substances at these surfaces also plays a role.

10.2.1
What Consequences Result from These Differences?

10.2.1.1 Applicability of GC
As already mentioned, GC is the method of choice for volatile and thermally stable (to 350–400 °C) compounds. The volatility depends on the size and polarity of the molecules. The larger the molecular weight and/or greater the polarity, the lower the volatility. Both factors must be considered, so large nonpolar molecules can possess greater volatility than small polar molecules. A polar group in a larger molecule has less influence on the polarity than a similar group in a smaller molecule. Compounds with several polar groups (sugars, amino acids, etc.) exhibit low volatility.

The domain for GC is, therefore, nonpolar and low polarity compounds, i.e. gaseous, liquid and solid compounds with less than 60 carbon atoms, a molecular weight under 600 and a boiling point under 500 °C. However, some high-boiling compounds, e.g. long-chain, saturated hydrocarbons (up to a C-number of 100) or triglycerides, are, due to their thermal stability, accessible by GC. An example of the GC of high-boiling compounds is shown in Fig. 10.1.

In the chromatogram of polybrominated diphenyl ethers (PBDE) a peak of the decabrominated diphenyl ether (last peak, nominal molecular mass 950!) can be seen, which is remarkable, because this substance has a tendency to decompose at active surfaces.

Through appropriate chemical reaction (predominantly silylation, alkylation, acylation), polar, less volatile and/or easily pyrolyzed substances can be transformed into stable volatile (less polar) derivatives that are accessible by GC. Through the introduction of more "detector-specific" groups such derivatization can lead, at the same time, to a more sensitive or selective detection. In addition, nonvolatile samples can be transformed, through the use of defined pyrolysis, into smaller, GC accessible molecules (pyrolysis-GC). Also, in numerous problems, it is not an analysis of the complete composition of the samples that is required but only the volatile materials content.

This necessary transfer of the analyte into the gaseous or vapor phase in GC offers an elegant possibility for linking the injection to an analyte/matrix separation and/or enrichment without use of a solvent (Table 10.1).

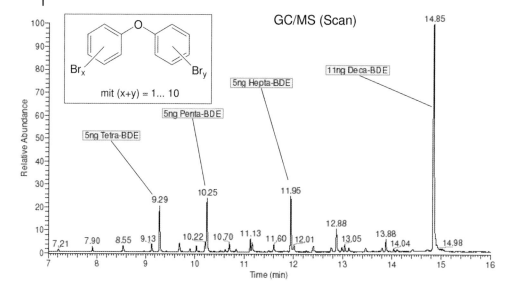

Figure 10.1 Gas chromatogram of a test mixture of polybrominated diphenyl ethers (flame protection agent) on a nonpolar column. Temperature-programmed function.

Table 10.1 Solvent-free (dry) extraction and injection techniques for the GC analysis of volatile components in heavy or non-volatile samples.

Technology	Principle
Headspace-GC (HS–GC)	
Head or gas space analysis	Sampling from the gas space over liquid or solid samples, static and dynamic variants
Thermal desorption	
Direct thermal desorption	Outgasing solid samples at a high temperature in the gas flow
Adsorptive enrichment/ thermal desorption	Trace enrichment from gases as adsorbents followed by heating in the gas flow
Solid phase micro-extraction	
Solid phase micro-extraction, SPME	Extraction from liquid samples or the gas space at an externally coated quartz fiber, followed by heating the fiber in the GC injector
Stir bar sorptive extraction, SBSE	Extraction at a coated stirrer ("Twister")
Solid phase dynamic extraction, SPDE (in tube SPME)	Extraction at an internally coated syringe needle

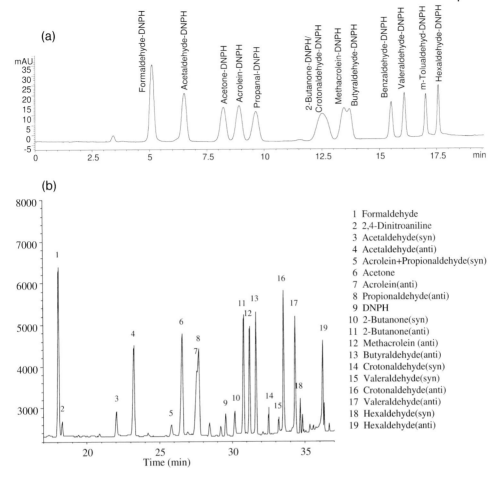

Figure 10.2 Separation of 2,4-DNPH-derivatives of low carbonyl compounds.
(a) HPLC; Column: Discovery RP Amide C16 (150 mm × 4.6 mm, 5 μm);
Flow: 1 mL min^{-1}, 30 °C; (b) GC; Column: Methyl silicone with 5% Phenyl (MDN-5),
(30 × 0.25 mm, 0.25 μm); Programmed temperature: 50 °C (1 min); 20 °C min^{-1} to
180 °C (5 min); 1 °C min^{-1} to 195 °C (1 min); 10 °C min^{-1} to 280 °C (50 min).
Injection: PTV, programmed temperature 50–280 °C at 12 °C s^{-1}. Carrier gas He (11 psi).

If one considers the various application possibilities of GC and HPLC, then both techniques complement each other ideally. However, some substances can also be examined with both techniques. Regarding the choice of method, it should be taken into consideration that HPLC offers mild conditions (separation at ambient temperature), whereas capillary GC offers significantly greater separation power: with HPLC columns one can separate up to 100 peaks in a run but with GC capillaries 500 to 1000 peaks can be separated.

The chromatographic analysis of carbonyl compounds in gaseous samples serves as an example for a comparison of the methods. This is of interest due to their

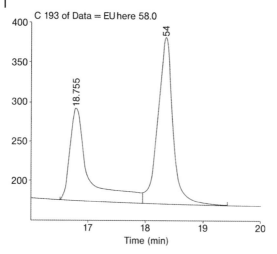

Figure 10.3 Isomerization on the column during the GC-separation of the syn- and anti-tautomers of 2,4-dinitrophenylhydrazone derivative of acetaldehyde (column and conditions as in Fig. 10.2b).

role in atmospheric chemistry and because of their toxicological relevance. Usually these compounds are determined, because of their high reactivity, not in the free form but after conversion into more stable and less reactive derivatives. The acid-catalyzed conversion with 2,4-dinitrophenylhydrazine (DNPH), combined with enrichment, and following HPLC of the formed hydrazone represents a standard method, although in the presence of ozone or nitrogen oxides interference can occur. These DNPH derivatives can also be separated, in principle, using GC.

Figure 10.2 presents the comparison of the chromatograms of a DNPH standard mixture obtained by means of HPLC (a) and GC (b). There is a substantial difference between the chromatograms in the separation of double peaks for some DNPH derivatives in GC. These double peaks result from the reaction of DNPH with asymmetrical carbonyl compounds generating syn- and anti-isomers of dinitrophenylhydrazone, which can be separated on the nonpolar capillary column but not by HPLC. For quantitative analysis the areas of both isomers must be added together, the higher resolution of the capillary GC is, in this case, a disadvantage. The isomer ratio depends on the reaction conditions of the derivatization (pH, solvent, temperature). Moreover, isomerization can take place even in the column during the chromatographic process, as can be seen in Fig. 10.3 for the signal between the two isomer peaks (the signal does not go back to the baseline but forms a plateau between the two peaks). This makes the quantification more difficult.

In the latter region of the gas chromatogram (Fig. 10.2b) asymmetrical and some additional peaks, related to decomposition of the aromatic DNPH derivatives, can be observed. In GC, because with the increasing sample number the peak form and resolution clearly worsen, HPLC remains the method of the choice for these substances.

10.2.1.2 **Sample Injection**

Because of their reliability as well as the electronic control of temperatures and gas flows, efficient autosampler and saving of analysis parameters, the modern gas chromatographs offer a high degree of control, which tempts some users to an uncritical approach with regard to the selection of parameters and methods. In particular, sample injection is, beside sample preparation, the most error-prone procedure in the GC analysis. The injection technology, and with it the applied parameters, have a large influence on the accuracy and the reproducibility of the quantitative results [2, 6, 7]. Here we will only discuss some aspects of this.

The most common type of sample introduction realizable with an autosampler is the syringe injection of more or less diluted solutions (extracts), with or without a flow splitting, cold-on-column-injection and different variants of the programmed-temperature injection (PTV, programmed temperature vaporizer). Besides the dimensions of the capillary columns and the applied injection technique, special requirements should be considered through appropriate choice of parameters, e.g. small sample amount, narrow starting range or a re-focusing of the analytes at the beginning of the capillary column.

In the so-called "hot" injection techniques the samples are introduced into the hot injector and volatilized. Besides possible dosing errors it should be noted that the volatilization is connected with a considerable volume increase. Depending on the polarity and boiling point of the solvent, as well as the temperature and the pressure in the injector, from a 1 μL liquid sample a vapor volume of between 0.1 and 0.5 mL develops, which must be considered during the dimensioning of the replaceable insert liner, to avoid memory effects as a consequence of back-flowing sample vapor. With the vapourizing injection the liner is the hottest surface in the sample path, therefore the use of well deactivated liners and their inspection each time when the septum is changed is strongly recommended. Also, the commonly used fillings, glass or quartz wool, should be thoroughly deactivated. A whole set of analytes (pesticides, medicines, acids, amines ...) tend to be adsorbed and decomposed on the active surfaces. This substance loss leads to smaller peaks (reduced "response") and worse reproducibility. The precipitates from such decompositions in the liner can catalyze further decomposition of subsequent samples. Other sources of contamination can be the sample itself (non-volatile residues) or the septum (when piercing with the syringe needle punches out particles). It is recommended that the activity of the chromatographic system is judged periodically on the basis of a chromatogram of indicator substances that tend to decompose (e.g. benzidine, 2,4-dinitrophenol, pentachlorophenol (PCP), endrin, dichlorodiphenyl-trichloroethane (DDT) or appropriate substances from your samples).

So the recommendations of the American Environmental Protection Agency, EPA, are that the decomposition of endrin to endrin aldehyde and endrin ketone and of DDT to DDD and DDE should not exceed 15%.

10.2.1.3 **Column**

For the GC-user there is today a whole arsenal of columns with different stationary phases and dimensions (length, inner diameter and film thickness) available.

The majority of separation problems can be overcome with a small number of film capillaries with different polarities, with others reserved for particular applications. Gas analysis is the domain of the so-called PLOT-columns with solid stationary phases. Other very important attributes of columns are their maximum working temperature, long-term stability, robustness, inertness and the so-called "bleed", which in recent years has been considerably improved by various measures taken by the column manufacturers (deactivation of the internal surface, use of extremely pure chemicals, immobilization of the polymer chains and new polymer structures) [7, 9, 13]. Column bleed refers to an unwanted contribution of the stationary phase to the underlying signal from a gradual volatilization of low-molecular weight components of the liquid polymer phases and from volatile products of thermal or chemical decomposition of the phases. In spite of their high stability, the most commonly applied polysiloxane phases undergo decomposition at high temperatures (above 250 °C), which leads to splitting off of cyclic siloxane, accelerated by smallest amounts of oxygen as well as inorganic acids, bases and salts. This column bleeding is greater with polar silicone phases than with nonpolar ones and is expressed as added noise (isothermal GC) and/or a rising baseline (programmed-temperature GC), an unfavorable signal-to-noise ratio as well as in "impure" mass spectra. It occurs in particular at high column temperatures (end of the temperature program) and when using very sensitive detectors in trace analysis. New columns are usually delivered by the manufacturer with a test chromatogram. In this, the bleeding is denoted as the difference in the FID signal between the upper temperature limit of the column and the isothermal test temperature (110 to 130 °C, "delta bleed" in [pA]). From the retention times, peak areas and the peak form of substances with different functional groups the column efficiency, information about column polarity and also adsorption activity (losses with polar analytes by irreversible adsorption and decomposition) are derived. It is recommended to test the efficiency of the used column occasionally with a commercially available or even specially prepared, problem-related, test mixture. Because experience shows that not only does the occasional introduction of air-containing samples but also the constant effect of oxygen increase the pyrolysis of silicone phases, the use of extremely pure or cleaned carrier gas and the prevention of leakages are very important in the analysis of trace components. Although linked and chemically bounded stationary phases are robust and tolerate larger amounts of organic solvents, and even water, it is recommended to avoid samples with inorganic acids and bases. In these instances, water also has an unfavorable effect. Obviously acid traces carried along in the samples are a major cause of the deterioration in the separation of DNPH derivatives discussed in Section 10.2.1.1. Over the course of time heavy volatile impurities from the samples (small amounts of impurities are found even in the cleanest samples!) are deposited at the beginning of the column, which changes the separation characteristics and implies the decomposition of the stationary phase and/or unstable sample components. By cutting off the first 10–20 centimeters of the column this interference can usually be eliminated.

10.2.1.4 **Detector**

Unlike the injector, the errors at the detector are less likely to be due to the inadequate instrumental technique and unsuitable operating parameters, but rather a result of wrong interpretation of the results.

Response

The signal produced by the detector is proportional to either the concentration or to the mass of the substance of interest. The appropriate proportionality factors are called material-specific response factors RF and must be determined by calibration:

$$RF = peak\ area/substance\ concentration\ or\ substance\ mass$$

One can differentiate between universal detectors, which result in a signal for all compounds, and selective detectors, which respond, depending upon the measurement principle, only to certain classes of substances. Quantification with an internal standard is usually performed using relative response factors, RRF, in which the response of the analyte is referred to a standard substance(s). The experimentally determined response factors contain all the inadequacies associated with the injection, separation and integration of the peak areas. Possible material losses by injection or adsorption cannot be neglected in the trace analysis. For these reasons, it is recommended that literature values should be used with care and that the concentration range for the experimental determination of the response factors should be selected, which is needed later for the analyses.

Linearity

For quantitative analysis it is important that the linear range of the detector is not exceeded. Note that the equipment manufacturers determine this detector characteristic directly, jas for the detector limit, thus in the absence of the injector and separation column. In reality, however, the detector is a part of the chromatographic system. Adsorption effects on the surfaces of the injector and column can lead to an apparently higher value for the detection limit than denoted in the detector specification and so reduce the usable linear range. If we assume a GC system using a detector with a detection limit of 10 pg, then, if in the column and the injector there is adsorption of polar components in amounts of approximately 50 pg, the detection limit of the method will be above this value. Thus for an injected amount of 100 pg of this component the material loss would be 50%, with 1 ng it would be only 5% and with still higher amounts of the substance, this effect can be neglected.

Time Constant and Data Acquisition

Of fundamental influence to the quality of the detector signal, and thus on the accuracy and the reproducibility of the chromatographic results, are the time constant and data acquisition rate, both of which must be chosen through close consideration of the width of the peaks [14].

The time constant (response time, rise time, detector filter time, see also Chapter 1) of a detector, including the related evaluation electronics (A/D transducer and noise filter), corresponds to the time taken from the creation of an input signal until 63.2% of this signal is indicated. If the time constant is too fast then the noise is increased; if it is too slow it leads to skewed peaks, which in turn lead to more difficulties in identifying the beginning and end of the peaks, a shift in the peak maximum and deterioration in the resolution of closely neighboring peaks. Optimal values are within approximately 10% of the peak width at half height of the narrowest peak in the chromatogram that can still be evaluated.

The data acquisition (sampling rate, sampling frequency) also has a fundamental influence on the peak form. For an accurate report, qualitatively and quantitatively, of the chromatographic peaks about 10 data points per peak width at half height are required, corresponding to about 20 points per peak (see Chapter 2). If a lower sampling rate is selected, the peaks are also skewed (recognizable by the "corners and edges"), which introduces the associated disadvantages already mentioned.

10.2.1.5 Fast Gas Chromatography

Fast gas chromatography [15, 16] always produces narrow peaks regardless in which way the reduction in the analysis times is achieved. As mentioned above, small peak widths demand a fast data acquisition (at least 8–10 data points per peak width at half height) and a short detector time constant (up to about 10% of the peak width at half height), which are only possible with the newer devices. For the various forms of fast GC different terms, such high speed GC, fast GC, ultra-fast GC or super-fast GC are common. In order to clarify the practical requirements for the detector time constant and the data acquisition it is common to consider the peak width at half height w_h for the various classifications of fast GC:

- Fast GC: Separation in a few minutes, w_h under 1 s
- Super-fast (very fast) GC: Separation within minutes, w_h under 0.5 s
- Ultra-fast GC: Separation in seconds, w_h less than 0.1 s

From the range possibilities for the reduction in the analysis times, two have proven successful: the use of shorter columns with smaller inner diameter (e.g. 0.05 or 0.1 mm in place of 0.2–0.3 mm in conventional columns) and extremely high heating rates with a programmed temperature mode of operation. Fast analyses must be accompanied by higher precision of the analysis conditions if the retention times are to be used for the identification of the components. Further requirements of the device concern, in particular, the fast injection of small amounts of sample at high column input pressure, precise regulation of small gas flows with high input pressure and the precise measurement of short retention times. During the method translation from "normal" to narrow columns, it must be realized that because of the decrease in the inner diameter, the film thickness must also be reduced in order to keep the phase ratio constant. So the sample capacity and the tolerance of the column to large differences in the concentration of the components in the sample is reduced. Extreme trace analysis at high separation speeds is therefore problematic. With high heating rates the "fine" structure of the chromatograms can be lost.

If the selectivity of a method is to be preserved through the reduction of the analysis time, a freely available "GC Method Translation Program" [17, 18] proves very helpful, provided that the same stationary phase is used.

From the values for the peak width at half height one can determine the instrument technique to be used. For a peak width within the seconds range, then maximum time constants of less than 100 ms and data acquisition rates of over 10 Hz are required, for ultra-fast GC the corresponding values are approximately 10 ms and over 100 Hz. The analytical instrument producer is increasingly addressing these challenges.

Fast GC makes particularly high demands on the mass spectrometric detection, because the data points in the scan mode, mentioned above, correspond to a complete mass spectrum, which must be an appropriate quality for the purpose of identification. Either very fast scanning mass spectrometers or time of flight mass spectrometers (TOF-MS) are needed, in association with deconvolution software.

10.3
Qualitative Analysis

10.3.1
Introductory Remarks

Use of the term identification must differentiate between two different analytical questions that require different strategies. On the one hand there is the identification of the materials present in unknown samples (non-target analysis) and on the other the specific search for selected components in samples as a condition for quantification (target analysis). For non-target analysis the combination of chromatographic separation with spectroscopic identification is the method of choice. From the possible coupling techniques GC-MS has become the most efficient combination for volatile analytes because of its high sensitivity, versatility (universal or mass-selective detection), substance-specific information and the extensive spectra libraries that are available. The coupling of GC with Fourier transform infrared spectroscopy (GC-FTIR) and with atomic emission spectrometry (GC-AED) supplements the MS structure information (see Table 10.2), but these techniques are presently only in use in a few laboratories.

In target analysis an allocation of the peaks to the target analytes is always necessary. In many cases pure substances or calibration standards are available but retention data are often used as decision criteria for the presence or absence of the target compounds.

10.3.1.1 Fingerprint Analysis
A further analytical question concerns the identity or similarity of complex compound mixtures.

Table 10.2 Capillary GC- coupling techniques.

	GC-AED	GC-MS (EI; LRMS)	GC-FTIR
Procedure	Atomization, irradiation emission	Ionization, depending structure-typical decomposition reaction	Stimulation of molecule oscillations and rotations
Spectrum	Emission lines (atoms, ions)	Molecule and fragment ion peaks	Interferograms → rotation and oscillation spectra
Information	Presence or absence of elements, Elementary composition, Element ratios	Molecular mass, Hetero atoms (Cl, Br, S), Isotopes, Basic structure: aliphatic, aromatic, chain branching	Functional groups, *cis/trans* isomers, Position isomers (aromatic compounds), Linkage of rings, Conformers
Molecule identity Spectrum comparison	–	☺ large spectra libraries	☹ small spectra libraries (gas phase spectra)
Sensitivity	☺	☺	☹ approx. 100 times less than MS, structurally dependent (strong, medium, weak)

By comparison of the peak patterns (position and intensity of the GC peaks) various conclusions can be reached, e.g. two samples are identical without identification of the individual contents. For this "fingerprint analysis" some programs (also recently including mass spectra) are available, which are based on the mathematical methods of pattern recognition.

10.3.1.2 Peak Purity

Even with the use of high resolution separating columns, co-elution cannot be avoided. From the shape of the peaks it can be recognized, in a few cases, whether they represent a pure component or a mixture. With very small differences in the retention times and/or large concentration differences it is particularly difficult to recognize unseparated components.

An indication of the peak purity can be obtained from a comparison of the peak width with neighboring peaks, by parallel detection with universal and selective detectors or by a comparison of mass spectra from different peak positions. Mathematical programs developed for the separation of overlapped signals (deconvolution, see later) present new possibilities for the detection and evaluation of co-elution, for the generation of cleaner spectra and for the quantitative evaluation of only partly separated peaks.

10.3.2
Comparison of Retention Times

Retention times taken from gas chromatograms cannot easily be used for comparison purposes, because they depend not only on the structure and amount of the stationary phase and the column temperature, but also on the column dimensions and speed of the mobile phase. A comparison of the retention times for a sample and suspected components (target analysis) must therefore take place under accurately identical experimental conditions. Any agreement in responses represents a necessary, but still insufficient, criterion for the presence of the substance concerned in the sample and should be repeated on another column with stationary phase with a clearly different polarity. If there is then a discrepancy one can immediately state with confidence that the substance of interest is not contained in the sample. If an identifying detector (MS) cannot be used then a two-column technology can prove to be favorable in some cases. In this method, after injection and volatilization, the sample is divided between two columns with different selectivities. Thus two chromatograms can be obtained per injection with different retention times. The retention times (recognition windows) of the target compounds must be determined with calibration runs. Only if a peak appears on both columns in the determined identification window, the substance of interest can be considered as being present. The peak areas and/or the quantification result should also be comparable on both columns. If this is not the case, then co-elution with another substance is probable occurring on one column and only the smaller concentration value should be indicated. Greater reliability in the qualitative and quantitative statements justifies the use of the more complex tandem GC.

10.3.3
Relative Retention Times

To minimize the influence of parameter fluctuations (mainly column temperature and carrier gas speed) as well as the amount of the stationary phase in the column on the measured values, the use of relative retention times, which are largely independent of these variations, has been suggested. Kovats (1958) established the retention index concept, which used as a reference not an individual compound but the homologous series of n-alkanes. He defined the retention index (I) of an n-alkane for any column at arbitrary temperature as 100 times its carbon number z (e.g. for n-octane: $I = 800$). Once determined, the isothermal and temperature-programmed GC assignable indices can be tabulated and consulted for comparison with experimental values.

At the predominantly applied programmed-temperature function (PTGC) with a linear temperature program the direct (brutto) retention time from the chromatogram can be used for the calculation:

$$I_x^{PT} = 100\, z + 100 \frac{t_{R(x)} - t_{R(z)}}{t_{R(z+1)} - t_{R(z)}} \tag{10.1}$$

where $t_{R(x)}$ is the retention time of the substance x and $t_{R(z)}$ is the retention time of the *n*-alkane with z C-atoms.

With the programmed-temperature retention index, often called the linear retention index (LRI), the temperature regime (heating rate, initial and final temperatures) should be indicated besides the stationary phase. With increasing application of GC-MS coupling, and the constant expansion of spectrum libraries over the past 20 years, the retention indices have lost much of their usefulness, because the mass spectra give substantially more information for structure detection than a retention index does. However, with compounds having very similar mass spectra a structure assignment from the mass spectra alone is not definitiv. Retention index values can therefore serve, especially with complex mixtures, as an additional criterion for peak identification, if the mass spectra of the compounds of interest differ only slightly, e.g. in hydrocarbon analysis or with smell and flavor materials. In [18] the linear retention index is included in the algorithm for spectra comparison. The inclusion of retention indices in the recent edition of the NIST spectrum library indicates a return to the use of retention values for peak allocation.

10.3.4
Retention Time Locking (RTL)

With the improvements in instrument technique, mainly the electronic regulation of column temperature and gas flows, as well as the improvement in column technology (closer tolerance and better reproducibility of column length, inner diameter and film thickness) a system was developed, one decade ago, with which the absolute retention times (total retention times in place of retention indices) can be used for identification with target analyses [19]. After the normal analytical procedure the dependence on input pressure of the retention time of a selected "adjustment substance" is experimentally determined in a few test runs. The electronics pressure control (EPC, Agilent Technologies) steered by the RTL software is then able to adjust small fluctuations or deviations in column temperature, input or output pressure and the column dimensions by variation of the variable input pressure and so keep the retention times constant for all compounds in the chromatogram. This fixation of the retention times leads to high reproducibility, which makes smaller retention time windows possible for peak detection. With RTL, chromatograms that were produced at different times or on different instruments and/or in different laboratories, on nominally the same columns but with different detectors, become comparable. The different column output pressures of FID (normal pressure), MS (negative pressure) or AED (moderate positive pressure) are compensated for. Also, after shortening the column or installation of a new column the same retention times can be obtained again. So the time-consuming correcting of the identification windows (e.g. with SIM analyses) is not required. On standardization of the column and operating conditions these fixed retention times can then be added to databases as criteria for identification.

10.4
GC-MS Coupling

The standard MS instrument suitable for coupling with GC is a bench top device with electron impact ionization (EI) using high energy and a quadrupole or ion trap as the mass analyzer. With these devices, a virtually constant resolution over the complete mass or nominal mass ranges (unit mass resolution or nominal mass resolution) is obtainable. However, a determination of signals with different accurate mass numbers but the same nominal mass is not possible with these low-resolution instruments. So, for example, the compounds

Benzamidine	$C_7H_8N_2$	exact molar mass:	120.069
Ethyltoluene	C_9H_{12}		120.090
Acetophenone	C_8H_8O		120.058

cannot be differentiated on the basis of their mole peaks because they all possess the same nominal mass of 120. (Another interesting number: there are more than 1500 well known compounds with the same nominal mass of 250.) In contrast, with high-resolution mass spectrometers very small mass differences Δmu can be differentiated, e.g. CH_4/O (Δmu = 0.036), CH_2/N (0.013) and O_2/S (0.018). For the determination of polychlorinated dioxins and similar compounds in complex environmental samples, therefore, the coupling of a GC with high-resolution mass spectrometers is recommended, but this is outside the scope of this chapter. As previously mentioned, the mass spectrometer can be utilized for both the identification of unknown substances and for the selective and extremely sensitive determination of target analytes.

10.4.1
MS as an Identifying Detector (Scan Mode)

With electron impact ionization the molecules leaving the GC column are ionized by bombarding with high-energy electrons. With high energy (70 eV, so-called hard ionization), depending on the molecular structure, fragmentation and rearrangement reactions occur, i.e. a part of the molecular ions formed disintegrates immediately with simultaneous formation of several smaller, more stable fragments (fragment ions). In the mass scan mode (full scan mode) the electrical fields of the mass analyzer are changed sequentially with time, i.e. the relevant mass range is scanned sequentially in a short time and all formed ions are thus detected, independently, whether a substance from the GC column arrives in the mass spectrometer or not (cyclic scan, mass run). In this way a very large number of mass spectra are recorded and stored. These line spectra contain the relative intensity (with corresponds to the abundance) of the formed ions as a function of their mass number. By addition of these intensities and plotting the total intensity versus the retention time or the number of scans, the so-called total ion current chromatogram (TIC) is shown on the monitor. This is very similar to a FID chromatogram. All compounds that are ionizable by electron impact are detected. With the MS in the scan mode it works, as with the FID, as a universal detector for organic substances. Owing to the different means of ion production (FID: hot flame; MS: bombarding with electrons) different signal intensities can result. If, for example, during the MS fragmenting of a compound in the ion source there is preferential formation of neutral molecules, which do not give a signal in the MS, these compounds will appear in the TIC with only a small intensity. At all times in the chromatogram the appropriate mass spectra are available for identification. With the mass scan mode there are two closely related selectable parameters that can affect the qualitative and quantitative result. There are the mass range and cycle time.

- Mass range: The complete EI mass spectrum of a substance ranges from small – relatively nonspecific – fragments with low mass numbers up to the molecular ions. Therefore the molecular mass of the expected compounds determines the upper end of the range (usually between 400 and 600, with polybromination compounds accordingly higher), while the cut-off weight should not be selected too low (if possible not under 40 Da), in order to keep the nonspecific background low.

- Cycle time: The cycle time or scan rate is a fundamental parameter, which depends on the peak width. As previously mentioned, for accurate reproduction of the chromatographic peaks at least 10 measuring points are needed or, better still, over 20. For example with a cycle time of 0.5 s (two scans per second) for a peak with a base width of 10 s, 20 data points are acquired. With a peak width of 1 s, only 2 data points are acquired, which is definitely insufficient. Narrow peaks (peaks at the beginning of the chromatograms or with fast GC) thus require substantially higher scanning rates in order to avoid peak distortions.

In the total ion current chromatogram each of these data points represents a complete mass spectrum. With a mass range of 500 amu (atomic mass units, e.g. of 50–550 amu) this corresponds, allowing for the reset time between the scans, to a residence time of less than 1 ms per mass unit. However, with reduction of the recording time the signal intensity is also reduced and hence for trace analysis the SIM (single ion monitoring) mode is preferred. Owing to the high vacuum the fragmentation occurs unimolecularly. No intermolecular reactions occur. The fragmenting paths are well known (fragmentation rules). Because the characteristic fragmentation pattern of the substance of interest, the amount of information contained in the EI-mass spectra is high and can be used for identification.

10.4.2
Use of Spectral Libraries

The high excitation energy of 70 eV from the electron impact ionization provides the precondition for a comparison of determined EI-mass spectra from different devices with those in comprehensive spectral libraries. Thus the recent version of the NIST/EPA/NIH Mass Spectral Library (NIST05) contains 190 825 mass spectra of 163 200 different chemical substances [20]. Using computer-controlled comparison of the determined spectra with the library spectra or reference spectra the GC peaks can either be assigned to the target analyte or suggestions for the structure of the unknown compounds can be derived. With the help of a search algorithm, the locations and intensities of the most important fragments in the determined mass spectrum are compared with the appropriate data in the library. The result of this comparison is a hit list of the substances with which the best agreement was found. As a measure of the agreement, a similarity index (SI or purity index, PURE) is indicated. The user is, however, well advised not to accept automatically and uncritically the compounds with the best agreement, but to regard this substance list only as what it actually represents, a ranking of suggestions! Critical examination of the possible structures with regard to their plausibility remains the task of the operator, because the computer cannot make a statement as to how meaningful its suggestions are on the basis of the similarity of the computations. Other factors involved are the range and quality of the library spectra (some libraries often still contain spectra which were recorded with impure substances or with devices from the early days of this work). Because the mass spectra resulting from the coupling consist of a summation of all signals of the substances present in the ion source, when co-elution is used a spectrum of a mixture will be obtained, as shown in Fig. 10.4. In addition, contaminated spectra can exhibit signals from chemical noise (usually column bleeding). In the algorithms for spectrum comparison there is, therefore, a background correction (subtraction of baseline spectra) as well as a smaller priority given to the low mass range. By spectrum comparison of a GC-peak (peak beginning, maximum, end)

TIC & Spectrum

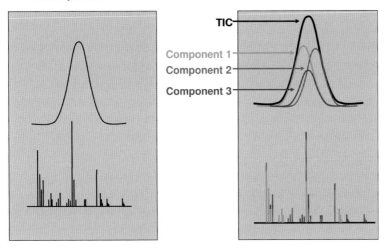

Figure 10.4 Compound mass spectrum with co-elution (from [21]).

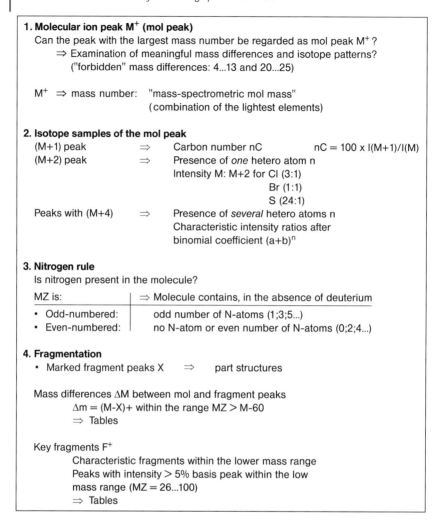

Figure 10.5 Scheme for the structural interpretation of EI-mass spectra
(*I* = Intensity; *a, b* relative intensity of the light or heavy isotopes).

from a pre-supposed co-elution, one can check the extend of overlapping provided that the mass spectra of the substances of interest will differ from each other.

As an important criterion with regard to plausibility, the retention time in gas chromatography should not be forgotten. The spectral library NIST05, mentioned above, contains for the first time the retention indices of over 25 000 compounds on nonpolar stationary phases [20]. Also a determination of the mass spectra (mol peak, isotope sample, characteristic fragments and key fragments) can help, without having to make a complete interpretation. Figure 10.5 shows the scheme for a systematic approach to the interpretation of mass spectra. The characteristic isotope pattern for the elements chlorine and bromine are shown in Fig. 10.6.

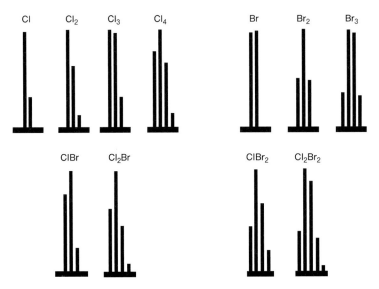

Figure 10.6 Isotope pattern of the elements chlorine and bromine.

10.4.3
Spectrum Deconvolution

In recent times mathematical techniques have been adopted for the fast and precise extraction of "clean mass spectra" from those overlapped with other components and/or high background signals [22], which leads, especially with complex samples, to a higher level of confidence of the identification by comparison with library spectra. Furthermore, in deconvoluted spectra overlap-free ion traces can be realized, which makes reliable quantification in the neighborhood of large peaks possible. The principle of deconvolution is clarified in Fig. 10.7. An example of its use is shown in Fig. 10.8. The differences in the mass spectra of the overlaid peak 3 before and after deconvolution can be clearly seen. For the detection and deconvolution of overlaid spectra, intensity conditions of all mass signals are constantly compared automatically with one another. For this a high sampling rate (scan rate) is necessary, with other conditions being the different mass spectra of the compounds of interest as well as a slight separation. Besides the freely available automated measure of spectral deconvolution and identification system (AMDIS), that can be linked with the search algorithms (Agilent Technologies [21]), there are further deconvolution programs on the market. There are major developments being made in this area at present.

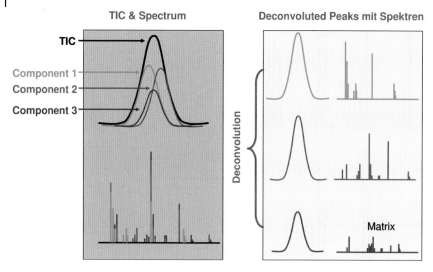

Figure 10.7 The principle of spectrum deconvolution (from [21]).

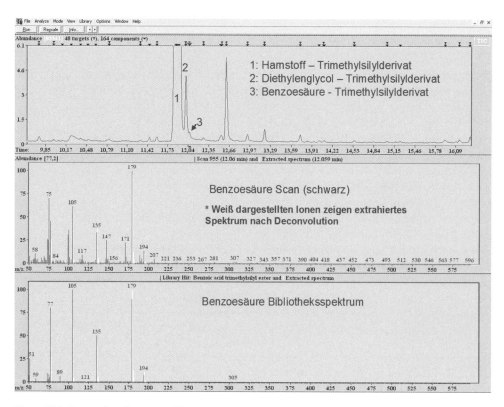

Figure 10.8 Data analysis by means of AMDIS [23].

10.4.4
MS as a Mass-Selective Detector

A selective presentation of certain components is achieved if, in place of the total ion current, only the processed signal for one or a few of the characteristic mass numbers is provided. In such an ion chromatogram only the compounds whose mass spectra exhibit the adjusted m/z values are registered. For mass-selective detection two variants exist.

10.4.4.1 Mass Fragmentography (Reconstructed or Extracted Ion Chromatogram)
The registration of selected ion traces after treatment of the scan data facilitates the recognition and evaluation of co-eluted compounds in complex mixtures. At the push button any mass numbers can be extracted from the stored data. An example is shown in Fig. 10.9 in which the total ion current chromatogram and two reconstructed ion chromatograms are presented with the mass numbers 220 and 236. It relates to the biodegregation of 4-n-nonylphenols [24]. The ion trace at 220 is the molecular ion of the nonylphenols, the hydroxylated nonylphenol formed on degradation studies is visible in the ion trace at m/z 236.

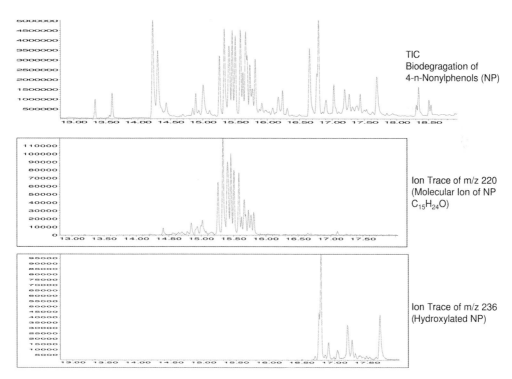

Figure 10.9 Total ion current chromatogram (TIC) and ion chromatograms with mass numbers 220 and 236 [24].

10.4.4.2 SIM (Single Ion Monitoring) or MID (Multiple Ion Recording)

In this method the coupling is used without the complete mass scan and thus without recording the complete mass spectra, only the intensities of one or a few ions that are characteristic of the determined compounds are followed. Apart from the high selectivity, this technology offers the advantage of longer recording times (retention time, dwell time) per mass number whereby, depending upon the type of device in relation to the scan mode, a sensitivity increase by a factor of 20 to 100

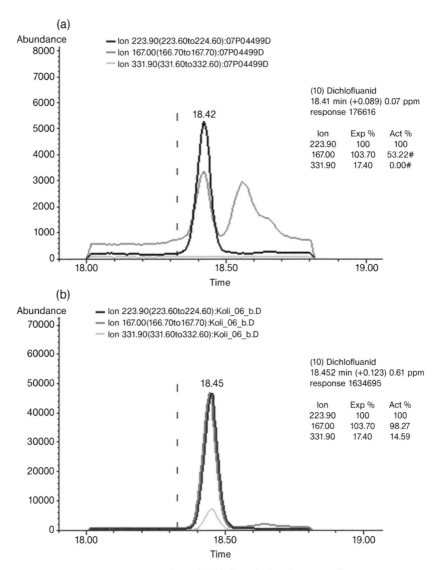

Figure 10.10 GC/MS-SIM-Trace analysis of dichlofluanid; identification on the basis of the retention times and the intensities of characteristic ions [25].

results. This technology is used with quadrupole MS in particular for the selective and sensitive trace determination of a target analyte in matrix-loaded samples, e.g. in residue analysis. The price for the increase in sensitivity and selectivity is, however, the loss of molecule-specific information, because the mass spectra are no longer available for identification purposes. Therefore the intensities of three characteristic ions per target analyte are recorded and from these the most intense is used for quantification and the others are used to confirm the identity (so-called qualifier ions).

Characteristic and high intensity ions should, if possible, be selected from the upper spectrum range and should occur only with the concerned target analytes. The identity of a compound with SIM analysis is considered secure if all selected SIM ions are present in the given retention window and the area ratios correspond to those of the calibration standard. Figure 10.10 shows this identity confirmation using an example from residue analysis of pesticides in agricultural products.

In the upper section of the chromatogram the fungicide dichlofluanid is not present although the qualifier ion (m/z 224) is present and the retention time is correct. The qualifier ion 167 does not lie in the expected range and the other qualifier ion with m/z 332 is completely missing. In the lower chromatogram dichlofluanid is positively identified, as all criteria are fulfilled. All ion traces are present, the retention time is correct and the intensities of the qualifier ions are in the expected range. During the elaboration of multi-component methods the chromatogram is appropriately divided into several time windows in which, in each case, different ions are registered.

Newer MS devices permit quasi-simultaneous full scan and SIM measurements in the same run and so can be used for structure confirmation as well as for the identification of other sample components (screening).

10.4.5
Chemical Ionization

Some compounds fragment so strongly under electron impact conditions that only nonspecific ions with a low intensity arise in the mass spectrum with no molecular ions or compound-characteristic ions. Here chemical ionization (CI), in the presence of a reactant gas (methane, isobutane, ammonia, methanol, water) proves helpful as only a small amount of energy is transferred. With this mild ionization technology very stable positive or negative ions, usually the quasi-molecular ions (MH^+, $(M-H)^+$ or adduct ions), are formed whereas fragments are hardly formed. With CI the molecular mass can be determined. Further structural information can be derived from MS/MS experiments, which are realized with ion trap devices or triple quadrupole systems. Because of their selectivity and the intense characteristic signals chemical ionization and MS/MS techniques are being used increasingly in ultra-trace analysis.

10.5
Quantitative Analysis

As mentioned previously several times, before quantitative evaluation can proceed it is necessary to have a confirmed identification of the peaks of interest, including their purity (co-elution), and to try to improve the resolution if necessary.

Correct and efficient calibration still proves to be the critical step in quantitative analysis. This applies particularly to the increasingly demanding multi-methods, with which as many analytes as possible are determined in *one* chromatographic run. The strategy used is strongly problem- and sample-oriented and dependent on the available equipment. In GC, as with HPLC, techniques for quantitative analysis include the use of: an internal standard (Area normalization), an external standard, an internal standard with a foreign substance and standard addition. Most often an internal standard is used, with which possible injection errors can be corrected. Provided a good separation, the quality parameters of gas chromatographic analysis methods depend significantly on the used detector and on the nature of the analyte (and the sample). Sensitive and robust detectors, e.g. the FID, permit, with an internal standard, the verification of organic compounds down to the ppb range with a relative standard deviation of 1–3%. Generally (including MS detection) measuring variances of approximately 5% are attainable. The measuring variance, as is well known, is the sum of the contributions of the individual steps. For the detection of these different contributions one works with several standards: the so-called "extraction standard" (surrogate) is added to the samples before the sample preparation steps and serves for the detection of errors during the sample preparation. As far as possible the analytes and standard compounds should exhibit similar structure and characteristics. With a mass selective detector, isotope-labeled standards (see Section 10.3.3) are preferred. The second standard is added directly before the GC injection. During the evaluation of recovery analyses and sequences strong fluctuations in the response of the surrogate standard indicate errors during the sample preparation. Fluctuation in both standards indicates errors with the measurement (injection).

10.5.1
Setting up an Analysis Sequence

When setting up analysis sequences one starts with blanks for the detection of interference signals caused by solvents, reagents, glass containers, septa and decomposition of the stationary phase. Calibration standards follow and then the individual samples, but not arranged in ascending or descending order of concentration (randomized). Between the samples, at regular intervals, (e.g. after every 10 to 20 samples) blanks are determined again for detection of carry-over effects (i.e. of substance carry over from preceding analyses) and calibration checks and repeat samples are run. The results of the calibration checks are compared with the initial calibration. In the case of clear deviations between the calibrations the measured values should no longer be trusted and possible errors should be investigated. As a limiting value for the permissible tolerance, following FDA regulations, a value of 15 or 20% is usually selected. In this connection it should be remembered that with linear regression the regression coefficient should not only be regarded alone, because at high values the deviation for the lowest measured value may amount to no more than 20% and within the middle range no more than 15%.

10.6
Isotope Dilution Analysis (IDA) or Stable Isotope Dilution Analysis (SIDA)

Mass spectrometric detection offers the possibility of using stable-isotope-labeled target compounds as internal standards. The labeling (insertion of heavy isotopes) is usually with deuterium or 13C-atoms. The stable-isotope-labeled standard substances possess nearly the same physical-chemical characteristics as the unlabeled substances, but they can be separated in the mass spectrometer by their higher mass numbers. If a defined quantity of an isotope-labeled internal standard is added to the sample, if possible at an early stage in the sample preparation, then the weight ratio of analyte to standard remains constant during the entire analytical procedure, as losses or irregularities affect the analyte and standard equally. With mass-selective detection (SIM or extracted ion chromatogram) interference free signals result for both due to their different molecular masses. Quantification is a result of the ratios between the ion intensities (peak areas) of the target analyte and the standard.

Generally the degree of labelling (number of heavier atoms in the molecule) should be greater than two, in order to avoid possible overlap with isotope peaks (M + 2) of the native analytes. Further, it should be noted that deuterium labeled compounds on a liquid stationary phase exhibit smaller retention values than unlabeled compounds. On nonpolar phases the reduction in retention value amounts to about 0.5 to 0.7 retention index units per deuterium atom. With several deuterium atoms in the molecule this effect in a total ion current chromatogram leads to a broader substance peak and/or a partial or incomplete separation (see Fig. 10.11). If no isotope-labelled standards are available, a substance not occurring in the sample but eluting in the proximity of the target analyte and possessing similar structure and characteristics can be used as an internal standard.

10.7
Matrix Effects in Trace Analysis

With trace analysis in strongly matrix-loaded samples (extracts from agricultural or animal products or medical samples) one frequently observes, despite the sample preparation, a higher or lower value for the determination of the target analyte, although the reproducibility of the measured values and the linearity of the linear calibration (external calibration) are reasonable. The possibility of material losses

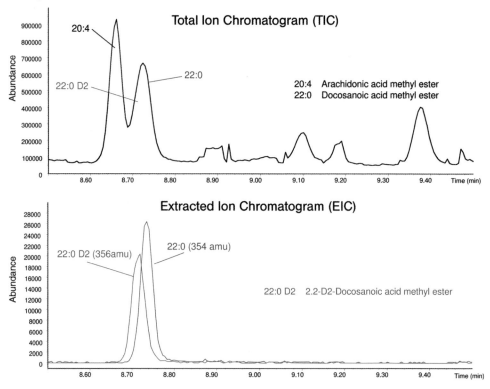

Figure 10.11 Recognition of deuterated compounds with the GC/MS on the basis of extracted ion chromatograms with the appropriate mass numbers [30].

has been referred to in several places in this chapter. Positive matrix effects, which lead to a signal enlargement, will usually be linked in gas chromatography with the covering of active surfaces by non-volatile matrix components in the samples.

If analyte-free sample standards are not available, then the use of stable-isotope-labeled standards is to be preferred (MS isotope dilution analysis). If these are not available either, then in the method development phase other standard additions experiments to test for the presence of matrix effects should be performed.

10.8
Headspace-GC

The problem of solvent-free extraction and injection techniques (see Table 10.1) can be addressed by static headspace technology, a sampling technology in GC analysis for volatile compounds with medium to high vapor pressure in less volatile samples [12, 26]. The liquid, solid or heterogeneous composite sample in a sealed container is heated for a set time at higher temperature, an equilibrium of the volatile sample components is formed between the sample and the gas volume and a sample from the gas region is passed to the GC for analysis. Quantitative evaluation can be carried out using internal or external calibration as well as standard addition. This elegant method of separation of the analyte from the matrix is, however, strongly matrix dependent as a consequence of the equilibrium balance. If a pure matrix is not available or cannot be simulated, then standard addition should be used. In addition, heterogeneous sample matrix adsorption affects the correctness and the precision of the measuring procedure. If the sample is a solid or contains solid particles (soil, mud, suspension), then the problem of inhomogenous distribution with the standard additions (spiking) comes along. On the outer surface adsorbed standard substances are available immediately but analyte from the inside must first migrate to the surface. For these reasons one should transfer inhomogenous or solid samples with a suitable solvent into a clear solution. With low-boiling analytes it is recommended to use a high-boiling solvent (e.g. dimethylsulfoxide, dimethylformamide, ethyleneglycol monomethyl ether), which also offers advantages when setting calibration standards. If it is not possible to get the sample into solution then one could try, by addition of a suspension agent (water) as well as salts, to decrease the matrix influence by a "displacement desorption". If this also is unsuccessful, multiple headspace extraction (MHE) must be used, a repeated sampling from the gaseous phase over the same sample with the same time intervals under the same conditions. There is an exponential concentration reduction with an increasing number of extraction steps. This gradual gas extraction in the same time intervals permits a matrix-independent quantification and is also feasible for the solid phase micro-extraction from the gaseous area (HS-SPME) [27].

10.9
Estimation of the Correction Factor with the FID

Chromatograms of multi-component mixtures consist of a variety of peaks, if they are obtained with a universal detector (FID, TIC). As such mixtures usually contain analytes of various classes of substances, these chromatograms cannot be regarded directly as an accurate image of the samples of interest because, by considering the individual response factors, other area ratios would result. Since, however, an experimental determination of all necessary correction factors is complex or impossible because of the lack of reference compounds, estimation from the molecular structure would be helpful.

Already in 1962 Sternberg et al. [28] had derived an equation (Eq. 10.2) for the prediction of relative response factors from the molecular structure which, more recently, with modern columns and devices, has been examined several times and confirmed [2, 29]. Deviations between computed and experimental factors of under 5% were found in most cases.

$$F_{w(x)} = \frac{MW_x}{MW_{Ref}} \cdot \frac{ECN_{Ref}}{ECN_x} \tag{10.2}$$

where $F_{w(x)}$ is the relative weight response factor of substance x, MW_x the molecular mass of substance x, MW_{Ref} the molecular mass of the reference compound, ECN_{Ref} the effective carbon number of the reference compound and ECN_x the effective carbon number of substance x.

The effective carbon number is the sum of the increments specified in Table 10.3, e.g.

for benzene $\quad\quad\quad 6 \times$ C aromatic = 6
for acetaldehyde $\quad\quad 1 \times$ C aliphatic = 1
$\quad\quad\quad\quad\quad\quad\quad 1 \times$ CO = 0

From this result the molecular response factor for acetaldehyde compared with benzene and by considering of the molecular mass ratio 44/78 a "weight factor" of 3.38 is obtained.

In spite of the potential usefulness of this concept it has so far found no general application.

With the electron capture detector (ECD) an estimation of the extremely substance-dependent response factors is not possible because, apart from the polarizability of the molecule, other factors play a role.

Table 10.3 Atom increments to the effective carbon number (ECN), after Sternberg.

Atom	Structural unit	ECN-increment
C	aliphatic	1
C	aromatic	1
C	olefinic	0.95
C	acetylenic	1.30
C	carbonyl	0
C	carboxyl	0
C	nitrile	0.3
O	ether	−1.0
O	primary alcohol	−0.5
O	secondary alcohol	−0.75
O	tertiary alcohol, ester	−0.25
N	amines	such as O in alcohols
Cl	two or more amaliphatic C	−0.12 per Cl atom
Cl	at olefinic C	+0.05

References

1 E. Leibnitz, H. G. Struppe (Eds.), *Handbuch der Gaschromatographie,* 3. Aufl., Geest & Portig, Leipzig, 1984.

2 G. Guiochon, C. Guillemin, *Quantitative Gas Chromatography,* Elsevier, Amsterdam, 1988.

3 G. Schomburg, *Gaschromatographie (Grundlagen, Praxis, Kapillartechnik),* VCH, Weinheim,1987.

4 W. Gottwald, *GC für Anwender,* VCH, Weinheim, 1995.

5 B. Baars, H. Schaller, *Fehlersuche in der Gaschromatographie (Diagnose aus dem Chromatogramm),* VCH, Weinheim, 1994.

6 K. Grob, *Einspritztechniken in der Kapillar-Gaschromatographie,* Hüthig-Verlag, Heidelberg, 1995.

7 H.-J. Hübschmann, *Handbuch der GC/MS,* VCH, Weinheim, 1996.

8 M. Oehme, Praktische Einführung in die GC/MS-Analytik mit Quadrupolen, Hüthig-Verlag, Heidelberg, 1996.

9 P. J. Baugh (Ed.), *Gaschromatographie – Eine Anwenderorientierte Darstellung* (Übersetzung aus dem Englischen), Vieweg-Verlag, Braunschweig/ Wiesbaden, 1997.

10 W. David, B. Kusserow, GC-Tipps, *Problemlösungen Rund um den Gaschromatographen,* Hoppenstedt-Verlag, Darmstadt, 1999.

11 B. Kolb, *Gaschromatographie in Bildern,* 2. Aufl., Wiley-VCH, Weinheim, 2003.

12 B. Kolb, L. S. Ettre, *Static Headspace-Gas Chromatography,* Wiley-VCH, Weinheim, 2006.

13 D. Rood, *The Troubleshooting and Maintenance Guide for Gas Chromatographers,* Wiley-VCH, Weinheim, 2007.

14 J. V. Hinshaw, *LC.GC Europe,* 2002, 15, 152–155.

15 A. van Es, *High Speed, Narrow-Bore Capillary Gas Chromatography,* Hüthig-Verlag, Heidelberg, 1992.

16 H.-U. Baier, L. Mondello, *Die schnelle Gaschromatographie (fast GC) in der Lebensmittelanalytik – Prinzip und Anwendung,* in W. Baltes, L. Kroh (Eds.), *Schnellmethoden zur Beurteilung von Lebensmitteln und ihren Rohstoffen,* Behr's Verlag, 2004.

17 www.agilent.com/chem.

18 L. Mondello, P. Dugo, A. Basile, G. Dugo, K. D. Bartle, *J. Microcol.* Sep. 1995, 7, 581–591.

19 L. M. Blumberg, M. S. Klee, *Anal. Chem.* 1998, 70, 3828–3839.

20 www.gcimage.com/nist.html.

21 B. Rothweiler (Agilent Technologies, Waldbronn) Analytische Trends und Möglichkeiten zur Überwachung der Lebensmittel-und Wasserqualität, Vortrag.

22 J. M. Halket et al., *Rapid Commun. Mass Spectrom.* 1999, 13, 279–289.

23 K. Dettmer, Regensburg (unpublished).

24 M. Möder, Leipzig (unpublished).

25 T. Knobloch, Leipzig (unpublished).

26 H. Hachenberg, K. Beringer, *Die Headspace- Gaschromatographie als Analysen- und Messmethode,* Vieweg-Verlag, Braunschweig/Wiesbaden, 1996.

27 S. A. Scheppers Wercinski (Ed.), *Solid Phase Microextraction – A Practical Guide,* Marcel Dekker, New York, 1999.

28 J. C. Sternberg, W. S. Gallaway, D. T. L. Jones, in *Gas Chromatography,* Academic Press, New York, pp. 231–267, 1962.

29 A. D. Jorgensen, K. C. Picel, V. C. Stamoudis, *Anal. Chem.* 1990, 62, 683–689.

11
Data Evaluation in LC-MS

Hartmut Kirchherr

11.1
Introduction

The development of robust ionization sources under atmospheric pressure, such as APCI (atmospheric pressure chemical ionization interface) and ESI (electrospray interface) led to a revolution in liquid chromatography and the routine application of LC-MS and the replacement of laborious and time-consuming separation procedures and detection methods with pre- or after-column derivatization.

In particular, by the application of tandem MS as an additional dimension to the separation, the sample preparation and chromatographic separation for quantification in complex matrices can be significantly simplified and shortened. Despite the undisputed success of this technique, a fact that cannot be ignored is that the ion sources are still the bottleneck in the coupling of HPLC and the mass spectrometer.

Whether a substance can be quantified with sufficient accuracy is dependent on the variable conditions of the apparatus, such as liquid and gas flow, ionization energy and temperature, arrangement and organization of component parts and on the ionization techniques in addition to the chemical structure of the analyte, side effects due to the matrix, including the mobile phase, cluster formation and the possibility of its avoidance, the surface conditions of the entire ion path and in particular the constant conditions during the ionization process. Alongside these are the additional technical requirements to get precise and accurate results, which is dependent on the coupling to the mass spectrometer and the number of analytes. These specific features will be discussed on the basis of certain examples. The multiple reaction monitoring (MRM) mode is particularly suitable for studying the quantification.

For a discussion of the basic general features of LC-MS please consult references [1–3].

Quantification in LC and GC: A Practical Guide to Good Chromatographic Data
Edited by Hans-Joachim Kuss and Stavros Kromidas
Copyright © 2009 WILEY-VCH Verlag GmbH & Co. KGaA, Weinheim
ISBN: 978-3-527-32301-2

11.2
Influence of the Matrix in Chromatography

One of the main problem in chromatography is the possible impact of the matrix on the result. Compared with a UV chromatogram, the chromatogram of an MRM transition often appears to be interference-free. However, in the background and not visible to the observer, there could possibly be some effects present that would affect the signals. There are different strategies to control the problem of such matrix effects:

1. Separation of the interfering matrix through sample preparation.
2. Dilution of the sample until no matrix effect is visible.
3. Chromatographic separation of the matrix components.
4. Use of isotope-labeled analogs as internal standards.

Matrix components that can cause an interference are mainly ionic substances such as proteins and salts in addition to nonpolar materials, for example lipids. The simplest method of sample preparation for a biological matrix – e.g. serum – is protein precipitation with acetonitrile/methanol or the addition of acid.

A mixture of nine parts acetonitrile and one part methanol is particularly suitable for protein precipitation in biological matrices. The danger of using pure acetonitrile is the possibility of analyte precipitation due to the very rapid precipitation process. If only methanol is used, a large volume is needed for complete precipitation. In the standard procedure 100 µL of serum are mixed with 300 µL of the acetonitrile/methanol mixture including the internal standard. After centrifugation the clear supernatant can be further diluted with mobile phase [4].

With this method of sample preparation drugs can be completely released from the protein binding. If only the free part of a drug is to be determined, separation from high-molecular weight substances through ultracentrifugation is possible.

If this method of sample preparation does not give the required analyte concentration in relation to the sensitivity of the mass spectrometer, analytes can alternatively be enriched and specimens purified with liquid–liquid extraction (LLE) or solid phase extraction (SPE). Solid phase extraction has the additional advantage that the process is easy to automate. With a multidimensional arrangement and colunm switching, serum can be used directly [5, 6]. Because of the low recovery, e.g. for polar analytes, this is not a universally applicable method of sample preparation.

Whether there is a matrix effect, and if so which region of the chromatogram is affected, is easy to find out with the assistance of the infusion method [7]. The mobile phase is flushed though the HPLC column. A solution of the analyte is simultaneously introduced using a syringe pump via a T-connection in the same way as in the tuning operation, resulting in a high and constant background signal. If the matrix (e.g. after sample preparation) is injected, a signal quenching effect can be observed as a reduction in the signal level. The analyte should not elute in these matrix-affected time zones.

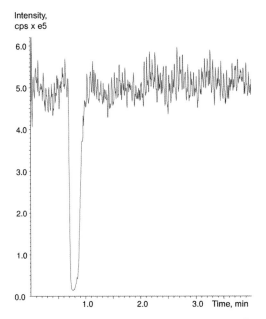

Intensity, cps x e5

Figure 11.1 API 4000, Infusion risperidone 10 ng mL^{-1}, 40 μL min^{-1}. Matrix effect after injection of 10 μL serum after protein precipitation (dilution factor 4) on Chromolith C18, 50 x 4.6 mm, mobile phase: CH_3CN : 5 mM acetate-buffer pH 3.9, 40:60 (v/v), 1 mL min^{-1}.

Figure 11.1 presents just such an infusion chromatogram with risperidone as the target substance and serum matrix after protein precipitation with a dilution factor of 4 on an API 4000 instrument (Applied Biosystems) using the turbospray source and MRM mode of detection. In this chromatogram with isocratic separation mode on a C18 reversed phase column it can be seen that the retention time of risperidone should not fall below 1 min to avoid any matrix effect and signal suppression. With additional sample dilution after protein precipitation, the level of signal suppression with respect to intensity and duration could be further reduced (Fig. 11.2). Even after SPE sample preparation this signal suppression could be observed around the 1 min retention time (Fig. 11.3).

If the elution of the analyte is within the critical time zone, a modification of the chromatographic conditions could result in a delayed elution. In this case the critical time zone of signal suppression was not affected (Fig. 11.4). With gradient-elution HPLC it must be taken into account that when using electrospray ionization, the signal intensity usually increases with a higher fraction of organic solvent in the mobile phase. A shift in the retention time might then lead to a change in the signal intensity. With high organic solvent content in the mobile phase, lipids can also be eluted, which then constitute another "forbidden zone" (Fig. 11.5). With isocratic chromatography these lipids are able to elute during series of samples and could produce a reduction in signal intensity. Flushing between series of samples could minimize this risk.

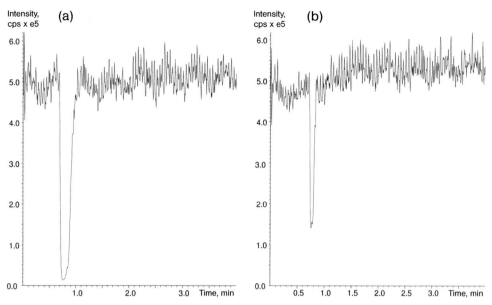

Figure 11.2 API 4000, Infusion risperidone 10 ng mL^{-1}, 40 µL min^{-1}.
Matrix effect after injection of 10 µL serum after protein precipitation.
(a) Dilution factor 40, (b) dilution factor 400 on Chromolith C18 50 x 4.6 mm,
mobile phase: CH$_3$CN : 5 mM acetate-buffer pH 3.9, 40:60 (v/v), 1 mL min^{-1}.

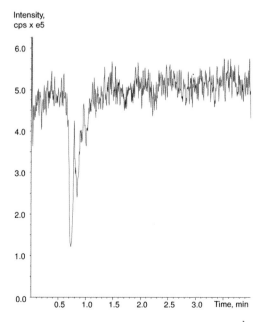

Figure 11.3 API 4000, Infusion risperidone 10 ng mL^{-1}, 40 µL min^{-1}.
Matrix effect after injection of 10 µL serum after C18-SPE (dilution factor 4)
on Chromolith C18 50 x 4.6 mm, mobile phase: CH$_3$CN : 5 mM acetate-
buffer pH 3.9, 40:60 (v/v), 1 mL min^{-1}.

Intensity,
cps x e5

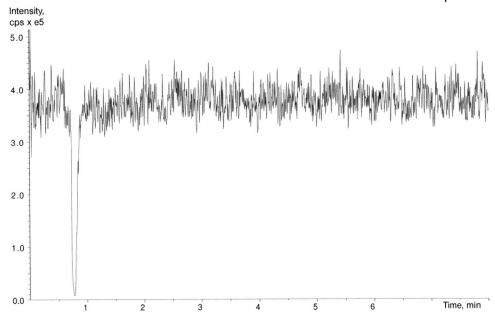

Figure 11.4 API 4000, Infusion risperidone 10 ng mL^{-1}, 40 µL min^{-1}.
Matrix effect after injection of 10 µL serum after protein precipitation
(dilution factor 4) on Chromolith C18 50 x 4.6 mm, mobile phase:
CH$_3$CN : 5 mM acetate-buffer pH 3.9, 20:80 (v/v), 1 mL min^{-1}.

Intensity,
cps x e5

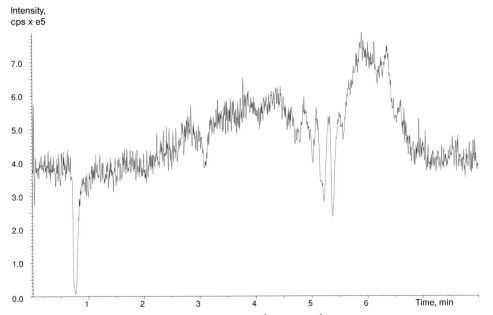

Figure 11.5 API 4000, Infusion risperidone 10 ng mL^{-1}, 40 µL min^{-1}.
Matrix effect after injection of 10 µL serum after protein precipitation
(dilution factor 4) on Chromolith C18 50 x 4.6 mm, Mobile phase:
gradient 20–70% CH$_3$CN in 4 min, until 5 min 70% CH$_3$CN,
until 8 min 20% CH$_3$CN : 5 mM AcOH pH 3.9, 1 mL min^{-1}.

It is also possible to proceed in a similar way with a less sensitive tandem mass spectrometer such as the API 2000 (Applied Biosystems). The spray hits the curtain plate in an oblique geometry, in contrast to the orthogonal arrangement of the API 4000 instrument. Flow rates of 0.2 mL min^{-1} and column inner diameters of 2 mm are typical. To obtain similar intensities during infusion as with the API 4000, the concentration of risperidone must be 1000 instead of 40 ng mL^{-1}. The matrix effect after protein precipitation of serum and a dilution factor of 4 is similar to that with the API 4000 (Fig. 11.6). With this method of sample preparation the sensitivity is not sufficient and therefore an extraction in order to enrich the analyte is necessary.

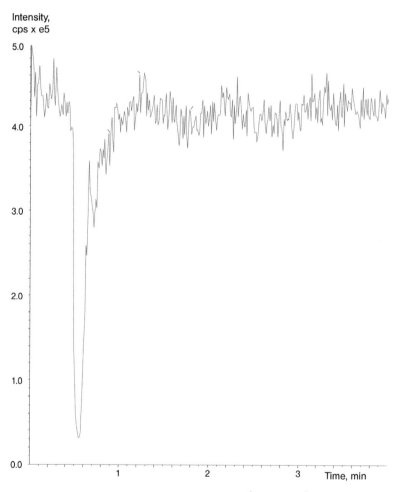

Figure 11.6 API 2000, Infusion risperidone 1000 ng mL^{-1}, 20 µL min^{-1}. Matrix effect after injection of 5 µL serum after protein precipitation (dilution factor 4) on LUNA C18 50 x 2 mm, mobile phase: CH$_3$CN : 5 mM acetate pH 3.9, 36:64 (v/v), 0.2 mL min^{-1}.

The further possibility of diluting the sample until no matrix effect is apparent, as compared with use of a standard without any matrix and recognizable by identical intensities, can of course only be used if either the concentration of the analyte is high enough or the mass spectrometer has adequate sensitivity. For a given set of chromatographic conditions, this comparison of intensities is suitable to establish the exclusion of a matrix effect, at least for the substance to be analyzed. If samples are extracted, the recovery of the extraction procedure must also be considered. A detailed description of this procedure was given by Matuszewski et al. [8]. Alternatively, a "relative" matrix effect can be calculated by comparing the standard line slopes for various samples of a biofluid [9]. Figure 11.7 shows the signal ratio of the substances with and without matrix, depending on dilution and the fraction of organic solvent in the mobile phase using electrospray ionization. The higher the dilution and the lower the organic fraction in the mobile phase, the lower the serum matrix effect is expected to be using risperidone as the test analyte. In practice, an elution shortly after the matrix suppression is sufficient to produce short retention times with narrow peaks and high intensities.

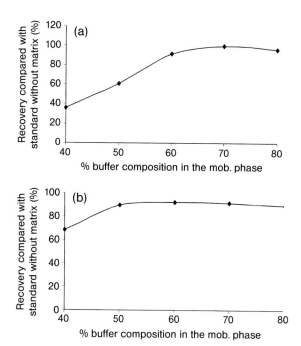

Figure 11.7 API 4000, matrix effect ESI – variation of mobile phase: Merck Chromolith C18 50 x 4.6 mm, CH_3CN : 5 mM AcOH pH 3.9 1 mL min^{-1}, Rt 0.88 min (40% buffer), Rt 0.90 min (50% buffer), Rt 0.97 min (60% buffer), Rt 1.28 min (70% buffer), Rt 3.25 min (80% buffer). 100% Intensity MRM (Area): 39 900. (a) Serum + 1 ng mL^{-1} risperidone, protein precipitation 1:4 diluted, 10 µL injected. (b) Serum + 1 ng mL^{-1} risperidone, protein precipitation 1:40 diluted, 10 µL injected.

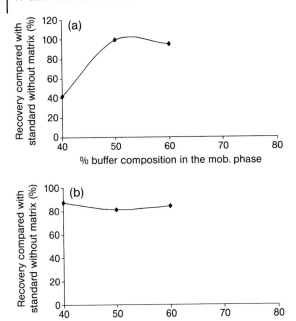

Figure 11.8 API 4000 Matrix effect APCI – Variation of mobile phase:
Merck Chromolith C18 50 x 4.6 mm, CH$_3$CN : 5 mM AcOH pH 3.9, 1 mL min^{-1},
Rt 0.88 min (40% buffer), Rt 0.90 min (50% buffer), Rt 0.97 min (60% buffer).
100% Intensity MRM (Area): 2810. (a) Serum + 1 ng mL^{-1} risperidone, protein
precipitation 1:4 diluted, 10 μL injected. (b) Serum + 1 ng mL^{-1} risperidone,
protein procipitation 1:40 diluted, 10 μL injected.

The APCI source clearly shows that there is less influence from the matrix. With
most polar analytes the present given chromatographic conditions a substantially
lower intensity of the signal and therefore the sensitivty attained will definitely be
inadequate (Fig. 11.8). This type of ionization is therefore mainly suited to less
polar analytes, such as fatty acids or steroids.

Some analytes readily form adducts or clusters. If non-optimized ionization
conditions are used, it is possible that inadequate sensitivities and even irreproduc-
ible results close to the determination limit could be obtained. A careful tuning
of the reference standard substances is still the requirement for further method
development. Formation of an adduct of immuno-suppressant agents such as
tacrolimus and sirolimus can be affected by the spray temperature. Normally these
analytes are quantified as their ammonia adducts, using for example ammonium
acetate buffer in the mobile phase.

However, as sodium is omnipresent when using glass bottles, at higher temperature of the turbo spray the sodium adduct will be formed preferentially and
the intensity of the ammonia adduct will be reduced (Figs. 11.9 and 11.10). It is
predominantly the most intense fragments that are selected for the MRM transi-

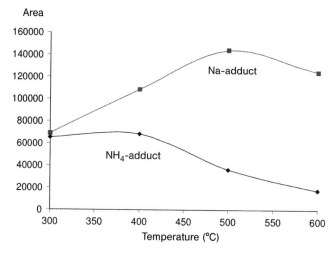

Figure 11.9 Influence of turbospray temperature on the Na- and NH_4-adduct formation with tacrolimus (API 4000).

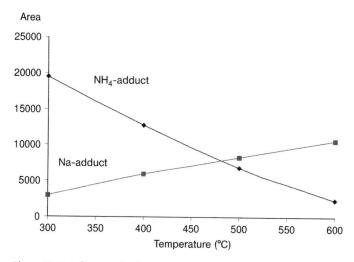

Figure 11.10 Influence of turbospray temperature on the Na- and NH_4-adduct formation with sirolimus (API 4000).

tions. In isolated cases such a fragment can be relatively nonspecific, if only water is split off, or a high background signal prevents the measurement at the trace level. In the ideal case, two intense and structure-specific fragments are obtained so that one fragment can be used as the quantifier and the other as qualifier ion.

11.3
Internal Standards

The use of a suitable internal standard for quantification purposes with MS/MS is especially important. In comparison with classic HPLC, the internal standard should not only compensate for mistakes during the sample preparation and injection, but it is also essential that it compensates for possible variations during the ionization process.

Isotope labeled analogs are particularly suitable for the internal standards. Of course stable isotopes have to be used, but it must be noted that an isotopic transfer, e.g. from deuterium with the mobile phase, has to be excluded. The difference between the mass of the analytes and that of the isotopic internal standards must be at least 2 in order to exclude possible interference with natural isotopes; for halogenated analytes it must be at least 3 mass units. Furthermore, it must be demonstrated whether the internal standard has any influence on the value of the analyte result as the available isotopes are often not completely 100%. If these preconditions are satisfied, an internal standard is able to compensate for the matrix effects within certain limits. Of course there must be enough intensity for a reliable quantification.

The situation becomes more difficult if – as in most cases – no isotopic internal standards are available. In those situations one has to strive for chromatography where there are no visible matrix effects on the analyte. Typically, substances with a similar chemical structure to the analyte are the most suitable internal standards. Use of internal standards that are used in conventional HPLC, is not always possible. For example, in the determination of terbinafine with HPLC-UV detection, clotrimazole was used as the internal standard. However, in contrast to terbinafine, clotrimazole is thermally unstable, and its signal intensity in electrospray ionization thus changes dramatically with spray temperature (Fig. 11.11).

Whether an internal standard is suitable for correct quantification or not can be easily tested by carrying out an analysis of the precision for a small set of samples (e.g. $n = 10$). The variance in the measurement should clearly be no worse than without internal standard. If it is worse, then opposing effects during the ionization process make this internal standard unsuitable for quantification. In practice one can define a tolerable upper limit for variation (e.g. a maximum of 5–10%). This upper limit should not be exceeded when using the internal standard.

Under ideal conditions, analyte and internal standard have the same retention times, but to force this by using an increasing steep gradient in the chromatography is not appropriate as co-elution of other substances could result in interferences similar to a matrix effect. More important is the control of signal stability within a long series of runs. The time it takes to obtain until stable signal intensities can vary, depending on the analyte, from 5 min to more than 2 h. If the conditions are changed at all, a long period of stabilization can be expected in some instances. If the analyte and internal standard react differently, errors due to beginning the analysis too early or drift in the signal intensities during a long series of runs are an inevitable consequence.

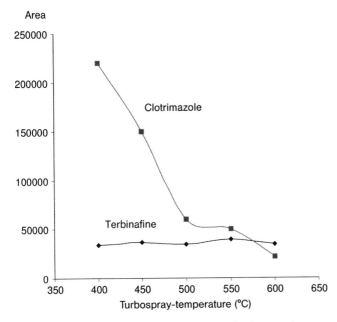

Figure 11.11 Influence of turbospray temperature for the example
terbinafine and internal standard clotrimazole (API 4000).

Table 11.1 Clonidine determination in a serum matrix with different internal standards.
Precision and deviation from reference value (n = 10).

Internal standards	(A) Protein precipitation		(B) C18 – SPE	
	CV (%)	Deviation (%)	CV (%)	Deviation (%)
d4-Clonidine	4.7	3.9	6.3	−11.4
Tizanidine	4.7	9.4	5.2	−19.2
Aciclovir	44.1	out of control	31.7	out of control
Without	5.0	6.7	5.1	−7.4

In Table 11.1 the influence of the internal standard on precision and accuracy
is given for the example of the determination of clonidine in serum after protein
precipitation and C18 SPE sample preparation. Nine different human serum
samples and one pig plasma sample were processed.

d4-Clonidine was, as expected, the most suitable internal standard. It appears
that with protein precipitation slightly better results are obtained with respect to
precision and accuracy than with C18 SPE sample preparation (CV 4.7%, deviation
+3.9% compared with CV 6.3% and deviation −11.4%). The pig plasma sample had
no recognizable negative influence on the results. The deviation is with reference
to a standard solution with no biological matrix.

Table 11.2 Clonidine determination in a urine matrix with different internal standards. Precision and deviation from reference value ($n = 10$).

Internal standards	(A) Dilution		(B) C18 – SPE	
	CV (%)	Deviation (%)	CV (%)	Deviation (%)
d4-Clonidine	4.6	1.1	3.8	1.1
Tizanidine	9.3	2.1	10.3	−3.7
Aciclovir	35.5	out of control	53.3	out of control
Without	15.0	−18.6	17.8	−18.2

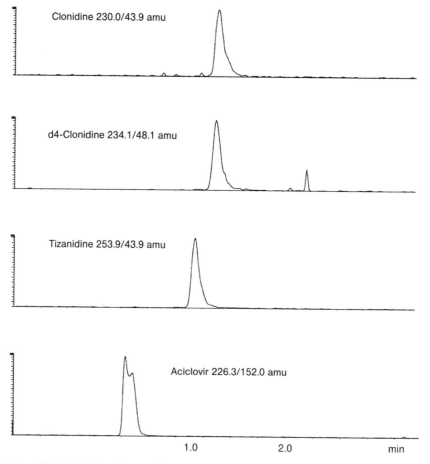

Figure 11.12 API 4000, selection of internal standards for use in the determination of clonidine.

Tizanidine is an imidazolamine derivative, and as such has a structural resemblance to clonidine. The retention time in the chromatogram is slightly less than clonidine (1.2 instead of 1.4 min). Precision and accuracy are in the same region as when using d4-clonidine as the internal standard. However, the retention time for aciclovir is 0.9 min under the given chromatographic conditions, just in the "prohibited area", i.e. matrix effects are expected after protein precipitation and after solid phase extraction. Under the given conditions is aciclovir therefore totally unsuitable as an internal standard for the determination of clonidine.

In some instances, when correction with the internal standard is not used, similarly good results with respect to precision and accuracy can be obtained. However, this observation should not tempt one into renouncing the use of an internal standard on principle. Control of the chromatographic process, and especially the conditions of ionization, can prevent the production of wrong results in isolated cases.

When the imprecisions and inaccuracies of ten different human urines were compared in relation to various internal standards (Table 11.2), there was a clear deterioration of these parameters if tizanidine was used instead of d4-clonidine as the internal standard. Aciclovir is again completely unsuitable. Without correction for the internal standard, essentially worse results were obtained than those for the serum samples alone.C18-SPE gives the best results when using d4-clonidine as an internal standard (CV 3.8%, deviation +1.1%). In Fig. 11.12 the chromatograms of the mass transitions of clonidine and the internal standards are described. In contrast to HPLC-UV chromatograms, the analyte and the internal standards are presented in individually extracted single ion chromatograms. The evaluation is made using the peak area. An evaluation using the peak height gives no improvement in the precision, which is different to the case for a UV-chromatogram.

11.4
Adjustments to the Mass Spectrometer

The advantage of tandem mass spectrometers over ion traps is that it is possible to include several analytes and internal standards in one MRM method. However, the scan time for a mass transition (dwell time) must be adapted to the number of MRM transitions. Even at low concentrations near the limit of quantification (LOQ) it is necessary to keep the dwell time high and to optimize the sensitivity. In Fig. 11.13 the dwell time is plotted against the signal-to-noise ratio using the determination of risperidone as an example. With a dwell time of 200 ms instead of 20 ms there is an 8-fold improvement in the signal-to-noise ratio. Thus the precision will be clearly better with an increase in the dwell time, but this is not relevant at higher concentrations, as shown by the example of 9-hydroxyrisperidone (Fig. 11.14).

If there are several MRM transitions in one method, the dwell time must be reduced in order for an adequate number of datapoints to be recorded over the peak. There should be about 10 to 15 datapoints for a sufficiently accurate quan-

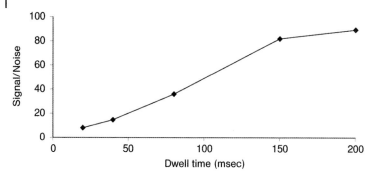

Figure 11.13 API 4000, relationship of dwell time to signal-to-noise for 10 ng mL^{-1} risperdione.

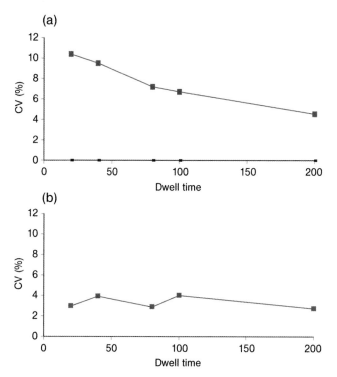

Figure 11.14 API 4000, effect of dwell time on the precision for
(a) 5 ng mL^{-1} risperdione and (b) 50 ng mL^{-1} OH-risperidone.

tification. In Table 11.3 the possible number of MRM transitions in a method in relation to the dwell time and the peak width is listed. At a peak width of 20 s, 50 MRM transitions can be fitted if a dwell time of 20 ms is chosen for the individual transitions. In this case 16 datapoints will be obtained. With 5 MRM transitions and the same peak width, a 200 ms dwell time can be set to receive 20 datapoints per peak.

Table 11.3 Relationship of the number of datapoints to the dwell time (DT) and the number of MRM transitions (minimum 15 datapoints per peak).

DT (ms)	Number of datapoints (peak width 20 s) 5 ms break between the MRM transitions					
	2 MRM	5 MRM	10 MRM	20 MRM	50 MRM	100 MRM
20	400	160	80	40	16	8
40	222	89	44	22	9	4
60	154	62	31	15	6	3
80	118	47	24	12	5	2
100	95	38	19	10	4	2
200	49	20	10	5	2	1
400	25	10	5	2.5	1	0.5

Calculation:

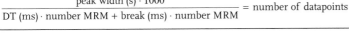

$$\frac{\text{peak width (s)} \cdot 1000}{\text{DT (ms)} \cdot \text{number MRM} + \text{break (ms)} \cdot \text{number MRM}} = \text{number of datapoints}$$

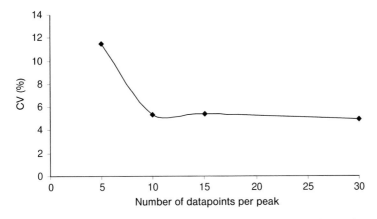

Figure 11.15 Precision in a run series ($n = 10$) for the determination of risperidone on API 4000. Number of recorded datapoints per peak in relation to the precision.

Too few datapoints have negative consequences for the reproducibility of the results. Figure 11.15 depicts the precision in a series of analyses ($n = 10$) related to the number of datapoints per peak. It becomes evident that 5 datapoints are definitely insufficient for an acceptable precision.

11.5
Evaluation Software

Modern evaluation systems are able to integrate chromatographically pure peaks reliably and without too many problems. This is the case with most MRM transitions. In contrast, the situation at low concentrations near the limit of quantification (LOQ) remains problematic. A control is absolutely necessary and a manual correction of the integration is often the quickest way to obtain better results. This process can be supported by a smoothing function, used for the whole series and not only for certain samples. In Fig. 11.16 it is shown that by application of the smoothing function several times, the signal-to-noise ratio for the determination of risperidone can be improved approximately two-fold.

With the evaluation software Analyst 1.4.1, the precisions for the determination of risperidone in serum ($n = 10$, protein precipitation, dilution factor 8, API 4000) at two concentrations (1.0 and 0.1 ng mL^{-1}) were calculated. The limit of quantification (signal-to-noise = 10:1) was found at 0.1 ng mL^{-1} for this method. With automated integration (smoothing factor 1, bunching factor 2) and without manual correction, CVs of 6.5 and 15.1% at 1.0 and 0.1 ng mL^{-1}, respectively, were found. After manual correction a minor improvement to 6.3% CV at 1.0 ng mL^{-1} and a clear improvement to 11.1% CV at 0.1 ng mL^{-1} were observed.

Using manual integration, the operator influence on the results was checked. Ten employees integrated a complete series independently of each other. For the determination of risperidone at 1.0 ng mL^{-1} the average of the precision was 5.2% with a range of 3.8 to 6.8%. At the LOQ (0.1 ng mL^{-1}) an average of the precision

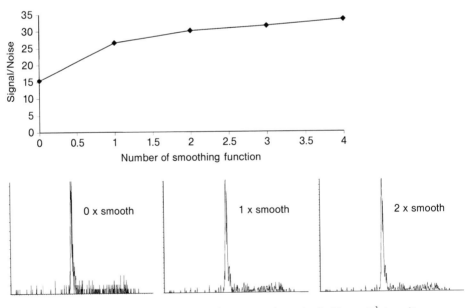

Figure 11.16 API 4000, relationship of smoothing to signal-to-noise for 10 ng mL^{-1} risperdione.

of 12.8% with a range from 6.9 to 16.0% was found. An operator influence was to be expected. If the employees were well trained the values in the region of the LOQ were rather better than with an automated integration. Manual interventions are – especially with automated sample preparation – not always allowable, or must be specifically documented.

References

1 Willoughby R. et al., *A Global View of LC-MS*, Global View Pub., **2002**.

2 McMaster M. C., LC-MS. *A Practical User's Guide*, Wiley-VCH, Weinheim, **2005**,

3 Traldi P. et al., *Quantitative Applications of Mass Spectrometry*, Wiley-VCH, Weinheim, **2006**.

4 Kirchherr H., Kühn-Velten K. N., *J. Chromatogr. B* **2006**, *843*, 100–113.

5 Boos K. S. et al., *Fresenius' J. Anal. Chem.* **2001**, *371*, 16–20.

6 Koal T. et al., Clin. Chem. Lab. Med. **2006**, *44*(3): 299–305.

7 Bonfiglio R. et al., *Rapid Commun. Mass Spectrom.* **1999**, *13*, 1175–1185.

8 Matuszewski B. K. et al., *Anal. Chem.* **2003**, *75*, 3019–3030.

9 Matuszewski B. K., *J. Chromatogr. B* **2006**, *830*, 293–300.

12
Evaluation of Chromatographic Data in Ion Chromatography

Heiko Herrmann and Detlef Jensen

12.1
Introduction

The use of ion exchange in analytical chemistry was initially reported in 1917 by Folin and Bell for the determination of ammonia in urine [1]. The first chromatographic determination of inorganic species was performed in 1937, when Schwab separated alkali and transition metals [2]. Modern ion chromatography (IC) as a form of liquid chromatography and is based on three different separation mechanisms, which also provide the basis for the nomenclature that is typically used: ion-exchange chromatography, ion-exclusion chromatography, and ion-pair chromatography.

The separation in ion-exchange chromatography is accomplished as a result of the analytes' interaction with the ion-exchange groups attached to the stationary phase. Depending on the properties of the analytes, the separation is performed using either anion- or cation-exchange chromatography. Ion-exclusion chromatography is used for the separation of weak inorganic and organic acids. The high-capacity stationary phase consists of a totally sulfonated polymeric cation exchanger. Separation is based on Donnan exclusion, steric exclusion, and adsorption as well as the formation of hydrogen bonds, depending on the stationary phase used.

Ion-exchange selectivity is mediated by both the mobile and stationary phases. In contrast, the selectivity of an ion-pair separation is determined primarily by the mobile phase. The two major components of the aqueous mobile phase are the ion-pair reagent and the organic solvent; the type and concentration of each component can be varied to achieve the desired separation. The ion-pair reagent is a large ionic molecule that carries a charge opposite to that of the analyte of interest. The large molecule usually has both a hydrophobic region to interact with the stationary phase and a charged region to interact with the analyte. Stationary phases used for ion-pair chromatography are neutral, hydrophobic resins such as polystyrene/divinylbenzene (PS/DVB) or bonded silica. A single stationary phase can be used for either anion or cation analysis. Although the retention mechanism

Quantification in LC and GC: A Practical Guide to Good Chromatographic Data
Edited by Hans-Joachim Kuss and Stavros Kromidas
Copyright © 2009 WILEY-VCH Verlag GmbH & Co. KGaA, Weinheim
ISBN: 978-3-527-32301-2

for ion-pair chromatography is not fully understood, three major theories have been proposed [3, 4]:

- ion pair formation,
- dynamic ion exchange,
- ion interaction.

Owing to higher sensitivity and greater flexibility (e.g. gradient elution), conductivity in combination with a continuously regenerated suppressor is the detection method most commonly used today. In addition, UV/VIS, amperometric, and fluorescence detection are also used.

This chapter's focus will be on the interpretation of data obtained by anion-exchange chromatography in conjunction with suppressed conductivity detection [3].

12.2
Eluents

12.2.1
Purity

Eluents most commonly used in modern IC consist of diluted acids or bases [5, 6], which can exhibit possible corrosive properties (see Section 12.4). They are conventionally prepared by diluting commercially available chemicals of suitable purity. Depending on the starting material, precautions have to be taken to prevent contamination of the final eluent. Diluted sodium hydroxide, for example, can be contaminated with carbonate through the uptake of CO_2 from the surrounding atmosphere. This uptake can be prevented by degassing the water with helium or by vacuum, and by keeping the final eluent under an inert gas. If carbonate is formed in the hydroxide solution, the composition of the eluent and the chromatographic properties will change. Contaminated hydroxide eluents also exhibit a higher conductivity, giving rise to unwanted baseline drifts in the case of gradient elution.

The introduction of Reagent-Free™ IC systems (RFIC™, Dionex Corp., Sunnyvale, CA), based on a concept developed by Small *et al.* [7, 8], has significantly improved automation and ease-of-use of IC. RFIC systems combine electrolytic generation of an eluent (i.e., potassium, sodium, or lithium hydroxide; sodium carbonate/bicarbonate; or methanesulfonic acid) from deionized water, an electrolytic continuously regenerated trap column, and electrolytic suppression. RFIC is a significant improvement over traditional assays, eliminating the handling of conventional reagents, as well as eliminating the need for degassing of the eluent. Electrolytic generation of the eluent allows precise control of eluent concentration through control of the current. The ability to accurately program the exact eluent concentration through software has allowed analysts to streamline the development process of many analytical methods, improve the precision of the methods, and provide better reproducibility between laboratories and operators.

Contamination of carbonate/bicarbonate-based eluents can also lead to a different chromatographic separation. A shifting ratio between carbonate and bicarbonate, due to the addition of NaOH or through the absorption of CO_2, leads to a different pH value, which will influence the peak retention times (e.g., phosphate). Because of the pH-dependent dissociation equilibrium

$$H_3PO_4$$
$$\downarrow -H^+$$
$$H_2PO_4^-$$
$$\downarrow -H^+$$
$$HPO_4^{2-}$$
$$\downarrow -H^+$$
$$PO_4^{3-}$$

a different chromatographic selectivity is obtained at first glance. The chemical explanation is that the P-oxoanion elutes from the ion-exchange column depending on the formal charge. The more negative the charge (the higher the pH of the eluent), the more the anion is retained. Therefore, shifting retention times can point to possible contamination of the eluent. RFIC also offers an easy-to-use solution for a continuous and online preparation of carbonate/bicarbonate eluents.

Eluent generators produce the eluents online in a closed system. Contamination with CO_2 from the surrounding atmosphere is consequently prevented, benefiting trace analytical determinations. Degassing of water used in the final eluent is then not necessary.

12.3
The Water Dip – the Solvent Peak in Ion Chromatography

Conductivity is the most commonly used method of detection in IC [9]. For a sensitive and specific detection of ions using their electrical conductivity, the use of a suppressor is required.

The suppressor (illustrated in Fig. 12.1) converts the high ionic strength eluent into a poorly conducting species (such as water) immediately prior to the conductivity detector. This allows low concentrations of ionic analytes to be detected following separation by ion-exchange chromatography. Counter ions are exchanged continuously in the suppressor with hydronium (anion determinations) or hydroxide (cation determinations) and removed from the eluent.

If NaOH is used as an eluent to separate chloride and sulfate by anion-exchange chromatography, the suppression reaction (in this case a cation-exchange process) for both the eluent and the analyte counter ions takes place prior to the conductivity cell. The corresponding acids of both the eluent and sample anions then enter the detection cell. As all cations are exchanged with hydronium ions, interferences with inorganic cations eluting near the void and possible co-elution of metals with inorganic anions are thus prevented. The suppression product of NaOH is water,

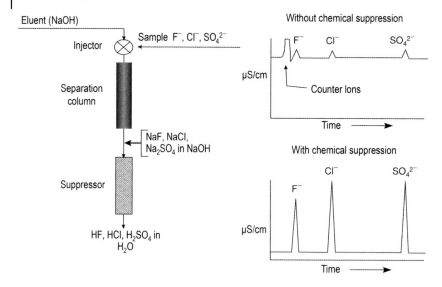

Figure 12.1 Schematic representation of suppressed conductivity in ion chromatography.

so that the background conductivity is lowered and the temperature dependence of conductivity detection is minimized. Suppression can also increase analyte conductivity. For example, NaCl and Na_2SO_4 are converted into the strongly conducting inorganic mineral acids HCl and H_2SO_4 (Fig. 12.1).

The lower background conductivity combined with the increased analyte signals additionally improves the signal-to-noise ratio, allowing MDLs in the low $\mu g\ L^{-1}$ range. Applying suppressed conductivity, both eluent and analyte ions are converted into their corresponding acids (anion analysis) or their corresponding bases (cation analysis), turning the conductivity detection into a quasi-analyte-specific detection scheme (Fig. 12.2).

Modern suppressors are continuously regenerated, resulting in a so-called dynamic ion-exchange capacity (suppression capacity). This capacity is expressed in $\mu mol\ min^{-1}$ and, depending on the specific suppressor properties, can reach values up to 350 $\mu mol\ min^{-1}\ Na^+$.

Hence, modern suppressors can be used with high capacity separation columns associated with high ionic strength eluents as well as concentration gradient elution. Due to the higher sample loading of such columns lower MDLs are obtained.

A specific feature of ion chromatography is the so-called "water dip". In conventional HPLC solvent peaks are known, which can be attributed to excessive sample solvent. In contrast to most solvent peaks in HPLC, the solvent peak in IC (the water dip) is a negative peak. It is due to the eluting sample solvent (water) diluting the eluent, resulting in a decrease in background conductivity. The size of the resulting peak depends on the background conductivity: the higher this background conductivity, the larger the negative peak.

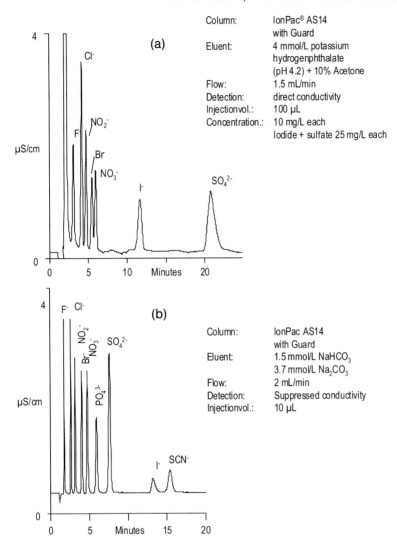

Figure 12.2 Comparison of conductivity detection.
(a) Direct conductivity without suppressor, (b) suppressed conductivity.
In both cases the conductivity cell was not thermally stabilized.

The integration of components eluting at the rising water dip flank, e.g. fluoride or acetate, etc., can be compromised. Specific integration algorithms – known as "Void-Volume Treatment" – simplify the evaluation. This algorithm prevents the integration of negative peaks (water dip) at the beginning of the chromatogram, facilitating the integration of the subsequent peaks (Fig. 12.3).

Alternatively, hydroxide eluents and gradient elution can be used. Owing to the lower background conductivity (e.g. of KOH eluents) and the higher peak

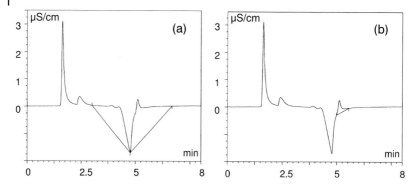

Figure 12.3 Example of "Void-Volume Treatment".
(a) Without Void-Volume Treatment, (b) with Void-Volume Treatment.
All other integration parameters were kept constant.

capacity resulting from gradient elution, quantitation of early eluting compounds is simplified. In these cases the use of specific integration algorithms is usually not necessary.

12.4
Contaminants

The constituent parts of modern ion chromatographs, namely the pump, capillaries, valves, injectors, etc., are all made from metal-free components. This is essential in preventing corrosion, which would lead to metal contamination of the separator columns and the suppressors. Metal contamination (e.g. iron) can negatively influence the recovery, peak shape, and retention times of certain ions and can increase the system pressure.

Iron and various other metals are present in many environmental samples. If injected directly, these metals can be retained in the separation columns and suppressors. When using a modern electrolytically regenerated suppressor, these processes can be monitored easily by controlling the electrical resistance. The consequence of contamination is an increased background signal, higher noise, and, in some cases (phosphate and fluoride), a change in the shape of the signal. Changes in column and suppressor properties can easily be diagnosed by regularly injecting control standards.

Cleaning instructions, e.g. rinsing the columns and suppressors with oxalic acid solutions, can be found in the particular instrument manual. In most cases, the initial performance can be restored [10].

Hydrophobic compounds and substances with a high affinity for the stationary phase used are retained first on the pre-columns. Fats, oils, surfactants, humic acids, and lignins, as well as cellulose, proteins, and other high-molecular weight compounds can spoil the separator column and therefore must be removed by

Table 12.1 Effects of possible contamination and remedies.

Observation	Possible cause	Action
Pressure increase	Metal precipitation	Exchanging the column frit
		Rinsing e.g. with oxalic acid solutions (column manual)
		Sample preparation
Peak broadening (e.g. due to fluoride or phosphate)	Contamination with iron	Rinsing e.g. with oxalic acid solutions (column manual)
Increase in the background conductivity	Metal contamination of the suppressor system	Cleaning of the suppressor (manual)
Increase in the noise level		

Table 12.2 Cleaning solutions for modern ion-exchange columns.

Contamination	Cleaning solution[a]
Organic compounds	In most cases 80% acetonitrile, to which, an electrolyte such as NaCl or HCl ($\sim$ 0.1 mol L^{-1}) is added, depending on the properties of the column
Metals	0.1 mol L^{-1} oxalic acid or HCl
Anions	10–100-fold concentrated eluent 0.1–0.5 mol L^{-1} NaOH 0.1–0.5 mol L^{-1} HCl

[a] The pH stability has to be checked in advance prior to using alkaline- or acid-based cleaning solutions. The respective manuals and instructions of the column producer have to be followed.

a sample preparation technique such as SPE prior to injection. The "poisoning" of the analytical column, generally recognized by a decrease in resolution and retention time, is thus prevented. The capacity of the pre-columns is limited, and they have to be rinsed or exchanged occasionally. The type of cleaning procedure depends on the properties of the stationary phase. As most modern ion-exchange columns are solvent compatible, rinsing with conventional HPLC solvents is an option.

In general the specific cleaning procedures supplied by the column manufacturer should be *exactly* followed. Once the cleaning steps have been completed, the columns are reconfigured and equilibrated with the eluent (Table 12.2).

12.5
Calibration Functions

In anion-exchange chromatography in conjunction with suppressed conductivity, linearity of the resulting calibration function is mainly affected by the mobile phase. As a result of the suppression reaction, the analyte ions and the eluting anions are converted into their corresponding acids. In the case of carbonate/bicarbonate eluents, the resulting, only formally discussed, carbonic acid, dissociates through the following reaction:

$$\text{"}H_2CO_3\text{"} + H_2O \leftrightarrow H_3O^+ + HCO_3^-$$

Eluting anions, e.g. halides, are converted into their corresponding, strongly dissociated acids. These strong acids influence the above equilibrium and direct it to the left-hand side of the equation. This change in equilibrium decreases the background conductivity underlying the analyte peak, decreasing peak response and typically yielding a quadratic calibration function.

In contrast, hydroxide-based eluents exhibit significantly lower background conductivity after the suppressor. The water formed is a much weaker acid than the carbonic acid, which results in a lower influence of strong acids on the autoprotolysis of the water. Calibration functions are linear with a steeper ascending slope (higher sensitivity).

Although high performing data analysis software for chromatography data is available, allowing analysis of calibration functions of higher order (quadratic, cubic ...), very often the linear calibration concept is favored. In many cases existing SOPs or other regulations force the user to apply linear calibration functions. Other, obvious advantages of linear correlations between the concentration and the detector signal are: the applicability of the standard additions calibration (spiking) ensuring a higher analytical quality and safety for complex samples as well as minimizing the calibration efforts in routine analysis.

Besides the properties of the eluents, the acid and base strengths of the analytes also affect the linearity of the conductivity detection signal. One example is a decrease in the dissociation of weak acids and bases with increasing concentration. The slope of the calibration curve levels out with increasing concentration, usually following a second-order function.

In cases where linear calibration functions are requested, direct conductivity could be applied. This, however, leads to lower specificity in addition to significantly lower sensitivity.

References

1 O. Folin, R. Bell, *J. Biol. Chem.* **29** (1917) 329.

2 G.-M. Schwab, *Naturwissenschaften* **44** (1937) 25.

3 J. Weiß, *Ionenchromatographie*, 3rd ed., Wiley-VCH, Weinheim, 2001.

4 *Methods Development Using Ion-Pair Chromatography with Suppressed Conductivity Detection*, Technical Note 12, Document No. LPN 0705-01, Dionex Corporation (2000). Dionex Reference Library or www.dionex.com.

5 ISO 10304 Teil 1 bis 4. Verschiedene ISO-Verfahren zur ionenchromatographischen Anionen-Bestimmung in unterschiedlichen Matrizes. Beuth-Verlag, Berlin.

6 ISO 14911: 1998. Bestimmung von Alkali- und Erdalkalimetallen sowie Ammonium in Wasser und Abwasser mit der IC. Beuth-Verlag, Berlin.

7 H. Small, J. Riviello, *Anal. Chem.* **70** (1998) 2205–2212.

8 H. Small, Y. Liu and N. Avdalovic, *Anal. Chem.* **70** (1998) 3629–3635.

9 J. R. Stillian, V. Barreto, K. Friedman, S. Rabin, M. Toofan, J. Statler and L. Takahashi: *Chemical Suppression in Ion Chromatography: 1975 to Present*, Vortrag anlässlich des International Ion Chromatography Symposium, Linz, Österreich, 1992.

10 Oxalic Acid Cleanup Procedure Notice: *Removal of Iron Contamination from Electrolytic Suppressors*. Document No. 031841-02. Dionex Corporation (2001). Dionex Reference Library.

13
Qualification of GPC/GFC/SEC Data and Results

Daniela Held and Peter Kilz

Gel permeation chromatography (GPC), also referred to as gel filtration chromatography (GFC) or size-exclusion chromatography (SEC), is a well established
liquid chromatographic method for the determination of the molecular properties
of synthetic and natural macromolecules and biopolymers in solution. The theoretical background and applications of this method have been described in many
review articles and books [1]. This chapter discusses the distinctive features of the
method and its specific data processing requirements, including their influence
on the results and the quality of the results.

13.1
Introduction

In contrast to other characterization methods, GPC/SEC allows the fast, simple, and
precise determination of macromolecular properties with just one easy measurement. Moreover, a single GPC/SEC experiment provides a complete view of the
characteristics of the properties. These are directly related to macroscopic characteristics of the product, as they depend on the physical (molar mass) and chemical
(composition) heterogeneity of the macromolecule. Therefore many laboratories
use this technique for the analysis of natural, synthetic and biopolymers, even
where there is little or no expertise in the characterization of macromolecules.

For readers that are more familiar with HPLC experiments, Table 13.1 and
Figure 13.1 compare the main differences between HPLC and GPC/SEC measurements.

In GPC/SEC data processing the peak retention volume and the peak shape
are evaluated with specific algorithms (see Section 13.2.1) to measure the molar
mass distribution and the molar mass averages. This is in contrast to HPLC and
GC data processing, where peak sequence and peak area are used for the data
evaluation.

A GPC/SEC separation is according to the molecular size in solution, and the
theoretical background of the separation mechanism is described in detail in the

Quantification in LC and GC: A Practical Guide to Good Chromatographic Data
Edited by Hans-Joachim Kuss and Stavros Kromidas
Copyright © 2009 WILEY-VCH Verlag GmbH & Co. KGaA, Weinheim
ISBN: 978-3-527-32301-2

Table 13.1 Comparison of HPLC vs. GPC/SEC with regards to the special requirements of GPC/SEC.

	HPLC	GPC/SEC
Chromatogram shape	many narrow peaks	(one) broad peak (event. impurity peaks visible in the RI signal)
Analysis goals	(a) qualitative analysis (b) quantitative analysis	(a) molar mass averages (b) molar mass distribution
Results based on	(a) peak order (b) peak area	(a) absolute peak position (b) peak shape
Calibration	detector signal → concentration	retention time → molar mass
Detection	single detector (UV, DAD)	multidetection (UV, RI, LS, viscometry, FTIR, etc.)

literature [1]. For ideal GPC/SEC separations, the retention volume of the eluting macromolecule depends only on its size in solution. There is no interaction with the stationary phase under ideal conditions.

However, molar masses cannot be measured directly by GPC/SEC. Their determination requires the calibration of the retention volume axis with polymer standards of known molecular weight (molar mass standards). This calibration allows one to relate the peak position to a molar mass. Since the mid-1980s molar mass sensitive detectors have been introduced into GPC/SEC systems. A typical molar mass sensitive detector is an on-line light scattering detector that measures the molar mass at every elution volume directly, if certain sample parameters are known (see Section 13.6.2). In this case the relationship molar mass *vs.* peak position (calibration curve) is measured on-line for this specific sample.

13.2
Principles of GPC/SEC Data Processing

GPC/SEC data processing requirements are, as briefly discussed in the Introduction, fundamentally different from those of other chromatographic techniques. Figure 13.2 shows the flow scheme for GPC/SEC data processing without the data acquisition part, which is the same as in any other real-time data capture process.

Meta data originate from several sources, such as automatically created process parameters, user input and user default settings, etc. GPC/SEC meta data contain sample related information, method parameters and all evaluation parameters required for correct and consistent GPC/SEC standard evaluation: e.g. baseline limits, integration regions and parameters for mass fraction evaluation as required by GPC/SEC guidelines.

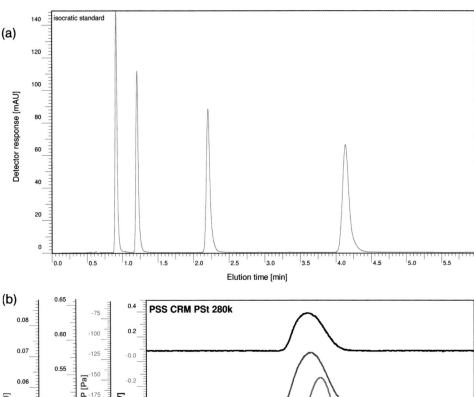

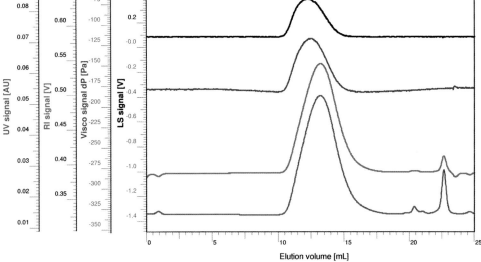

Figure 13.1 Comparison of HPLC and GPC chromatograms (raw data).
(a) Isocratic HPLC separation with one DAD signal.
(b) GPC/SEC chromatogram of a certified reference material (CRM) with a
broad molecular weight distribution with UV (red), RI (green), viscosity (blue),
and light scattering (black) detection (Source: CRM: PSS).

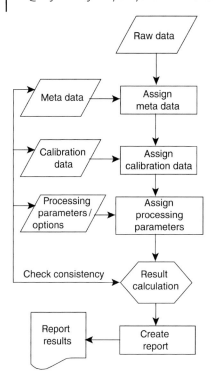

Figure 13.2 Flow diagram for GPC/SEC data evaluation indicating evaluation requirements and user input.

GPC/SEC calibration data are a specific type of meta data and can be seen as a combination of raw data and meta data. The calibration data box summarizes retention and intensity calibration (optional) as well as parameters documenting the type of calibration, the reason for calibration and the used reference values.

The box "processing parameters/options" contains the parameters used to obtain the final results. This includes all sample independent system and infrastructure parameters and defines the data processing schemes. It also includes the parameters used for molar mass calculation and other evaluation options.

13.2.1
Calculation of Molar Mass Averages

The calculation of the molecular weight averages nowadays uses the slice method [1d, 2], where the eluted peak is separated into equidistant time or, more accurately, volume slices. Data acquisition with flow meters is now obsolete as modern chromatographic pumps provide the high precision and reproducibility required to convert the elution time into elution volume. The elution volume is then converted, by use of the calibration, into the molecular mass.

When calculating the molecular averages, the slice concentrations, c_i, must be corrected according to the slope of the calibration curve. This is necessary because the data recording occurs linearly with respect to time, the molecular mass, however, does not change in a linear fashion. This means, in practical terms, at the same concentrations, the number of polymer chains with a defined molecular weight on the high molecular weight part of the elugram is much smaller than on the low molecular weight part. In many chromatography software programs this correction is not included. The errors caused by this will increase with the width of the sample and decrease with the data recording frequency. The correction can only be neglected in the case of strictly linear calibration curves, which commercial linear mixed bed columns do not usually exhibit.

The macroscopic properties of macromolecular products can be derived from the molar mass distribution, $w(M)$, which in turn can be calculated from the signal heights, $h(V)$, in the chromatogram. The molar mass distributions describe, in contrast to the molar mass averages, the overall sample characteristics. Molar mass averages are reduced information, in that two samples can have the same molar mass averages but still have very different molar mass distributions and, therefore, macroscopic behavior.

The differential distribution, $w(M)$, of the molar mass M is defined as

$$w(M) = \frac{dm}{dM} \tag{13.1}$$

where dm/dM is the mass fraction of polymer in the dM interval.

By simple transformation $w(M)$ can be expressed by measured quantities:

$$w(M) = \frac{h(V)}{M(V)\,\sigma(V)} \tag{13.2}$$

where $\sigma(V) = \dfrac{d\log M}{dV}$, $h(V)$ is the detector signal and $\sigma(V)$ is the slope of the calibration curve.

The above correction that is introduced due to the gradient of the calibration curve can now be allocated by the mathematical derivative of the calibration curve.

The cumulative distribution, $I(M)$, will be used as a normalization condition resulting from:

$$I(M) = \int_0^\infty w(M)\,dM \tag{13.3}$$

The molecular weight averages can be calculated from the moments, μ_i, of the molar mass distribution:

$$\mu_i = \int_0^\infty M^i\,w(M)\,dM \tag{13.4}$$

where μ_i is the i-th moment of the molar mass distribution.

The molar mass averages are defined and calculated by:

- number average molecular weight:

$$\overline{M}_n = \frac{\mu_0}{\mu_{-1}} = \frac{\sum\limits_i h_i}{\sum\limits_i \frac{h_i}{M_i}} \tag{13.5}$$

- weight average molecular weight:

$$\overline{M}_w = \frac{\mu_1}{\mu_0} = \frac{\sum\limits_i h_i M_i}{\sum\limits_i h_i} \tag{13.6}$$

- z-average molecular weight:

$$\overline{M}_z = \frac{\mu_2}{\mu_1} = \frac{\sum\limits_i h_i M_i^2}{\sum\limits_i h_i M_i} \tag{13.7}$$

- viscosity average molecular weight:

$$\overline{M}_v = \left(\frac{\mu_v}{\mu_0}\right)^{\frac{1}{a}} \qquad \mu_v = \int\limits_0^\infty M^a \, w(M) \, dM \tag{13.8}$$

The width of the molar mass distribution can be described by the polydispersity, D, or the polydispersity index, PDI:

$$D = \frac{\overline{M}_w}{\overline{M}_n} \tag{13.9}$$

A list of additional GPC/SEC results is summarized below:

Value   description:
$[\eta]$ intrinsic viscosity either measured using an on-line viscometer or
            calculated using the Mark-Houwink coefficients
V_p elution volume at peak maximum of the elugram
M_p molecular weight at peak maximum of the elugram
A peak area
$w(\%)$ mass fraction within defined molecular weight limits
$I(M)$ molecular weight at defined mass fraction

13.3
Guidelines, Standards and Requirements for GPC/SEC Data Processing

In many application areas where scientific software is applied, it is necessary to demonstrate the correct functioning and reliability of the system(s) being used. Additionally, certification and accreditation bodies require comprehensive traceability of data and results [3]. A general proof for GPC/SEC specific data processing has not yet been devised so every single system needs to be validated. This is best done in close cooperation with the vendor, who enables the user to optimize this process with respect to time and efficiency.

Currently only one vendor of GPC/SEC products offers a commercial tool for vendor independent validation of GPC/SEC instrumentation and software (Easy-Valid Validation Kit for GPC/SEC systems, PSS GmbH, Mainz, Germany). The authors do not know of any general tools for validating GPC/SEC software. Most GPC/SEC software products do not have standardized software validation and verification options, thus for this task, the users are left on their own. Besides the validation and verification tools for end users, the vendors need comprehensive software life-cycle processes with revision management, extensive documentation, and strict and independently audited quality control processes (see Table 13.2).

Additional requirements for GPC/SEC software in regulated laboratories consist of general guidelines and policies, as exemplified by e.g. ICH, GxP, GAMP 4,

Table 13.2 Requirements for GPC/SEC software vendors for users working in a regulated environment (ICH, GMP, GAMP 4 and CFR guidelines).

General	Certification, reference projects
Quality management system	QM handbook, SOPs, audits (internal, external), review of vendors
System development life cycle	Project management, design-/test phases, documentation
Conformity	ICH, GxP, GAMP 4, ISO 9000, CFR; usability tests (ISO/DIN 9241-10), German ordinance for work with visual display units; GPC/SEC specific standards
Documentation	Guidelines, change control, test plans, approvals, validation processes
Configuration and risk management	SOPs, version control, traceability, risk management for change
Security aspects	Retrieval processes, security procedures, archiving
Support/warranty	Installation service, support plans, revision service, escalation management, customer specific support

Table 13.3 Requirements of GPC/SEC standards and guidelines for data analysis and evaluation.

Requirement	ISO 13885-x	ISO 16014-x	ASTM D 5296-05	DIN 55672-x
Plate count	> 20000/m	n/s	> 13100	> 20000/m
Asymmetry	1.00 ± 0.15	n/s	n/a	1.00 ± 0.15
Resolution	> 2.5	n/s	> 1.7	> 2.5
Separation efficiency	> 6.0	n/a	n/a	> 6.0
Temperature constancy	±1 °C	n/s	< 3 °C	±1 °C
Signal/noise ratio	> 1:100	> 1:200	< 1:50	> 1:100
Signal drift	< 1% I_{max}	< 20% I_{max}	< 2% I_{max}	< 1% I_{max}
Data points per peak	> 60 data points	n/a	> 40 data points	> 60 data points
Resolution peak height	> 9 bit	n/a	n/a	> 9 bit
Calibration	> 5 calibration points	n/a	n/a	> 5 calibration points
	> 2/decade	n/a	> 3/decade	> 2/decade
Decision criteria calibration	residues and derivative	n/a	n/a	residues and derivative
Baseline	linear	linear	linear	linear
Integration limits	separate	separate	n/a	separate
Flow accuracy	RSD(V_p) < 0.3%	RSD(V_p) < 0.2%	RSD(V_p) < 0.3%	RSD(V_p) < 0.3%
Molar mass determination	cf. Chapter 13.2	cf. Chapter 13.2	cf. Chapter 13.2	cf. Chapter 13.2
MMD determination	cf. Chapter 13.2	cf. Chapter 13.2	cf. Chapter 13.2	cf. Chapter 13.2
$I(M)$ determination	cf. Chapter 13.2	cf. Chapter 13.2	cf. Chapter 13.2	cf. Chapter 13.2

n/s Not specified: mentioned in standard without detailed requirements.
n/a Not applicable: not mentioned in standard.

ISO 9000, 21 CFR Part 11 [4]. Moreover, user oriented ergonomics standards and directives for software usability (ISO/DIN 9241-10) and the German Ordinance for Work with Visual Display Units and should be met.

Additionally, conformity to specific national and international GPC/SEC standards (ISO 13885, ISO 16014, ASTM D 5296-05, DIN 55672, etc.) has to be followed. They provide guidelines for accurate data treatment, robust data evaluation, result calculation, and traceability. Table 13.3 summarizes the most important requirements discussed in these standards. Very detailed and strict procedures can be found in the ISO 13885 and DIN/EN 55672 standards. Other GPC/SEC standards are less detailed and less instructive, which can be a drawback when assessing inter-laboratory results based on those guidelines.

13.4
Validation and Tests for GPC/SEC Data Evaluation

The molar mass distribution (MMD), $w(M)$, is *the* comprehensive description of a macromolecular sample and allows correlation with macroscopic properties. Today GPC/SEC is the method of choice for measuring the molar mass distribution. However, the primary information (raw data) of GPC/SEC measurements is the apparent concentration distribution (chromatogram, $c(V)$), which is a convolution of sample related parameters and experimental conditions. The task during GPC/SEC data processing is to separate the instrument specific parameters (e.g. flow rate, eluent, number and type of separation columns, detection type) from the sample specific parameters, to translate the chromatogram into the molar mass distribution and to calculate the resulting molar mass averages that are widely used to describe the samples.

The difference between a molar mass distribution and a chromatogram can be easily understood with the following example. Two different laboratories measure the same sample on different instruments with a different number of columns with different column lengths. This results in two different chromatograms (primary information). Without additional information it is not possible to decide whether these chromatograms show the same sample or not. However, this can be easily decided when the molar mass distribution is calculated from the raw data because, for the molar mass distribution, all instrument related parameters have been eliminated. Therefore only correctly calculated molar mass distributions allow direct inter-laboratory and long-term comparison of samples and sample properties.

13.4.1
Description of a General Verification Procedure for GPC/SEC Software

Kilz *et al.* [5] have developed a method for the validation of GPC/SEC data processing algorithms. Reference data have been generated from a theoretical molar mass distribution (Flory-Schulz and Wesslau distributions with various molar masses, polydispersities and different signal-to-noise ratios) [6]. This novel approach is independent of any preconditions or influence from the system. Therefore this route avoids vicious circles in the validation process.

Without the necessity of any additional assumptions the method allows the independent validation of
- *all* software algorithms,
- the *complete* data flow path within the data management system,
- the in-house installation.

One vendor of a macromolecular chromatography data system (MCDS) offers this procedure as an automated software tool so that users can check and prove the correctness of the installation and the evaluation process on the dedicated workstation(s) [7].

If the software validation is successful and discrepancies occur in the results, the chromatographic conditions and instrument performance need to be checked. Dedicated GPC/SEC validation kits (e.g. PSS EasyValid Validation Kit, PSS, Mainz, Germany) are used for the operational qualification during instrument qualification and to verify whether the problem is system or application specific.

The validation procedure consists of three steps:

1. Using the theoretical molar mass distributions (with well-known M_n, M_w and M_z, $I(M)$, $[\eta]$) the raw data, $c(V)$, $P(V)$, $I(V)$ and $M(V)$ are generated according to equations published in the literature.
2. The theoretical raw data, with well-known molecular parameters, are imported into the MCDS and treated as newly acquired data within the software. This ensures that identical data processing and data presentation steps are being used for validation and evaluation of unknown samples.
3. The results of the theoretical molar mass distribution and the software-calculated experimental results are compared as numerical values and distribution curves. The results must be identical (taking numerical errors into account) for the test to be successful.

13.4.2
Validation of the Molar Mass Distribution Calculation

The procedure for the calculation of the molar mass distribution is described in Section 13.2.1. This section discusses the difference between the true molar mass distribution and the molar mass diagram calculated by many HPLC programs with GPC/SEC add-ons or software modules.

As already shown in the previous section, the chromatogram presents an apparent concentration distribution where the molar mass decreases with increasing retention time (elution volume).

The first step for the transformation of a chromatogram into a molar mass distribution is to replace the retention volume at the x-axis by the corresponding molecular weight from the calibration curve. In the second step the y-axis needs to be converted into mass fractions (within a molar mass interval), $w(\log M)$ [8].

This is necessary because chromatogram signals are recorded at constant time intervals while the molar mass distribution requires the mass fraction to be in constant molar mass intervals.

Sometimes it is difficult to understand whether a commercial chromatography data system shows a true molar mass distribution or only a molar mass diagram where the y-axis shows just signal intensity and not the mass fraction, $w(\log(M))$.

A simple experiment can show whether true molar mass distributions are obtained or not. A mixture of polymer standards with different molecular weights but equal concentrations is injected onto a single-porosity column/column combination. A calibration curve with a nonlinear fit function (e.g. cubic fit function or any other polynomial function of higher degree, see Fig. 13.3) is used to evaluate the data and to display the molar mass distribution for the sample mixture. If the peak heights and widths in the chromatogram and the molar mass distribution do not vary, the data system shows a molar mass diagram and not the molar mass distribution (see Fig. 13.4a). As discussed in the Introduction, inter-laboratory comparisons can be extremely difficult when different GPC/SEC software packages are used that might not generate the correct molar mass distributions and results.

Figure 13.4a shows in detail that in molar mass diagrams the peak position (molar mass at peak maximum) as well as the peak width (polydispersity) can be wrong.

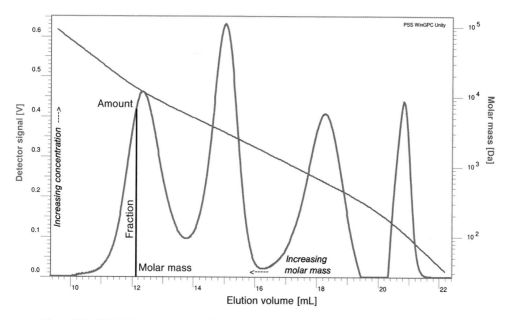

Figure 13.3 GPC/SEC chromatogram of a sample mixture (green) containing four different molar mass calibration standards (PSS ReadyCal green) with overlaid calibration curve (red).

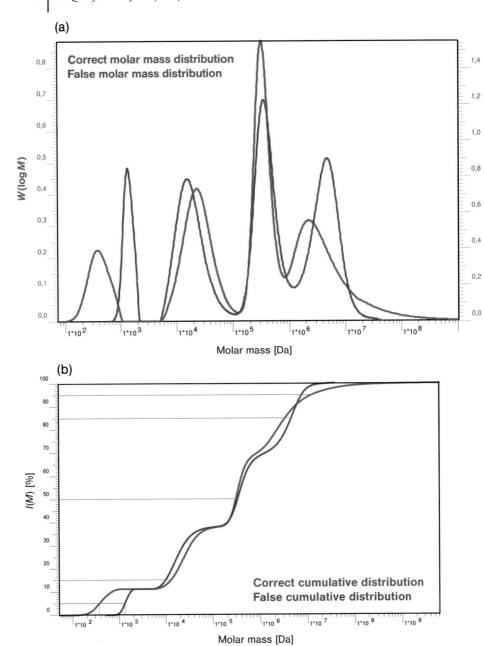

Figure 13.4 Overlay of a true molar mass distribution (green) obtained
from a GPC/SEC evaluation software and a false (apparent) molar mass
distribution calculated by HPLC data system (red).
(a) Differential molar mass distribution, (b) cumulative molar mass
distribution.

The question that arises here is: will the molar mass averages calculated by these data systems also be incorrect?

In general, this is not the case because most CDS vendors calculate the molar mass averages from the sum equations and not directly from the molar mass distribution. However, Table 13.4 shows that peak molecular weights and polydispersities can be significantly wrong if the data processing is not compliant with GPC/SEC standards.

Moreover, Table 13.5 shows that other important GPC/SEC results, such as the mass fractions above and below certain molecular weight limits (e.g. 500 Da or 1000 Da) will also be affected (cf. Fig. 13.4b). Such information is often required for the registration of products with federal bodies or for an improved quality control of products that have a broad molecular weight distribution.

Table 13.4 Comparison of peak molar mass and polydispersity index for correct and false molar mass determination. Source: data in Fig. 13.4.

Sample	M_p [kDa]			M_w/M_n		
	Correct	False	Deviation (%)	Correct	False	Deviation (%)
Peak 1	3685	4764	29	12.10	1.40	−88.4
Peak 2	319	352	10	1.13	1.22	8.0
Peak 3	21	15	27	1.41	1.38	−2.1
Peak 4	0	1	250	1.21	1.04	−14.0

Table 13.5 Comparison of additional GPC/SEC results for correct and false molar mass determination. Source: data in Fig. 13.4.

Property	Correct	False	Deviation (%)
Mass < 500 Da	7.85%	0.00%	−
Mass > 10 000 kDa	5.88%	2.11%	64
Molar mass at 5% (kDa)	0.37	1.31	254
Molar mass at 15% (kDa)	12.8	9.87	23
Molar mass at 50% (kDa)	304	331	9
Molar mass at 85% (kDa)	3325	4510	36
Molar mass at 95% (kDa)	12254	7520	39

13.5
Influence of Data Processing, Calibration Methods and Signal Quality on Accuracy and Precision of GPC/SEC Results

In the next sections the deviations of GPC/SEC results will be evaluated, based on theoretical molar mass distributions, applying various GPC/SEC techniques. Additionally, the influence of signal quality and signal stability on GPC/SEC results will be investigated.

13.5.1
Validation of GPC/SEC Evaluation with External Molar Mass Calibration

As described in Section 13.4.1, the deviation caused by the evaluation process may be tested by comparing the calculated results with theoretical results. This test analyzes the influence of the data processing procedure on the quality of GPC/SEC results. In this example, the GPC/SEC instrument is equipped only with a Refractive Index (RI) concentration detector. The molar masses are obtained from a calibration curve established with polystyrene molar mass standards with narrow distributions.

Table 13.6 shows the results obtained through a dedicated macromolecular chromatography data system (MCDS) in comparison with the theoretical results according to Schulz–Flory. The minor deviations are caused by the limited number of data points being compared (in this case 52).

Table 13.6 Comparison of results and reference values for conventional data evaluation for one concentration detector. The reference values are based on a Schulz–Flory distribution.

	Reference	Calculated by MCDS	Deviation (%)
M_n (Da)	150 000	150 300	0.2
M_w (Da)	300 000	300 070	0.023
M_z (Da)	450 000	450 010	0.002
D	2	1.997	0.15
M_v (Da)	280 869	280 230	0.228
$[\eta]$ (mL g^{-1})	105.68	105.69	0.009
c (g L^{-1})	4.343	4.343	0

13.5.2
Validation of GPC/SEC Evaluation with Viscometry Set-ups

Similar to Section 13.5.1, this study investigates the influence of advanced data processing methods in GPC/SEC instruments consisting of an on-line viscometer (type: 4 capillary viscometer measuring delta pressure and inlet pressure) and using RI as the concentration detector. It is assumed that the refractive index increment, dn/dc, is constant.

In GPC/SEC–viscometry systems the molar mass is calculated from a universal calibration curve ($\log [\mu]M$ vs. V) and the measured intrinsic viscosities at every elution volume. Results from a theoretical Schulz–Flory distribution for a polystyrene sample were compared with the results from a MCDS obtained with a universal calibration curve established with poly(methyl methacrylate) molar mass standards with a narrow distribution.

Table 13.7 summarizes the results and shows the very good agreement between theoretical and experimental data obtained with the MCDS. The small deviations (< 1%) are caused by the limited number of theoretical data points and by numerical rounding errors using transcendent functions.

Table 13.7 Accuracy of results for GPC/SEC–viscometry data analysis in comparison with a Schulz–Flory distribution.

	Reference	Calculated by MCDS	Deviation (%)
M_n (Da)	150 000	150 110	0.073
M_w (Da)	300 000	298 230	0.59
M_z (Da)	450 000	447 220	0.618
D	2	1.987	0.65
M_v (Da)	278 997	278 510	0.175
$[\eta]$ (mL g^{-1})	105.68	105.69	0.001
K (mL g^{-1})	0.0136	0.0137	0.514
a	0.714	0.714	0
c (g L^{-1})	4.343	4.343	0

13.5.3
Influence of Detector Noise on the Accuracy of GPC/SEC Results

The theoretical data used in the previous sections were obtained from raw data with infinite signal-to-noise ratios, S/N. In real experiments the raw data always show more or less signal noise. It is critical for any GPC/SEC data processing software to handle noisy signals correctly in order to obtain valid results even

Table 13.8 Influence of statistical noise (5% of signal maximum) on the results from various evaluation techniques for a sample with a Schulz–Flory distribution.

	Reference	Conventional	Viscometry	Light scattering	Triple detection
M_n (Da)	150 000	131 630	150 200	169 120	151 040
M_w (Da)	300 000	303 310	306 260	299 670	294 420
M_z (Da)	450 000	477 670	478 210	427 110	415 290
D	2	2.304	2.039	1.772	1.949
M_v (Da) (PS)	280 869	282 240	285 670	284 030	276 170
$[\eta]$ (mL g^{-1})	105.68	106.79	106.86	105.9	104.59
K (mL g^{-1})	0.01363	n/a	0.01457	n/a	0.01295
a	0.714	n/a	0.708	n/a	0.719
c (g L^{-1})	4.343	4.327	4.276	4.284	4.296

under difficult circumstances. The following experiments were pereformed with noisy data so that the ruggedness of the data processing software could also be tested (see Fig. 13.5).

All raw data in this "worst case" scenario show a 5% statistical noise to test the influence of noisy data. This corresponds to a signal-to-noise ratio of 20:1 while in laboratory experiments S/N levels > 100 are usually obtained. The results shown in Table 13.8 were obtained without any smoothing, FFT or despike algorithms.

The simulation shows the potential deviations for different GPC/SEC calculation methods caused by signal noise. In general, the deviations caused by the poor S/N ratios increase when more detector signals have to be used to obtain the results. As expected, additional signal noise in conventional GPC/SEC leads to lower number average molar masses, M_n, and z-average, M_z, being overestimated (approx. 10%). Larger deviations are seen using universal calibration, although the Mark–Houwink coefficients K and a are still close to the expected values. The largest deviations are observed for GPC/SEC light scattering results: M_n is clearly overestimated and the polydispersity index, PDI, is underestimated because the light scattering signals at lower masses are lost in the baseline noise.

A clear improvement in these results can be achieved if the integration limits in the elugram are changed in such a way that only significant data with an S/N > 3 are used for subsequent calculations. In our example these values are at 10.0 and 17.0 ml. This corresponds to a reduction in the calculated injected mass of about 1%. This shows that particular care is needed in the selection of the baselines when dealing with noisy detector data. The independent setting of baseline and integration limits, as required by the ISO GPC/SEC standard [8] and discussed in detail in Section 13.6.1.3 is, therefore, an important aspect in obtaining reliable results. With these caveats taken into account, accurate and precise molar mass

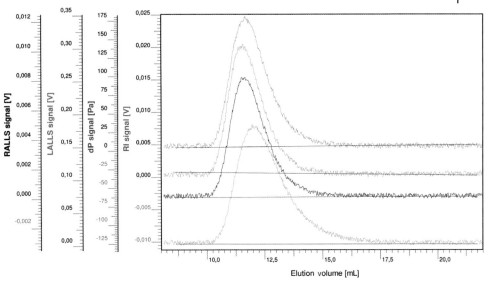

Figure 13.5 LALLS, RALLS, viscosity and RI raw data with baselines. All signals have been overlaid with a 5% statistical noise to test the robustness of the GPC/SEC algorithms (S/N: 20:1).

averages can be obtained even with noisy data. Despite the high detector noise surprisingly accurate intrinsic viscosities and Mark–Houwink coefficients can be measured when an additional on-line viscometer is used.

13.5.4
Influence of the Detector Drift on the Accuracy of GPC/SEC Results

Owing to the various types of possible detector drifts a comprehensive discussion of the influence of detector drift is not possible in this chapter. A constant drift can normally be handled easily within a (M)CDS, if the interpolation of data points is done correctly [8]. As this cannot be guaranteed and many vendors do not publish their interpolation procedures, only the maximum deviation for the most common case of detector drift, the linear drift, will be investigated in this section. A linear drift means that the signal, starting from the injection point, will drift linearly and constantly with the elution. The drift ratio is related to the maximum signal intensity of the peak (corrected with the interpolated baseline value). This means that if a signal shows a 1% drift, the last data point will be 1% of the peak maximum above the value for a non-drifting baseline.

Table 13.9 summarizes the influence of the detector drift on typical GPC/SEC results for a certified reference material with a broad molecular weight distribution (CRM source: PSS, Mainz, Germany). It is clearly visible that the central moments of the distribution, M_w and here also M_v, are more stable against detector drift than the averages M_n and M_z (see also Section 13.2.1). These results are also shown in Fig. 13.6.

Table 13.9 Influence of a constant signal drift (x% from signal maximum) on important GPC/SEC results studied for a certified reference material (CRM source: PSS BAM P005).

	CRM reference	1% Drift	5% Drift	10% Drift
M_n (Da)	115 000	98 700	72 500	57 100
M_w (Da)	330 000	329 000	328 000	327 000
M_z (Da)	612 000	622 000	662 000	691 000
D	2.88	3.34	4.53	5.72
M_v (Da)	299 000	298 000	295 000	291 000
$[\eta]$ (mL g^{-1})	110.44	110.06	109.29	108.27

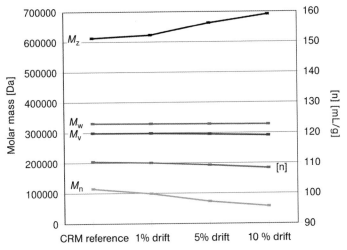

Figure 13.6 Influence of the detector drift on the results for GPC/SEC data analysis based on identical raw data for a certified reference material (Source: PSS BAM P005). The molar masses are displayed on the left y-axis and the intrinsic viscosity is displayed on the right y-axis.

13.5.5
Influence of the Number of Data Points on the Accuracy of GPC/SEC Results

In contrast to other analytical methods, the influence of the number of data points on typical GPC/SEC results, e.g. the molar mass averages, is limited. Only if a drastically reduced number of data points are used will the results change significantly. This is caused by the averaging process that reduces the information content of the molar mass distribution when molar mass averages are calculated by a mathematical integration.

Table 13.10 shows the influence on the results when the number of data points is reduced by a factor of 64 for a certified reference material with a broad molecular weight distribution (CRM source: PSS, Mainz, Germany, sample BAM P005) which is used as a model of typical polymer samples. Significant deviations can only be observed if the number of data points is below 50. This is also illustrated in Fig. 13.7. For samples with narrow molecular weight distributions this limit will be reached earlier if the same data capture rate is used.

Table 13.10 Influence of the number of data points on the accuracy of the molar mass averages and the polydispersity index for a certified reference material (CRM source: PSS BAM P005).

Points	1024	128	64	32	16
M_n (Da)	121 000	120 000	118 000	110 000	106 000
M_w (Da)	330 000	330 000	330 000	330 000	332 000
M_z (Da)	616 000	617 000	621 000	624 000	630 000
D	2.72	2.75	2.81	2.99	3.15
M_v (Da) (PS)	299 000	299 000	299 000	299 000	300 000
$[\eta]$ (mL g^{-1})	110.4	110.4	110.4	110.4	110.8
% < 10 kDa	0.57	0.57	0.58	0.59	0.74
% > 1000 kDa	3.27	3.29	3.31	3.61	4.57

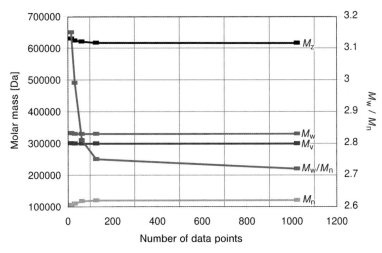

Figure 13.7 Influence of the number of data points on the accuracy of GPC/SEC results based on identical raw data for a certified reference material (Source: PSS BAM P005). The molar masses are displayed on the left y-axis and the polydispersity index PDI is displayed on the right y-axis.

13.6
Influence of GPC/SEC Specific Parameters on the Precision and Accuracy of GPC/SEC Results

This section discusses the influence of parameters other than the detection and data acquisition parameters that were addressed in the previous section. As shown in the Introduction and Table 13.1 the absolute position (elution volume, retention time) of the chromatographic peak is necessary to obtain the GPC/SEC results. All parameters that influence the peak position consequently influence the results. The most important parameter is the precision of the flow provided by the pump. A flow rate correction can be employed if the pump is consistently operating at a different flow rate for samples and for the calibration samples. However, if the flow rate changes continuously or step-wise during a run, a correction to obtain better precision is not possible. The peak position can also be influenced by overloading the separation columns due to too high an amount of the injected sample. The optimum injected mass depends on the molar mass of the sample and its polydispersity. For samples with a narrow molar mass distribution (e.g. molar mass calibration standards with polydispersities less than 1.1) the injected mass needs to be much lower than for samples with a broad molar mass distribution. Recommendations for the use of appropriate concentrations and injection volumes can be found in the ISO standard 13885 [8].

The precision and accuracy of the results is also influenced by the GPC/SEC calibration data and by additional evaluation parameters. For a detailed discussion of these influences it is important to distinguish between measurements with external molar mass calibration and measurements with on-line calibration with light scattering detectors. External molar mass calibration is done by establishing a calibration curve from the separately measured elution volumes of molar mass calibration standards. This set-up uses RI (refractive index), UV, ELS (evaporating light scattering) and on-line viscometer detection (and combinations thereof). In GPC/SEC measurements with on-line light scattering detection, the relationship of the molar mass to the elution volume axis is measured on-line with each sample. This set-up uses RI (refractive index) or UV detectors and an on-line light scattering detector. Both set-ups have their own dedicated requirements for achieving precise and accurate results.

13.6.1
Influence of Parameters for GPC/SEC with External Molar Mass Calibration

13.6.1.1 Influence of the Calibration Curve
As expected for all methods using an external calibration, the accuracy and precision of GPC/SEC measurements with external molar mass calibration depends strongly on the accuracy of the calibration itself. Table 13.11 shows an overview of typical GPC/SEC calibration procedures with their advantages and limitations.

The most commonly used calibration procedure is based on molar mass calibration standards with narrow molar mass distribution. If the calibration standards

Table 13.11 Overview of GPC/SEC calibration methods with molar mass calibration standards.

Method	Advantages	Disadvantages
Calibration with molar mass calibration standards with narrow distribution	• Easy and straightforward • Accurate method for chemical and structural identical samples and calibration standards	• Not all monomers can be polymerized with a narrow distribution • Calibration is only valid for chemically and structurally identical substances
Broad calibration	• Easy and accurate method	• Only a limited number of standards commercially available • Calibration is only valid for chemically and structurally identical substances
Integral calibration	• Easy and accurate method	• Only a limited number of standards commercially available • Calibration is only valid for chemically and structurally identical substances
Universal calibration Calibration with molar mass calibration standards with narrow distribution with additional on-line viscometer detection	• Easy and accurate method • One calibration curve valid for all types of samples	• Increased experimental complexity • Additional experimental error from concentration dependence, band broadening and detector delay

and the measured samples are chemically and/or structurally different, only apparent molecular weights related to the calibration standards are obtained. The deviation for the molar mass averages and the molar mass distribution can easily be in the region of 20% up to several 100%. As the results for different samples can still be compared with each other and the method is robust and easy to use, many laboratories apply such procedures for quality control and sample comparison as well as for applications where the absolute molar mass is not required.

The number of required molar mass standards depends on the molar mass range that has to be calibrated. The ISO standard recommends at least 3 molar mass standards for every molar mass decade [8]. After measuring and plotting the elution volumes for the molar mass calibration standards a fit function describing the molar mass/elution volume relationship needs to be selected. Unfortunately, GPC/SEC calibration curves are not linear over their complete separation range, even when so-called linear or mixed-bed columns are used. GPC/SEC calibration curves normally show a strongly sigmoidal behavior. Therefore the most commonly used fit functions are polynomial functions or specifically developed

GPC/SEC fit functions (PSS fit functions). The optimum fit function depends on the column or column combination in use and needs to be selected by the user.

Several criteria that support the user in the selection process are discussed in the literature and provided by software programs:

1. Regression coefficient, R^2.
2. Deviation of every calibration point from the fitted value (e.g. average deviation).
3. The slope of the calibration curve.

Table 13.12 shows the regression coefficients for identical raw data (Figs. 13.8 and 13.9) with different fit functions and the average deviation for all data points. The average deviation and the deviation of every single calibration point are correlated with all errors that can occur during characterization, analysis and evaluation of the molar mass calibration standards (e.g. wrong elution volume, inaccurate reference molar mass).

It is obvious from Table 13.12 that the regression coefficient is not a suitable parameter to select the best calibration fit function. Even for a regression coefficient close to 1 very large average deviations can be observed. If the data evaluation software provides the regression coefficient as the only selection criterion for the fit function, a value of 0.999 should be achieved for GPC/SEC results with the lowest possible error.

The regression coefficient and the average deviation decrease when selecting a polynomial function of higher degree. However, it is not physically meaningful to use the function with the highest degree that generates the lowest average deviation. More important than small deviations is that the shape of the calibration curve is in general agreement with the separation mechanism. This means that a lower molar mass (hydrodynamic volume) is always correlated with a higher elution volume. A good measure of this principle is the first derivative of the calibration curve (slope).

Figure 13.8 shows an ideal first derivative for a GPC/SEC calibration curve. The slope changes only near the exclusion limit and the total permeation volume and is constant for the optimum separation range. If a polynomial fit function of the seventh degree is chosen (compare Fig. 13.9) the slope is not constant and maxima and minima (that lack any physical significance) appear.

Table 13.12 Influence of the calibration fit function on the regression coefficient.

Fit function	R^2	Average deviation (%)
Linear (square)	0.9925	30.20
Polynomial 3 (cubic)	0.9986	10.44
Polynomial 5	0.9995	7.35
Polynomial 7[a]	0.9999	3.57
PSS Polynomial 7	0.9998	4.92

[a] First derivative is not uniformly continuous; this function should not be used.

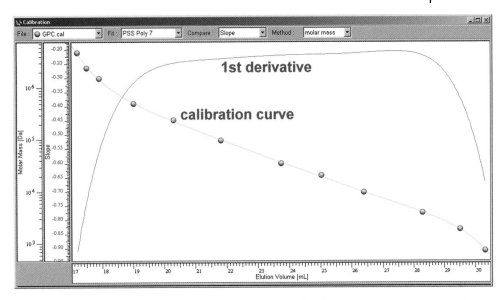

Figure 13.8 GPC/SEC calibration curve with physically meaningful first derivative.

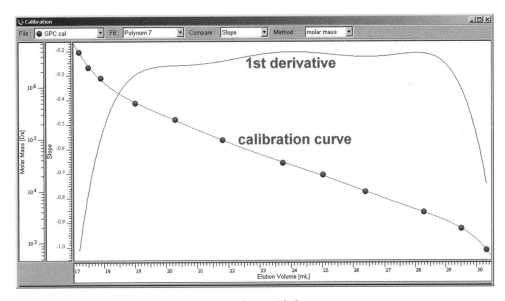

Figure 13.9 Calibration curve with seventh degree polynomial fit function.
The degree of the chosen function is high leading to a first derivative
with physically meaningless maxima and minima.

Table 13.13 Influence of the selected fit function on the accuracy of the determined molar mass: example with a conventional cubic fit function and a fit function developed specifically for GPC/SEC calibration curves.

Polystyrene standard	Results with PSS calibration fit function			Results with polynomial fit function 3rd order		
Reference M_w	M_w	M_w/M_n	ΔM_w (%)	M_w	M_w/M_n	ΔM_w (%)
985 000	1 006 000	1.05	+2	1 507 000	1.06	+53
325 000	316 000	1.02	−3	393 000	1.04	+21
98 000	96 300	1.05	−2	84 500	1.07	−16
34 000	31 500	1.05	+3	26 400	1.05	−29
10 000	9 803	1.09	−2	7 800	1.06	−28
8 100	8 080	1.08	0	6 650	1.04	−22

The influence of the fit function on the molar mass results is shown in Table 13.13. It compares the results for identical raw data and calibration points obtained with a conventional cubic fit function (third order polynomial) and a specially developed GPC/SEC fit function based on a polynomial fit. The PSS fit functions (PSS poly 3 to 7) are ideally suited to describe the sigmoidal behavior and the points of discontinuity at the exclusion limit and the total permeation limit. This allows one to make efficient use of the complete molar mass separation range of the column/column set, saves analysis time and reduces costs for additional columns.

13.6.1.2 Influence of Instrument Parameters (Instrument Performance)

As shown in Table 13.1, GPC/SEC results are obtained from the absolute peak position. All parameters that influence the peak position will influence the accuracy of the final results. The reproducibility of the analytical conditions during calibration and measurement is of the utmost importance for accurate results. One of the most important requirements for GPC/SEC is therefore reproducible and stable flow rates. Although modern HPLC pumps provide stable flow rates, the influence of this effect should not be neglected. Table 13.14 shows the influence of flow rate variations during calibration and analysis for typical GPC/SEC results.

The deviations in Table 13.14 depend on the column or column combination in use and the position of the investigated peaks. The steeper the slope of the calibration curve, the higher the deviation. This means that the deviations for so-called linear or mixed bed columns or for columns with larger particle size (> 10 µm) are even higher than those shown in this example.

Working with an internal flow marker (internal standard) helps to overcome this problem and to improve the accuracy and long-term repeatability of GPC/SEC.

Table 13.14 Influence of flow rate deviations on the accuracy of the molar mass results.

Deviation from calibration flow rate (%)	Measured molar mass M_w (Da)	M_w deviation (%)
2.1	46 500	36.8
1.31	41 900	23.2
0.61	37 500	10.3
0	34 000	0
−0.8	29 800	−12.4
−1.5	26 400	−22.4
−2.2	23 300	−31.5

A low molecular weight compound is added to the sample preparation solvent, so that an additional peak will appear in the low molecular weight region for all calibration standards and unknown samples. Every single measurement will then be corrected with respect to this peak. This ensures that the same chromatographic conditions during calibration and measurement of the unknowns are achieved (see Fig. 13.10) and avoids experimental errors as shown in Table 13.14. However, this approach works only for a constant change in the flow rate. There are no correction tools available for continuous flow variations or stepwise flow changes during the analysis. A well maintained pump should therefore be a prerequisite for every GPC/SEC user.

13.6.1.3 Influence of Evaluation Parameters

Sample Concentration

Precise knowledge about the sample concentration is not required for GPC/SEC with concentration detectors (RI, UV, ELSD). The concentration is not needed for the determination of the molar mass distribution and the molar mass averages (compare equations in Section 13.2.1).

This is no longer true if a viscometer or a light scattering detector is used in the experimental set-up. These methods require precise knowledge of the overall and slice concentrations (e.g. the injected mass). Table 13.15 shows the resulting deviations for the determined molar masses and typical results in GPC/SEC viscometry with inaccurate concentrations.

Every molar mass sensitive detector (on-line viscometer or light scattering detector) needs a concentration detector (RI, UV, ELSD) to measure the slice concentration. The volume difference between the two detectors, the inter-detector delay, has to be corrected so that the measured slice viscosity and the signal of the concentration detector match properly. A wrong inter-detector delay will influence the results. This is also shown in Table 13.15.

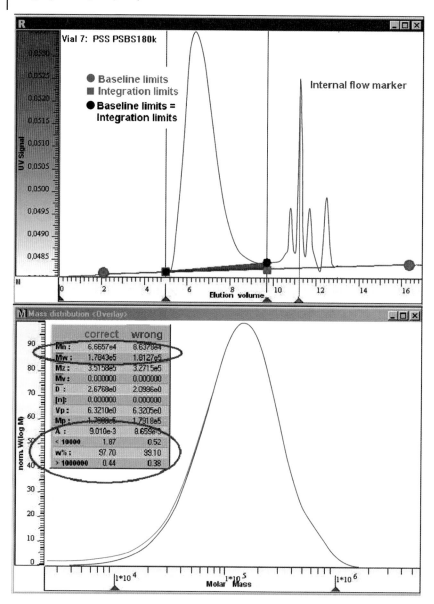

Figure 13.10 Influence of the baseline limits on the molar mass distribution and the molar mass averages. This figure compares the results of a one-step (only baseline limits) and two-step (baseline and integration limits) approach, as required by GPC/SEC guidelines.

Table 13.15 Influence of wrong concentration and inter-detector delay on the accuracy of molar masses for GPC/SEC with on-line viscometry detection.

	Variation (%)	M_w (Da)	Error (%)	M_n (Da)	Error (%)	$[\eta]$	Error (%)	a	Error (%)
Reference value	–	384 000	–	115 800	–	42.01	–	0.545	–
Concentration	–5	364 800	–5.21	110 000	–5	44.22	–5.26	0.545	0
Inter-detector delay	–5	385 900	0.49	112 700	–2.68	41.96[a]	–0.12	0.526	–3.49

[a] The inter-detector delay was changed for the viscometer, so that the calibration curve was still valid. Normally the bulk intrinsic viscosity $[\eta]$, is not influenced when the inter-detector delay is modified. In this case it changed as the integration limits were not adapted and a small part of the sample was therefore neglected.

Baseline limits

It is fairly common in GPC/SEC that for a polymer with a broad molar mass distribution the signal does not return to the original baseline but interferes with additional low molecular weight compounds. Therefore a two-step evaluation procedure is recommended in GPC/SEC. In the first step baseline limits will be defined, so that the baseline starts in the exclusion limit region and ends after the last system peak (for RI signals often a negative peak), when the signal returns to the original baseline. In the second step the integration limits are defined independently. Then the baseline is subtracted according to the limits in step 1, while the molar mass averages and the distribution are determined within the integration limits in step 2. This procedure allows a user independent and robust evaluation over the complete sample range, as is shown in Fig. 13.10 for a certified reference material (sample source: PSS, Mainz, Germany).

The vertical lines indicate the integration limits: the baseline limits from the two-step procedure are shown in green, while the red line shows a single step evaluation procedure where baseline and integration limits are not independent. The red area is the part of the sample that is not evaluated when only baseline limits are employed. In that case, the molar mass distribution is offset and the obtained molar mass averages are too high (cf. Fig. 13.10, bottom).

13.6.2
Influence of Parameters for GPC/SEC with Light Scattering Detection

Several on-line light scattering detectors are used in GPC/SEC. The main difference between these instruments is the number of detector angles used to measure the scattering intensity and the cell design. Table 13.16 summarizes the different techniques for on-line static light scattering detectors used in GPC/SEC.

The major advantage of light scattering detectors is that they measure the molar mass at every elution volume on-line with each sample. Therefore it is not nec-

Table 13.16 Overview of static light scattering techniques used in commercially available on-line light scattering detectors.

Type	Technique	Applicable for	Limitations	Prerequisites
LALLS Low angle laser light scattering	Molar mass determination without extrapolation	• MWD • High molecular weight samples	• Often noisy signals • High maintenance technique • R_g only in combination with viscometers	• Extremely clean system without particles or dust
RALLS Right angle laser light scattering	Molar mass determination without angular correction	• Only MWD of low molecular weight samples (isotropic scatterers) • Comparative data analysis	• No angular correction • R_g only in combination with viscometers	• $M < 200\,000$ Da • Sample properties known
TALLS Two/three angle laser light scattering	Molar mass determination with 2(3) point extrapolation	• MWD • High- and low molecular weight samples	• Often noisy signals for small angles • Limited angular correction • R_g accuracy low	• Coil statistics should be known
MALLS Multi-angle laser light scattering	Molar mass and radius of gyration determination with extrapolation	• Accurate MWD • Accurate R_g • Branching • Structure determination		

essary to establish a calibration curve with molar mass standards. However, for accurate molar mass determination several sample and instrument parameters need to be known independently for the instrument being used:

- refractive index increment, dn/dc;
- slice concentration, c_i (either calculated from the injected mass or by intensity calibration of the concentration detector);
- inter-detector delay between light scattering and concentration detector (RI, UV);
- refractive index of the eluent, n_{Eluent};
- detector constant of the light scattering detector (instrument calibration with toluene or by a molar mass standard);
- normalization coefficients for detector angles, $N(q)$ (MALLS detectors only).

13.6.2.1 Influence of Instrument Calibration

Although light scattering is an absolute method, the determination of a specific detector constant is required. This can either be done with toluene or with the molar mass calibration standard that is required to determine the normalization coefficients and the inter-detector delay volume. The instrument constant is determined with the Rayleigh ratio from the literature (toluene) or with the known M_w of the molar mass standard (certificate of analysis by the manufacturer). However, there are several literature values for the Rayleigh ratio of toluene that differ by about 10% and the M_w of the molar mass standard also has an uncertainty. Both errors will directly influence the results obtained for unknown samples. When using light scattering detection the results from two different laboratories can easily vary by approximately 20%: about 10% from the error in the light scattering measurement itself and about 10% resulting from the error in instrument calibration.

13.6.2.2 Influence of Instrument Performance

As the relationship between molar mass and elution volume is measured on-line during GPC/SEC with light scattering detection, the influence of different flow rates can be neglected. However, it is still important that the pump flow is reproducible and stable.

In contrast to GPC/SEC with concentration detectors the reproducibility of the injected mass plays a dominant role. Therefore precise and accurate manual or automated injection is required. The influence of an inaccurate injection volume is comparable to the influence of the wrong sample concentration (see Section 13.6.2.3).

13.6.2.3 Influence of Evaluation Parameters

The data of the theoretical Schulz–Flory distribution (see Section 13.4.1) have been used to investigate the influence of the evaluation parameters on the final results. Inaccurate evaluation parameters (e.g. refractive index increment dn/dc or concentration c) have been used in the data evaluation software to calculate the molar mass averages and other typical results. Table 13.17 summarizes the investigated parameters and the results obtained.

The highest impact on the results is from the refractive index increment and the concentration of the sample. Figure 13.11 illustrates the influence of a wrong dn/dc or concentration on the molar mass averages.

13.6.3
Repeatability and Reproducibility of GPC/SEC Analyses

Extensive round-robin tests have shown that accurate, precise and reproducible GPC results can be achieved if the requirements discussed above are met. In such tests performed by the German Institute for Standardization (DIN) for the establishment of GPC/SEC standards the repeatability and reproducibility of GPC/SEC analyses was determined [8]. These results are based on the analysis of complex commercial products and are summarized in Table 13.18. The results for the

Table 13.17 Influence of evaluation parameters on the accuracy of molar mass averages for GPC/SEC with on-line light scattering.

	Variation (%)	M_w (Da)	Error (%)	M_n (Da)	Error (%)	$<R_g>_z$ (nm)	Error (%)
Expected value	–	300 000	–	150 000	–	21.96	–
dn/dc	–5	331 000	10.3	165 600	10.4	21.96	0
A2 correction		326 000	8.67	155 500	3.67	23.05	4.96
Concentration	–5	315 800	5.27	158 000	5.33	21.96	0
Normalization	–5, –4, –3	289 800	–3.40	144 900	–3.40	22.10	0.64
	–5, 4, –3	296 300	–1.23	148 100	–1.26	22.03	0.32
Inter-detector delay	5	295 100	–1.63	159 500	6.33	21.96	0
n (Eluent)	–0.004	297 500	0.83	148 800	–0.83	21.96	0

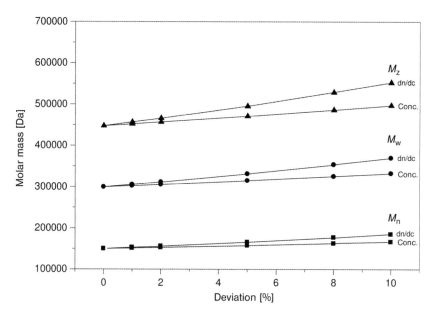

Figure 13.11 Influence of the refractive index increment dn/dc and the concentration on the molar mass averages obtained with GPC/SEC-light scattering.

repeatability were obtained in the same laboratory (intra-lab) under similar conditions using in-house standard operating procedures which were more elaborated than the round-robin test procedures. The results for reproducibility represent the precision for GPC/SEC results that can be obtained if samples are analyzed in

Table 13.18 Repeatability and reproducibility of GPC/SEC analyses of technical samples in various eluents; results based on standardized statistically valid round-robin tests according to DIN 5725-1 using complex samples (Source: ISO 13885 GPC standard [8]).

	Repeatability (%)			Reproducibility[a] (%)		
	THF	DMA	H_2O	THF	DMA	H_2O
M_n	3	2	2	15	15	15
M_w	2	2	2	10	15	15
M_z	3	3	3	15	24	24
M_w/M_n	3	3	3	15	24	24

[a] Reproducibility determined with complex commercial samples in round-robin test.

different laboratories (inter-lab) using different equipment by staff with different expertise, based on the specifications in the GPC/SEC standard.

13.7
Summary

The requirements of GPC/SEC/GFC data processing are fundamentally different from other chromatographic data analysis processes. Multiple parameter boxes intervene in these complex data processing procedures and can influence the final results in many ways. Additional factors that influence the final results and data integrity are:

- the choice of the data system itself,
- the calibration method,
- the type and selection criteria for the calibration fit,
- the type and settings of the baseline,
- the definition of the integration range,
- the chromatographic reproducibility of calibration runs and the analysis of unknowns,
- the instrumental, method-dependent and sample-related parameters in molar mass sensitive detection.

Accurate, precise, reproducible and operator-independent GPC/SEC results can only be obtained if the specific GPC/SEC data processing conditions have been obeyed. International and national GPC/SEC standards can be a perfect foundation to meet the most important pre-conditions for successful implementation and traceable GPC/SEC data analysis in every laboratory.

It might be advisable for laboratories without specific GPC/SEC expertise to hire external consultants or use expert contract laboratories if the analytical results are very important and/or are to be used in legal proceedings.

References

1 (a) P. Kilz, Optimization of GPC/SEC Separations by Proper Selection of Stationary Phase and Detection in: *HPLC Made to Measure. A Practical Handbook for Optimization*, S. Kromidas (Ed.), Wiley-VCH, Weinheim, 2006; (b) P. Kilz, H. Pasch, Coupled techniques in chromatography, in: *Encyclopedia of Analytical Chemistry*, R. A. Meyers (Ed.), Wiley, Chichester, 2000; (c) S. Mori, H. G. Barth, *Size Exclusion Chromatography*, Springer, Berlin, 1999; (d) W. W. Yau, J. J. Kirkland, D. D. Bly: *Modern Size-Exclusion Liquid Chromatography*, Wiley, New York, 1979.

2 E. Schröder, G. Müller, K.-F. Arndt: *Polymer Characterization*, Hanser, Munich, 1998.

3 (a) R. D. McDowall, *LC·GC Europe*, **12**, 568–576 (1999); (b) R. D. McDowall, *LC·GC Europe*, **12**, 774–781 (1999); (c) W. Winter, L. Huber: *Implementing 21CFR Part 11 – Electronic Signatures and Records in Analytical Laboratories*, BioPharm, Jan. 2000.

4 (a) Technical Requirements for the Registration of Pharmaceuticals for Human Use, Q2B: Validation of analytical procedures: Methodology, ICH-Q2B, Geneva, 1996; (b) Commission of the Council of the European Community: directives 87/18/EEC, 1987; 88/320/EEC, 1988 and 90/18/EEC 1990; (c) US FDA: Code of Federal Regulations, Title 21, Part 11, "Electronic Records; Electronic Signatures; Final Rule; Federal Register 62 (54), 13429/13466, 1997; (d) ISO/IEC Guide 25: General requirements for the competence of calibration and testing laboratories, 3rd ed., ICH, Geneva, 1990.

5 P. Kilz, G. Reinhold, Proceedings International GPC Symposium, San Diego 1996, Boston, 1997.

6 H. G. Elias, *Macromolecules*, Hüthig, Basel, 1992.

7 PSS, User documentation PSS WinGPC macromolecular chromatography data system, Mainz, 2006.

8 ISO 13885-1: Gel permeation chromatography (GPC) – Part 1: Tetrahydrofurane (THF) as Eluent, International Organization for Standardization, Geneva, 1998.

Part 3
Requirements for Chromatographic Data Analysis
from the Viewpoint of Organisations and Public Authorities

Quantification in LC and GC: A Practical Guide to Good Chromatographic Data
Edited by Hans-Joachim Kuss and Stavros Kromidas
Copyright © 2009 WILEY-VCH Verlag GmbH & Co. KGaA, Weinheim
ISBN: 978-3-527-32301-2

14
The Science Behind the Pharmaceutical Regulatory Chromatographic Procedures

Linda Ng

14.1
Introduction

The common chromatographic techniques are high pressure liquid chromatography (HPLC), gas chromatography (GC) and thin-layer chromatography (TLC). These techniques have expanded through the years to include new technologies for instruments, columns and detectors that provide improvements for better resolution, shorter run time, reduced solvent volume and monitoring chiral compounds, to name a few. Although HPL chromatographic procedures with UV detection are discussed for illustration purposes in this chapter, the principles can be extended to all types of chromatographic procedures and detection techniques. The regulatory comments apply mainly to the analytical procedures approved in the drug substance and drug product specifications. Here again, the principles could be extrapolated to testing procedures used in process manufacturing or cleaning validation.

Reliable, robust and reproducible analytical procedures are essential to provide accurate data on the quality of a pharmaceutical product, which subsequently benefits the patients. Quality analytical procedures can be achieved through thoughtful design, systematic development and appropriate validation, as shown in Fig. 14.1. Design is the thought process or foundation of a quality analytical procedure.

Figure 14.1 Designing a quality analytical procedure.

Quantification in LC and GC: A Practical Guide to Good Chromatographic Data
Edited by Hans-Joachim Kuss and Stavros Kromidas
Copyright © 2009 WILEY-VCH Verlag GmbH & Co. KGaA, Weinheim
ISBN: 978-3-527-32301-2

Studies at the development and validation stages may overlap. The validation stage includes the qualification of the equipment and validation of the analytical procedure. Though the design and development of an analytical procedure, and qualification of the equipment may be briefly mentioned, this chapter will cover only validation of chromatographic procedures and system suitability testing, planned during development and implemented during use, of the procedure.

Regulatory methods validation is the process of confirming that the analytical procedures are effective for their intended use and are suitable for regulatory purposes. To achieve this objective, the roles of various parties are outlined. For the applicant, this includes the qualification of instruments, validation of the analytical procedures during method development or changes, and system suitability testing for use during analysis. The review chemist evaluates the analytical procedures and validation data submitted in the New Drug Application (NDA) or Abbreviated New Drug Application (ANDA). As needed, an FDA (Food and Drug Administration) laboratory analyst demonstrates that the analytical procedures are reproducible by laboratory testing. The review chemist and laboratory analyst determine the suitability of the analytical procedures for regulatory purposes. The FDA investigator inspects the analytical testing site to ensure that the equipment and analytical procedures used for testing comply with current good manufacturing practices (cGMPs) (21 CFR part 211) or good laboratory practices (GLPs) (21 CFR part 58), as appropriate.

14.2
Instrument Qualification

An instrument is generally qualified before an analytical procedure is validated. This is performed by the instrument manufacturer before delivery, by the laboratory after receipt of the instrument, and by the laboratory during the development of a procedure.

The current instrument qualification practice [1] is divided into four parts: (1) Design Qualification (DQ), which is traditionally vendor driven; (2) Installation Qualification (IQ), user/vendor driven; (3) Operational Qualification (OQ), user/vendor driven; and (4) Performance Qualification (PQ), user driven. Independent of whether it is user or vendor driven, the user is responsible for documenting all levels of the qualification. Specifically, for HPLC systems, the DQ is to set instrument specification and selection criteria, the IQ is to ensure the instrument is received as specified and properly installed, the OQ is to verify the key performance attributes of the modular, not method-related aspects, and the PQ is to demonstrate the suitability of the "entire" HPLC system for routine use. PQ overlaps with the testing performed by analysts just before sample data are collected.

The purpose of the instrument qualification is to ensure that the criteria, as specified by the manufacturer, are met for the instrument. All instruments are thus qualified for assurance of reliable operations, and generation of accurate and reproducible data. PQ and system suitability tests (SST) have common properties even though both are performed for a different purpose. PQ is part of the qualification testing for the instrument whereas the SST is performed each time samples are analyzed. PQ tests can be built into SST to ensure evidence of satisfactory precision and linearity over the desired range.

PQ and SST are based on the concept that the equipment, electronics, analytical operations and samples to be analyzed constitute an integral system. Although not a guarantee for error-free chromatography, they ensure that the independent and objective qualification of critical system components (detector, pump and autosampler) will result in robust and transferable procedures. They verify that the resolution and reproducibility of the chromatographic system are adequate for the analysis to be done. The purpose is to accommodate for variations that should be kept to a minimum. Hence, SST will confirm that the chromatographic system functions correctly independent of the environmental conditions. SST, in the generic sense, covers all tests performed to confirm that the instrument will operate properly during use.

14.3
Chromatographic Procedures

Chromatographic procedures are normally used for assay, impurities and for quantitation in dissolution or rate release testing. The chromatographic analytical procedure should be described adequately so that any qualified analyst can follow the instructions in the procedure. The analytical procedure may be used in the developer's laboratory, the quality control laboratory of the manufacturer or may be transferred to other sites, e.g. contract laboratories.

Each procedure should be assigned a unique test procedure number that can be used to track changes. The list of changes, with dates or version number, should be included in the procedure. The procedure should include the scientific principle behind the procedure, and the sampling methodology, e.g. 20 tablets are triturated and aliquots are accurately weighed. The list of reagents, equipment (e.g. instrument type, detector, column type, dimensions) and equipment parameters (e.g. flow rate, temperatures, run time, wavelength settings) should be described. Appropriate system suitability tests should be proposed. How the analytical procedure is carried out should be described, including the standard and sample preparations, the calculation formulae and how to report the results. A representative chromatogram with labeled peak(s) should be included in the procedure. A quantitative analytical procedure should use well-characterized reference materials with documented purity. For an impurities procedure, the level of detectability and quantitation reliability should be addressed in the procedure design, development and validation.

All chromatographic analytical procedures should include system suitability testing and criteria. Setting SST criteria is performed during the development of the chromatographic procedure. Parameters typically used in system suitability evaluations are defined and discussed in the CDER Reviewer Guidance on Validation of Chromatographic Methods [2] and USP Chapter <621> [3]. The amount of testing will depend on the purpose or type of procedure.

The injection repeatability of a system suitability test for an RSD of < 1% is desirable for the analyte, that is, the drug substance, preservative, etc., but will be higher for impurities. This test is recommended to be conducted at the beginning, during and at the end of the analysis, as appropriate. However, with interspersed standards, a higher RSD may be reasonable if all standard values are pooled. Their use depends on the number of samples analyzed and the length of each chromatographic run. The purpose is to evaluate the reproducibility of the area count in the system suitability standard and provide information on the quality of the sample data.

The system suitability test for tailing is to evaluate the peak symmetry. Integration of the peak area relies on the changing detection signal at the beginning and end of the peak. Increased tailing or fronting makes the beginning or end of the peak more difficult to detect for integration purposes. A symmetrical peak having a front and tail of 1 is optimized from an integration viewpoint. A tailing factor of < 2 is desirable.

The system suitability test for theoretical plate number, N, is a measure of the column efficiency. For the same peak, the shape will increase in height and decrease in width with a decrease in elution time. With an increase in N, an increase in the number of possible peaks that can be detected would be helpful for the impurities procedures. Many factors can influence the capacity factor.

A system suitability test for an impurities procedure should address the issue that the peaks of interest are well resolved from each other and the void volume. With multiple peaks in an isocratic chromatographic analysis, the use of the resolution factor, applied to the two components that elute closest to each other, helps to ensure that the peaks are resolved thus resulting in reliable quantitation. In other cases, e.g. gradient analysis, more than one resolution factor may be needed. It is desirable that the resolution factor between the two peaks is greater than 2. With the evolution of the equipment technology, e.g. from HPLC to μHPLC, and columns, the resolution of peaks has improved dramatically [4, 5].

In addition, it is recommended to include a diluted standard at the quantitation limit of either an impurity or the drug substance, e.g. 0.1% in the drug product, to ensure detectability of impurities at that level during the analysis. Detector lamps could age with time, resulting in decreased sensitivity, or differences can occur among detectors from different manufacturers.

The system suitability tests selected should be adequate and appropriate for the analysis being performed. For example, the detectability limit test is not needed for an assay procedure. This discussion is to make the reader aware that useful SST and their respective criteria, which are critical, should be selected and adopted in the procedure.

14.4
Method Design, Development and Validation

Each NDA and ANDA should include the analytical procedures necessary to ensure the identity, strength, quality, purity, and potency, as appropriate, of the drug substance and the drug product. Data should be available to establish that the analytical procedures used in testing meet appropriate standards of accuracy and reliability (21 CFR 314.50(e)(2)(i) and 21 CFR 211.165(e)). Information should be submitted on the appropriate validation characteristics of the proposed analytical procedures as described in the ICH and FDA guidances [2, 6, 7] and USP Pharmacopoeia [3].

The development history of the procedure does not need to be submitted to the NDA/ANDA. Nonetheless, a summary of the procedure development process for the data supporting variations allowed for system suitability changes should be submitted. The summary data will help investigators and reviewers to understand the capability of the analytical procedure for future system suitability adjustments.

All validation characteristics should be addressed in the NDA/ANDA submission. All relevant validation data and formula used in the calculations should be included. Validation should not be a one-time operation to fulfill the agency's filing requirements. It should be performed during the development of the analytical procedure and with procedure changes. However, the amount of validation will depend on the drug product development stage, i.e. Investigational New Drug (IND) vs. NDA/ANDA. All analytical procedures are of equal importance from a validation perspective. In general, validated analytical procedures should be used, irrespective of whether they are for in-process, release, acceptance, or stability testing. Each quantitative analytical procedure should be designed to minimize any variation in the results.

The method validation parameters are described in ICH Q2(R1) [6]. Although not all of the validation characteristics are needed for every type of test, typical validation characteristics are: accuracy, precision (repeatability and intermediate precision), specificity, detection limit, quantitation limit, linearity, range, and robustness. Appropriate validation parameters will address how to generate reliable quantitative data and meet the procedure's claims.

The current method validation practices are to ensure that each procedure is designed to perform as expected during development and with procedure changes, and demonstrate robustness and reliability. In addition, the method validation is to ensure that the procedure is reproducible throughout the life of the drug substance or drug product provided the same procedure is maintained. The extent of the validation needed depends on the intended use for the procedure. Partial validation may be performed with analytical procedure changes.

Accuracy in a procedure is determined by how close the results are to the true value and the precision is a measure of how tightly grouped the results are around the mean result. The variations in accuracy, intermediate precision and reproducibility in the drug product are a sum of the procedure measurement and of the product manufacturing process. As a homogeneous sample is used for repeatability testing, the product manufacturing process variation is excluded. The sources of variation during a chromatographic procedure measurement could be from the sample preparation, dilutions and chromatographic assay. Sources of variation in the manufacturing process could include: the content uniformity of each dosage unit, especially for complex dosage forms such as transdermal and metered dose inhalers; evenness of mixing for solids in the tablet or capsule formulation; and distribution of solids in a liquid suspension.

Precision testing is split into three levels: (1) repeatability or intra-assay variation that evaluates a homogenous sample performed by a single person on a single piece of equipment, (2) intermediate precision or intra-laboratory variation that evaluates different analysts on different days using different chromatographs, and (3) reproducibility or inter-laboratory variation that evaluates multiple laboratories. Reproducibility is evaluated by companies with more than one laboratory location. Precision testing is readily achievable for all dosage forms.

On the other hand, except for solution formulation, it is harder to confirm accuracy or recovery because of the difficulty of reproducing the manufacturing process in a laboratory. Spiking a placebo with the drug substance is a common practice. Accuracy or recovery could also be inferred after establishing the precision, linearity and specificity of the procedure.

Although it is common for assay and impurities to be combined into a single chromatographic procedure, it is not always feasible. To address the specificity characteristic of the drug product procedure, various approaches are available. If the degradation pathway of the drug substance is known, spiking studies could be performed if the individual impurities are available. If impurities are not available, degradation/stress studies (e.g., acid and base hydrolysis, thermal degradation, photolysis, oxidation) [7–9] will support the specificity of the procedure. Peak purity tests (e.g. diode array, mass spectrometry) are tests that will support the purity of the peak. A combination of a range of analytical procedures based on different principles of separation, orthogonal methods, could also be used. ICH Q2(R1) defines specificity as the ability to access unequivocally the analyte in the presence of components that may be expected to be present. Though it is not always feasible to achieve specificity, the objective is to attempt to achieve the ultimate goal of complete baseline separation of the peaks. Hence the chromatographic procedure should be designed for resolution of the peak(s) of interest from each other, the solvent front and the excipients.

Linearity is validated to ensure that the compounds of interest are not out of the linear range of the detector. This is because external standards are commonly used for pharmaceutical procedures. Even though detector technology is constantly improving with time, this test is performed to confirm that the range expected for the compounds or its impurities is valid.

Quantitation and detection limits are of no concern in assay tests. An assay procedure will not need a sensitivity test at the quantitation level. However, for limit or impurities tests, knowledge of the sensitivity of the procedure is critical for reliable quantitation. Hence, the use of a standard at the quantitation limit will provide assurance of detectability of impurities down to that level.

It is recommended that reliability should be demonstrated with deliberate variations in chromatographic conditions. These studies, termed robustness, are performed in the development and validation stages. This will establish variations to support adjustments in system suitability testing during the life of the procedure. For these reasons, SST criteria should be meaningfully established to detect critical HPLC system performance problems for each procedure. The type of tests for SST will depend on the purpose of the procedure.

14.5
Compendial Procedures

Compendial procedures vary from quantitative determinations to subjective evaluation of the attributes. Compendial or published methods as per 21 CFR 211.194(a) (2), states that "the suitability of all testing methods used shall be verified under actual conditions of use". This assumes that published methods are validated by the developer of the method. However, the user of the procedure should confirm that the developed procedure works for the user's application.

The user should evaluate that the performance characteristics meet the requirements for the intended analytical applications. The recommended parameters required to evaluate such quantitative chromatographic procedures are specificity, intermediate precision, and stability of the sample solution.

The impurities profile will vary depending on the synthetic route of the drug substance. The composition of the drug product may differ for different manufacturers. For these reasons, specificity, as a validation parameter, should be addressed. In addition, the intermediate precision and stability of the sample solution should be evaluated to ensure reliable quantitation and assurance of reproducibility in the hands of the user. Intermediate precision takes the human factor into consideration. Use of equipment from various manufacturers by different analysts on different days confirms the precision of the procedure. The composition or properties, such as pH range of the compendial product, may differ. The analysis time includes sample preparation and multiple injections for multiple samples. It is not uncommon for the HPLC autosampler to collect data overnight. Thus, a stability study of the sample solution, for the period that the solutions are assayed at least, is needed. This stability study is of course also appropriate for non-compendial procedures.

14.6
Conclusions

In summary, the method developer should have an excellent, in-depth understanding of the purpose of the analysis, and should design the procedure thoughtfully to meet the claims of the procedure. The equipment should be qualified and the procedures should include supportive validation data. System suitability tests should be included with each procedure, and systemic tracking of the data collection, with either paper or electronic trails, should be maintained. Reliable analytical procedures, be they off-line, in-line, on-line or at-line, and an understanding of the manufacturing process will contribute to the improvement in quality of the analytical procedures for pharmaceutical products.

The reader is challenged to think through the requirements of the procedure. Consider the discussions in this chapter as examples to design and select appropriate critical components for a quality procedure.

Disclaimer

The views expressed above are the personal views of the author. The content of this article does not necessarily reflect the views or policies of the Food and Drug Administration, nor does the mention of trade names, commercial products, or organizations imply endorsement by the U.S. government.

References

1 W. B. Furman, T. P. Layloff and R. F. Tetzkaff, *J. AOAC Int. 77* (1994) 1314.
2 FDA Reviewer Guidance – Validation of Chromatographic Methods, 1994.
3 U.S. Pharmacopeia/National Formulary (Chapters <621> and <1225>).
4 M. W. Dong, LC-GC 25/7 (2007) 656.
5 J. W. Dolan, LC-GC 25/10 (2007) 1014.
6 ICH-Q2(R1) Validation of Analytical Procedures: Text and Methodology, 2005.
7 FDA Draft Guidance for Industry – Analytical Procedures and Methods Validation, 2000.
8 FDA Guideline for Submitting Samples and Analytical Data for Methods Validation, 1987.
9 ICH-Q1B Photostability Testing of New Drug Substances and Products, 1998.

Copies of the FDA and ICH guidances may be requested either as hard copies or as electronic copies. Hard copies can be obtained from the US Food and Drug Administration, Division of Drug Information, WO 51-2201, 10903 New Hampshire Ave, Silver Spring, MD 20993. Telephone: (301)-796-3400. Electronic copies are available through the CDER Home Page at http://www.fda.gov.cder. guidance. ICH guidances may also be obtained through the ICH Home Page at http:/www.ich.org.

15
Interpretation of Chromatographic Data According to the Pharmacopoeias – Control of Impurities

Ulrich Rose

15.1
Outline

The Pharmacopoeias are a collection of general and special texts (monographs), which serve to define generally applicable norms for the qualitative and quantitative analysis of drug substances, excipients and finished products. In the European Pharmacopoeia [1] these norms represent the obligatory standards in the 36 member states of the European Pharmacopoeia Convention.

Essential components of the Pharmacopoeias are the general sections which include the General Chapters describing analytical techniques and the special section containing the monographs on drug substances and excipients. The United States Pharmacopoeia (USP) [2] also contains monographs on finished products.

The special monographs describe the control of impurities as usually being carried out employing chromatographic methods, HPLC being the method of choice. In order to define a framework for the performance of the chromatography, the Pharmacopoeias provide definitions, formulae for the calculation of general parameters and generally applicable requirements for system suitability tests in the corresponding General Chapters. These are Chapter 621 "Chromatography" in the United States Pharmacopoeia (USP) [3] and Chapter 2.2.46 "Chromatographic Separation Techniques" in the European Pharmacopoeia (Ph. Eur.) [4], which was revised and adopted at the 130th session of the European Pharmacopoeia Commission in March 2008 [5]. In particular, the recommendations and guidelines for performance of chromatography and interpretation of chromatographic data given in this chapter are explained herein.

15.2
Interpretation of Qualitative Data

The definition of a suitable and robust chromatographic system represents a major challenge when a generally applicable pharmacopoeial purity test that uses HPLC

Quantification in LC and GC: A Practical Guide to Good Chromatographic Data
Edited by Hans-Joachim Kuss and Stavros Kromidas
Copyright © 2009 WILEY-VCH Verlag GmbH & Co. KGaA, Weinheim
ISBN: 978-3-527-32301-2

is being elaborated. The difficulties observed are often due to the availability of numerous, commercially available, stationary phases, particularly reversed phase materials, where not only different types but also batch-to-batch variations of the same phase can influence the selectivity and/or elution order. A standardization with the objective of obtaining qualitatively and quantitatively reproducible results is therefore the main objective in the elaboration of a pharmacopoeial method.

In Chapter 2.2.46 "Chromatographic Separation Techniques" of the European Pharmacopoeia there are qualitative chromatographic parameters listed that help to describe the retention data: these are, for instance, retention time and retention volume, hold-up time and hold-up volume, relative retention, retention factor (mass distribution ratio, capacity factor) and distribution coefficient. An important factor to describe the peak form is the symmetry factor, which is calculated as follows: $A_s = W_{0.05}/2\,d$, where $W_{0.05}$ is the peak width at 1/20 of the peak height and d is the distance between a perpendicular dropped from the peak maximum and the leading edge of the peak at 1/20 of the peak height. A value of 1.0 means ideal symmetry. Further, in more recent monographs a less often described parameter for the characterization of the column performance is the plate number (also referred to as the number of theoretical plates), which is calculated as $N = 5.54 \times (t_R/W_h)^2$, where t_R is the retention time of the peak corresponding to the component and W_h is the width of the peak at half-height.

Nowadays, criteria that are often for the characterization of chromatographic separations of two substances are the resolution or the so-called "peak-to-valley" ratio. According to Ph. Eur. the resolution between two components is calculated according to $R_s = 1.18 \times (t_{R2} - t_{R1})/W_{h1} + W_{h2})$. The previous requirement of Ph. Eur. that these peaks must be of similar height is no longer valid in the most recent version of this chapter. This allows the author of a monograph to describe a system suitability test that represents a "real life" situation, e.g. the separation between the substance to be examined and an impurity which are both contained in a reference solution in a realistic concentration ratio of e.g. 100:1. A pre-condition for this is that the main component is present in a concentration which does not cause an overload of the detector, because the measurement of the peak width at half height is not, or only approximately, possible otherwise. In the USP the calculation of the resolution is different insofar as the width of the peak is not measured at half height but at the base and a multiplication factor of 2 is used instead of 1.18. As a consequence, this leads to a result that is the same order of magnitude as the calculation according to Ph. Eur. Alternatively, USP offers the following formula for the use of electronic integrators: $2 \times (t_2 - t_1)/1.70 \times (W_{1,h/2} + W_{2,h/2})$. Although expressed differently, this is exactly the same formula as that given in the European Pharmacopoeia.

In those cases where the impurity and the main peak elute close to each other and are not baseline separated, the resolution cannot always be calculated as described above. In such a case monographs of the European Pharmacopoeia may describe a "peak-to-valley-ratio" as a system suitability test, a test which is not used in the USP in this form. Here the ratio between peak height of the impurity peak (H_p) and the lowest point ("valley") of the curve separating this peak from

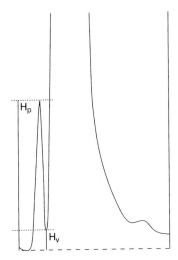

Figure 15.1 Chromatogram of an active substance with an incompletely separated impurity.
H_p = height above the extrapolated baseline of the minor peak;
H_v = height above the extrapolated baseline at the lowest point of the curve separating the minor and major peaks.

the peak due to the main component ($p/v = H_p/H_v$) is used as a system suitability criterion (Fig. 15.1).

An important issue concerning the qualitative interpretation of chromatographic data is the identification of components in the chromatographic system. The indication of retention times may be helpful but is also insufficient, because, depending on the stationary phase used, they may vary. Of more exact benefit is the indication of the relative retention, as described in Ph. Eur. and USP, which is calculated as $r = (t_{Ri} - t_M)/(t_{Rst} - t_M)$. As t_M, the retention time of a substance that is not retained (hold-up time), is often not experimentally determined, Ph. Eur. uses as a simplified method, the calculation of the "unadjusted relative retention r_G" which is calculated as the ratio of t_{Ri} (retention time of the peak of interest) and t_{Rst} (retention time of the reference peak, usually the peak corresponding to the substance to be examined). Experience shows, however, that relative retention values are also not always reproducible and should not be the only means of identifying impurity peaks in chromatographic systems. According to the Technical Guide [6] of the European Pharmacopoeia, the preferred option to identify impurity peaks is the use of reference substances (chemical reference substance, CRS) [7]. These reference substances can be the substance to be examined as well as impurities and mixtures of the impurities to be identified and the main compound (system suitability – or peak identification mixtures). Owing to the limited availability of impurities as isolated substances, the use of these mixtures, which are supplied together with a typical chromatogram, is the preferred option for identifying impurities and for carrying out the system suitability test in the more recent monographs of the Ph. Eur. (Fig. 15.2).

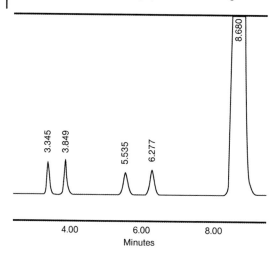

Figure 15.2 Chromatogram of an active substance with four impurities.
Requirement: minimum resolution of the corresponding pairs of impurities: $R_s > 2.5$.
Found: each pair $R_s = 2.9$.

An interesting alternative to the use of reference substances for identification of impurities in the chromatographic system is the method of "*in situ* degradation", where a specific impurity is produced under controlled conditions by degradation of the test substance by, for example, oxidation or hydrolysis. The reference solution thus obtained can be used directly in the chromatography without isolation of the substance [8]. This principle can also be usefully employed as a system suitability test.

15.2.1
System Suitability Test

In order to ensure an adequate performance of the chromatographic system throughout the analytical procedure, a system suitability test should be carried out not only at the beginning, but also during the test procedure. For this purpose, it is recommended that the above mentioned parameter is verified. Principally, the following requirements should be fulfilled according to Chapter 2.2.46 of the European Pharmacopoeia:

- The peak symmetry in the chromatogram of a reference solution should be between 0.8 and 1.5, unless otherwise described in the monograph.
- The LOQ (limit of quantification) should be equal to or less than the disregard limit (corresponding to ICH reporting threshold).
- The requirements for a minimum peak separation given in the monograph, e.g. by determination of the resolution or of the "peak-to-valley-ratio" must be fulfilled. More recent monographs in which several impurities must be identified and separated from each other often include more than one resolution criterion.

For example, the revised monograph for norfloxacin [9] describes a minimum resolution of 3.0 between impurities A and H and a minimum peak-to-valley ratio of 5 for the separation of impurity E from norfloxacin. These types of tests are used particularly for critical separations, i.e. for such impurities which elute close to each other and where there is a risk of co-elution. As a result of this test it should be ensured that all specified impurities are separated from each other and from the principal peak.

- Retention time and relative retention: these are usually given for information in the Pharmacopoeias, but not as a system suitability criterion.
- System repeatability: the repeatability of response is expressed as an estimated percentage relative standard deviation of a consecutive series of measurements for not fewer than three injections or applications of a reference solution. For the assay, maximum permitted relative standard deviations are given in the General Chapter, but not for the related substances test.
- Adaptation of chromatographic conditions: Often the analyst is confronted with the question as to how far the chromatographic conditions may be adapted in order to fulfil the system suitability requirements. In this case the European Pharmacopoeia gives clear guidance. It is surprising that the given tolerance is rarely used by pharmaceutical manufacturers. According to the revised version of Chapter 2.2.46 the allowed modifications particularly refer to isocratic systems.
- Composition of the mobile phase: the amount of the minor solvent component may be adjusted by ±30% relative or ±2% absolute, whichever is the larger. No other component is altered by more than 10% absolute.
- pH of the aqueous component of the mobile phase: ±0.2 unless otherwise prescribed or ±1.0 pH when non-ionizable substances are to be examined.
- Concentration of salts in the buffer component of the mobile phase: ±10%.
- Detection wavelength: no adjustment permitted.
- Stationary phase: no change to the identity of the substituent of the stationary phase permitted (e.g. no replacement of C18 by C8); particle size: maximum reduction of 50%, no increase.
- Column dimensions: length ±70%, internal diameter ±25%, when column dimensions are changed, the flow rate must be adjusted as necessary using an equation given in Chapter 2.2.46.
- Flow rate: ±50%, a larger adjustment is acceptable when changing the column dimensions
- Column temperature: ±10 °C, where the operating temperature is specified, unless otherwise specified.

The modification of the above-mentioned parameters is not acceptable for gradient systems. A modification of the chromatographic conditions is here more critical than in isocratic systems because it may lead to a shift of the peaks to other gradient steps and, therefore, the chromatographic profile may change; this may render the correct attribution of peaks more difficult or even impossible. Moreover, there is a risk that late eluting impurities will no longer be detected within the gradient system. For isocratic systems this is less problematic as the chromatographic run time here is usually expressed as a function of the retention time of the main substance (e.g. twice the retention time of the principal peak) and thus, even when retention times are prolonged, no impurity peak is "lost". In the revised General Chapter 2.2.46 of the European Pharmacopoeia it is recommended to apply only minor adjustments of the mobile phase and the gradient and this only under the following conditions:

- The system suitability requirements are fulfilled.
- The principal peak elutes within ±15% of the indicated retention time.
- The final composition of the mobile phase is not weaker in elution power than the prescribed composition.

Where the system suitability requirements can still not be achieved it is recommended to consider the dwell volume or to change the column [10].

A formula for the correction of the gradient times as a function of the dwell volume is given:

$t_c = t - (D - D_0)/F$, where D = dwell volume, in mL; D_0 = dwell volume, in mL, used for the development of the method; F = flow rate, in mL min^{-1}; t = time points stated in the gradient table; t_c = adapted time points.

The Technical Guide of the European Pharmacopoeia recommends that system suitability tests are carried out at each critical step of the gradient. In such a case preferably the above mentioned "system suitability mixtures" are used which should contain components eluting at the beginning, in the middle and at the end of the gradient in order to ensure selectivity as well as a complete elution of all impurities to be controlled.

15.3
Interpretation of Quantitative Data

The objective of the chromatographic impurity analysis described in the Pharmacopoeias is the quantitative determination or limitation of impurities in the substance to be examined. For this purpose, the above-mentioned qualitative requirements are described in order to ensure a sufficient separation and identification of the impurities to be determined. Apart from the qualitative system suitability criteria the European Pharmacopoeia describes in Chapter 2.2.46 a quantitative criterion [11] which must be fulfilled: the limit of quantification (LOQ) should be equal to or less than the "disregard limit" (reporting threshold). Impurities detected at lower levels shall not be taken into account. In most cases this disregard limit is 0.05% in relation to the test substance, exceptions may be described in the individual monographs. In the USP such a criterion is missing but will be introduced, as reported recently by Pappa *et al.* [12].

In the literature there are numerous methods described for calculating the limit of quantification [13, 14], also in the paragraph on validation of analytical methods in the Technical Guide of Ph. Eur. possible methods of calculation are given. The method described in Chapter 2.2.46 is based on the calculation of the signal-to-noise ratio, which should be at least $10 : 1$ in order to quantify impurity peaks correctly and reproducibly. According to Chapter 2.2.46 this ratio is calculated as follows: $S/N = 2\ H/h$, where H is the height of the peak corresponding to the component concerned in the chromatogram with the prescribed reference solution, and h is the range of the noise in a chromatogram obtained after injection or application of the blank, observed over a distance equal to at least 5 times the width at half height of the peak in the chromatogram obtained with the prescribed reference solution and, if possible, situated equally around the place where this peak would be found. The term "if possible", leaves the analyst a certain liberty in the measurement and stresses that an analytical judgement is required in all pharmacopoeial tests and sometimes individual solutions must be considered.

At present, the way the quantification is carried out in the European Pharmacopoeia is more of a limit test for individual impurities and a quantitative test for the sum of impurities. It is foreseen that this impurity test will become a real quantitative test in future editions. In principle, there are two methods that can be used for the determination of impurities: normalization or the method of external standard. Applying the less often used method of peak area normalization, the area of the impurity peak in the chromatogram obtained with the test solution is directly compared with the area of the principal peak and related to the latter. On the one hand this is a simple, quickly carried out procedure, which, on the other hand, requires a linearity of the detector response over the complete range, i.e. from concentrations of around 0.1 to 100% of the test substance. This linearity is often not given, as in the test for related substances concentrated test solutions are used in order to allow the detection of impurities at low levels; this may lead to a saturation of the detector, depending on the type of detector used.

The European Pharmacopoeia, therefore, prefers to recommend the method of external standardization [6], where the linearity must only be guaranteed in the range of 50 to 120% of the specification of the impurity concerned. This must be verified during the validation of the analytical method. In general, for the application of this method, diluted solutions of the impurities to be determined or a diluted solution of the test substance may be used as an external standard. The concentration of these solutions should preferably be in the range of the specification of the impurity, i.e. often between 0.1 and 1.0% in relation to the test solution. The area of the impurity peak in the chromatogram obtained with the test solution is then directly compared with the peak in the chromatogram of the reference solution. In the ideal case, a solution of a reference standard of the impurity to be determined is used, because here any problem of a possibly different detector response of the main substance and the impurity is excluded. As mentioned above, this approach cannot always be applied, therefore a dilution of the test solution is often used as the external standard. During the validation of the method it must be ensured that the impurity shows a similar detector response to that of the substance to be examined. Whilst the USP does not give further instructions on this subject but discusses an approach to the policy of Ph. Eur. [15], the European Pharmacopoeia allows, in Chapter 2.2.46, a range of

0.8 to 1.2 in which a different detector response of an impurity can be neglected, i.e. is considered equal to 1. Outside this range the monograph must indicate a response or correction factor. As well explained in [15], the expressions response factor and correction factor are often used in a different sense. According to Ph. Eur. the expression "response factor" means the ratio of the detector response of the impurity to the detector response of the main substance; the correction factor given in the monograph is then the reciprocal value of the response factor: the peak area of the impurity in the chromatogram of the test solution must be multiplied by this correction factor. In the Pharmacopoeias there is no guidance on how response factors should be determined. In each instance a verification of the linearity in the range of the specification of the impurity should be done. It is the opinion of the author that it is also necessary to examine the impurity level, water content and residual solvents of the impurity in order to determine the "content" of the impurity. Principally, the same tests should be carried out as for the establishment of a CRS [16]; the same is true for the substance to be examined. It is evident that, for instance, a high water content of the impurity, which is not considered, may lead to wrong values in the calculation of the response factor. As the amount of impurity available is often limited it is left to the discretion of the analyst to choose methods that use only small amounts of substance and still deliver scientifically valid results.

References

1 *European Pharmacopoeia*, 6th ed., Council of Europe, Strasbourg, **2008**.

2 *United States Pharmacopeia (USP) 29*, Rockville, MD, **2006**.

3 *United States Pharmacopeia (USP) 29*, General Chapter 621 Chromatography, Rockville, MD, **2006**.

4 Chapter 2.2.46, Chromatographic Separation Techniques, *European Pharmacopoeia*, 6th ed. (6.0, Vol. 1), **2008**.

5 Chapter 2.2.46 Chromatographic Separation Techniques, *Pharmeuropa* **2006**, *18.3*, 410–416, adopted at the 130th session of the Ph. Eur. Commission in a modified version.

6 *Technical Guide for the Elaboration of Monographs (European Pharmacopoeia)*, 4th ed., **2005**, p. 31

7 J. H. McB. Miller, A. Artiges, U. Rose, V. Egloff, E. Charton: Reference substances and spectra for pharmaceutical analysis, in *Reference Materials for Chemical Analysis*, Eds. M. Stoeppler, W. R. Wolf, P. J. Jenks, Wiley VCH, Weinheim, **2001**, pp. 172–195.

8 U. Rose, *J. Pharm. Biol. Anal.* **1998**, *18*, 1–14.

9 Monograph 1246, Norfloxacin, *European Pharmacopoeia 6.2*, **2008**.

10 Chapter 2.2.46 Chromatographic Separation Techniques, *Pharmeuropa* **2006**, *18.3*, pp. 413–414, adopted at the 130th session of the Ph. Eur. Commission in a modified version

11 Chapter 2.2.46 Chromatographic Separation Techniques, *Pharmeuropa* **2006**, *18.3*, 414, adopted at the 130th session of the Ph. Eur. Commission in a modified version.

12 T. J. DiFeo, O. A. Quattrocchi, H. Pappa, *Pharmacopoeial Forum* **2006**, *32 (6)*, 1862–1864.

13 J. P. Foley, J. G. Dorsey, *Chromatographia*, **1984**, *9*, 503–511.

14 M. Zorn, R. Gibbons, W. Sonzogni, *Eniron. Sci. Technol.* **1999**, *33*, 2291–2295.

15 L. Bhattacharyya, H. Pappa, K. A. Russo, E. Sheinin, R. L. Williams, *Pharmacopoeial Forum*, **2005**, *31 (3)*, 960–966.

16 Chapter 5.12, Reference Standards, *European Pharmacopoeia*, 6th ed., 6.0, Vol. 1, **2008**.

16
Requirements of (Chromatographic) Data in Pharmaceutical Analysis

Joachim Ermer

Compulsory compendial requirements with respect to treatment and interpretation of data are established to verify the suitability of analytical procedures including the equipment used, the so-called system suitability tests. Chromatographic and system suitability parameters of the European and the United States Pharmacopoeia will be discussed in Section 16.1.

In pharmaceutical quality control, the acceptance or rejection of materials is evaluated by determining the conformity to established acceptance limits (of the specification). Provided that efficacy and patient safety is guaranteed, these limits must include the variability of both manufacturing and the analytical procedure. In Section 16.2, statistical models are discussed to verify this compatibility between variability and limits.

Section 16.3 provides information on the General Information Chapter <1010> of the US Pharmacopoeia "Interpretation and treatment of analytical data".

16.1
System Suitability Tests

In pharmaceutical analysis, verification of the appropriate performance of the overall system composed of both the instrument and the method is regarded as an integral part of the analytical procedure [1, 2]. Details with respect to parameters of these system suitability tests are described in compendia [3, 4] and in guidelines [5]. As distinct from (basic) validation which must demonstrate the general suitability of the analytical procedure, the objective of the system suitability tests is to verify the actual suitability, each time the method is performed. Therefore, they can be regarded as an important component in ensuring the quality of analytical data (Fig. 16.1) and of the validation life cycle [6, 7].

In pharmaceutical analysis (quality control), the matrix of the respective analyte is usually fairly constant. Therefore, the suitability tests are focused on the analytical instrument and method. Changes within the matrix, such as in the composition of the drug, are subject to strict change control. Implications for the

Quantification in LC and GC: A Practical Guide to Good Chromatographic Data
Edited by Hans-Joachim Kuss and Stavros Kromidas
Copyright © 2009 WILEY-VCH Verlag GmbH & Co. KGaA, Weinheim
ISBN: 978-3-527-32301-2

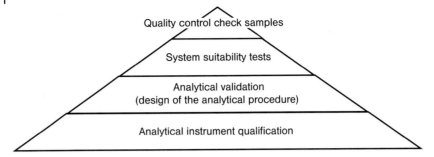

Figure 16.1 Components of data quality ([9], supplemented).

analytical procedure have to be evaluated and, if required, must be investigated and re-validated.

Analytical results are only valid if the defined system suitability criteria are fulfilled. Often, questions on the time and repetition of system suitability tests are raised. The compendia leave details to the user, but they make very clear that the suitability of the system must be maintained over the whole course of the analysis [4, 8]. It is common understanding that system suitability test solutions are injected at the beginning of the analytical series in order to verify the initial suitability. But how often, when and which tests must be repeated? The flexibility provided by the compendia is very sensible, because this depends strongly on the analytical procedure in question, for example on its robustness, run time, number of analyses, etc. Repetitions should also focus on tests on performance parameters that potentially change gradually with time, such as resolution, peak symmetry, or signal-to-noise ratio. These parameters should be repeated at least at the end of a series and, depending on the overall run time, perhaps also within the series. Concerning system precision, repetition is not necessary as short-term changes are not to be expected. Singular failure of the injection system must not be considered, because it may also happen during sample injections, can clearly be identified and not be avoided by a successful system suitability test. Of course, this must remain a rare event, otherwise an appropriate system suitability is also not guaranteed. The calculation of system precision from standard injections dispersed throughout the series (bracketing) is also acceptable [4]. However, this may impair the modular evaluation of sub-series and, in the worst case, the whole series must be repeated. The calculation of the system suitability test parameters does not matter, it may be performed either before the start of the sample injections, or after the conclusion of the series. The latter bears the risk of the whole series having to be repeated in the case of failure.

16.1.1
European Pharmacopoeia (EP)

In the EP-Chapter 2.2.46 "Chromatographic Separation Techniques" [3] – which was revised [8] – chromatographic parameters are discussed and compulsory system suitability tests are defined, in so far as the individual monographs do not prescribe otherwise.

A separate section "Adjustment of chromatographic conditions" lists possibilities for adjustments of method parameters, in order to meet the system suitability test requirements (see Table 16.1). This is essential for compendial monographs where there is no specific description of chromatographic columns, but can also be used as orientation for in-house methods to achieve (at least some) regulatory flexibility. In the revision, a separate treatment of isocratic and gradient methods is proposed, which takes the different complexity into account. When changing the column dimensions, the flow rate must be adjusted [8].

Table 16.1 Ranges for adjustments of chromatographic methods [3].

	Method parameter	Maximum permitted range for adjustments
1	LC, isocratic elution	
	Mobile phases	
1A	Composition	
1A1	• Minor solvent component	±30% relative
1A2		or ±2% absolute[a]
1A3	• Other components	±10% absolute
1B1	pH of aqueous component	±0.2 pH
1B2	for neutral substances	±1.0 pH
1C	Concentration of salts in buffer	±10%
1D	Detector wavelength	No adjustment permitted
1E	Stationary phase	
1E1	• Column length	±70%
1E2	• Inner column diameter	±25%
1E3	• Particle size	Maximum reduction 50%, no increase
1E4	• Identity of the substituent	No change (e.g. C8 by C18)
1F	Flow rate	±50%; Adjustment according to Eq. (16.1), if the column dimensions have been changed.
1G	Temperature	±10 °C
1H	Injection volume	May be decreased, as far as is acceptable for detection and precision; no increase
2	LC, gradient elution[b]	
2A	Mobile phases	Minor adjustments, provided that: • System suitability test requirements are fulfilled • ±15% of the principle peak retention time • Final mobile phase composition not weaker
2A1	pH of aqueous component	No adjustment permitted
2A2	Concentration of salts in buffer	No adjustment permitted

Table 16.1 (continued)

	Method parameter	Maximum permitted range for adjustments
2B	Dwell volume	Adjustment of the isocratic step before the start of the gradient to compensate for different dwell volumes. In the monographs, volumes up to 1 mL are considered.
2C	Flow rate	Adjustment according to Eq. (16.1), if the column dimensions have been changed.
2D	Detector wavelength	No adjustment permitted
2E	Stationary phase	
2E1	• Column length	±70%
2E2	• Inner column diameter	±25%
2E3	• Particle size	Maximum reduction 50%, no increase
2E4	• Identity of the substituent	No change (e.g. C8 by C18)
2F	Temperature	±5 °C
2G	Injection volume	May be decreased, as far as is acceptable for detection and precision; no increase
3	Gas chromatography	
3A	Stationary phase	
3A1	Column length	±70%
3A2	Inner column diameter	±50%
3A3	Particle size	Maximum reduction 50%, no increase
3A4	Film thickness	−50% to + 100%
3B	Flow rate	±50%
3C	Temperature	±10%
3E	Injection and split volume	May be adjusted, as far as is acceptable for detection and precision
4	Supercritical fluid chromatography	
4A	Composition of mobile phase	
4A1	Packed columns	
	• Minor solvent component	±30% relative or ±2% absolute[a]
4A2	Capillary columns	No adjustment permitted
4B	Detector wavelength	No adjustment permitted
4C	Stationary phase	
4C1	• Column length	±70%
4C2	• Inner column diameter	±25% (packed columns)
4C3		±50% (capillary columns)
	• Particle size	Maximum reduction 50%, no increase (packed columns)
4D	Flow rate	±50%
4E	Temperature	±5 °C[b]
4F	Injection volume	May be adjusted, as far as is acceptable for detection and precision; no increase

a) Whichever is the larger, b) Revision [8]

Adjustment of the flow rate:

$$f_2 = f_1 \frac{l_2 \, d_2^2}{l_1 \, d_1^2} \qquad (16.1)$$

where f_2 = adjusted flow rate in mL min^{-1}; f_1, l_1, d_1 = flow rate, column length, and inner column diameter indicated in the monograph, in mL min^{-1}, mm, and mm, respectively; l_2, d_2 = column length and inner column diameter of the column used, in mm.

Outside the described adjustment ranges, the analytical procedure is formally changed and a re-validation is required.

16.1.1.1 Chromatographic Parameters

Parameters of retention (time and volume, retention factor, distribution coefficient), of the chromatography (symmetry factor (Eq. 16.2) and apparent number of theoretical plates), and of separation (resolution (Eq. 16.3), peak-to-valley ratio (Fig. 16.2), relative retention) are discussed.

$$\text{Symmetry factor:} \qquad A_S = \frac{w_{0.05}}{2 \, d} \qquad (16.2)$$

$$\text{Resolution:} \qquad R_S = \frac{1.18 \, (t_2 - t_1)}{w_{1,0.5} - w_{2,0.5}} \qquad (16.3)$$

with $w_{0.05/0.5}$ = peak width at 5%/50% of peak height; d = distance between a perpendicular dropped from the peak maximum and the leading edge of the peak at 5% of the peak height; $t_{1,2}$ = retention times of two adjacent peaks with $t_2 > t_1$.

In most of the monographs with LC methods, a minimum resolution factor is prescribed; in a few cases where the baseline is not resolved a peak-to-valley ratio is

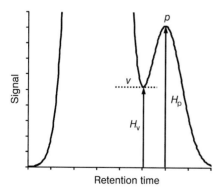

Figure 16.2 Peak-to-valley ratio (p/v) for peaks not baseline separated [3]. $p/v = H_p/H_v$ with H = height above the extrapolated baseline of the minor peak (p) and at the lowest point of the curve separating the minor and major peaks (valley, v).

given. Of the general system suitability test requirements which must be fulfilled, in addition to the monograph-specific ones, a symmetry factor between 0.8 and 1.5, determined with the reference solution for quantification is defined [8]. This is an important clarification for the determination of related substances because sometimes the shape of the main peak in the test solution is impaired by overloading.

Other parameters are only used in special cases, for example the plate number in a SEC application of the monograph for Alteplase.

16.1.1.2 Signal-to-Noise Ratio

The signal-to-noise ratio (S/N) is linked to the precision of the quantification and therefore is an important parameter for the determination of related substances (Eq. 16.4). The quantitation and from detection limit of an analyte correspond to concentrations with S/N of 10 and 2 to 3 [1], respectively. In the general system suitability test requirements, it is prescribed that the quantitation limit must be not more than the disregard limit defined in the respective monograph, usually 0.05%.

$$\text{Signal-to-noise ratio:} \quad S/N = \frac{2\,H}{h} \tag{16.4}$$

where H = height of the peak corresponding to the component concerned, in the chromatogram obtained with the prescribed reference solution, measured from the maximum of the peak to the extrapolated baseline of the signal observed over a distance equal to 20 times (5 times [8]) the width at half-height. h = range of the background noise in a chromatogram obtained after injection or application of a blank, observed over a distance equal to 20 times (5 times [8]) the width at half-height and, if possible, situated equally around the place where this peak would be found.

Nowadays, out of the various types of noise (e.g. short- and long-term, baseline irregularities and drifts, spikes, etc. [10]), only the short-term noise is relevant to the quantification. Noise can only impair peak integration if its frequency ("peak width") is similar to that of the analyte itself. In contrast, the long term noise can be "filtered" by means of an extrapolated baseline. Of course, some of the aforementioned types of noise can also affect the integration. However, they are easily recognised as discrete incidents and do not pose the risk of an unnoticed gradually decreasing performance. The determination of the relevant short-term noise is described in an ASTM standard for qualification of LC detectors [10]. The chromatogram is divided into segments of 0.5 to 1.0 min over about 15 min, in which the maximum vertical distances of the baseline signals are determined (see Fig. 16.3). The mean is calculated from all segments. Even in the rather moderate example shown in Fig. 16.3, a clear overestimation of the noise according to EP can be seen. Usually, the peak width at half-height ranges between 0.2 and 0.5 min, whereas 20 times the peak width would result in an interval of 4 to 10 min, which cannot really be claimed as "short term"! However, in the revision, the distance is reduced to 5 times the width at half-height [8], a better approach to short-term noise.

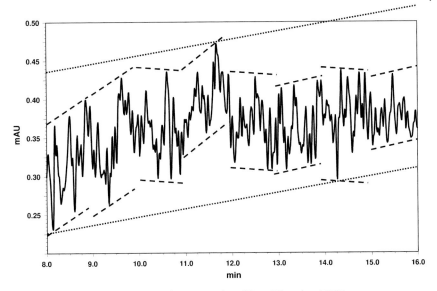

Figure 16.3 Noise estimation according to EP (dotted lines [3]) and to ASTM (dashed lines [10]). The result according to EP is 0.21 mAU, according to ASTM 0.13 mAU, with individual segments between 0.10 and 0.16 mAU.

16.1.1.3 System Precision

A maximum acceptable relative standard deviation of repeatedly injections is defined for assay determinations, as a dependence on their number and on the upper specification limit (Eq. 16.5 and Table 16.2). The difference B between the upper limit and 100% corresponds to the part of the specification range that can be attributed to the analytical variability (because the true content of a drug substance cannot exceed 100%). It is up to the user to reduce the number of injections. However, in order to maintain the same uncertainty, the acceptance limits must be tighter to compensate for the larger uncertainty in the standard deviations at lower degrees of freedom (see Fig. 8.2b). This strict limitation of system precision is appropriate, taking the increasing overall variability through the precision levels into account (see Section 8.3).

Table 16.2 Requirements for system precision according to EP [3, 8]. For discussion see text.

B (%)	Maximum permitted relative standard deviation			
	n = 3	n = 4	n = 5	n = 6
2.0	0.41	0.59	0.73	0.85
2.5	0.52	0.74	0.92	1.06
3.0	0.62	0.89	1.10	1.27

The term "system repeatability" will be used [8] to avoid confusion with the other precision levels. As discussed in Section 16.3, this may lead to the wrong interpretations and conclusions.

$$\text{Maximum injection precision [3]: } RSD_{max} = \frac{KB \sqrt{n}}{t_{90\%,\, n-1}} \tag{16.5}$$

where K = constant ($0.349 = \frac{0.6}{\sqrt{2}} \times \frac{t_{90\%,\, 5}}{\sqrt{6}}$), in which the first term represents the required relative standard deviation after 6 injections for $B = 1.0$; B = upper limit given in the definition of the individual monograph minus 100%, assuming that the upper limit is set according to the reproducibility of the method; n = number of replicate injections of the reference solution (between 3 and 6).

$t_{90\%,\, n-1}$ = Student's t at the 90% probability level (two-sided) with $n - 1$ degrees of freedom.

System Precision for Impurities?
In the draft up-date, system precision was also discussed for impurity testing. At concentrations of the reference solution larger than 0.2% (with respect to the active ingredient in the test solution), the relative standard deviation should be less than 5.0%. If concentrations are equal to or less than 0.2%, a maximum relative standard deviation of 10.0% is defined. It was claimed that the precision of the analytical system also needs to be controlled for impurity determinations. However, this objective cannot be met by the proposed approach. Within the concentration range of the assay, the injection variability is the dominating variance contribution (see Fig. 8.10b). Therefore, the performance of the injection system can be controlled with an appropriate statistical power and reliability, as described in Table 16.2. In contrast, within the concentration ranges described for impurities, the detection and integration variance is dominant (see Fig. 8.10b), which does not allow appropriate control of the suitability of the injection system. The former variability is already sufficiently addressed with the signal-to-noise ratio. Fortunately, in the final revised version, the system precision requirement for impurities is dropped [8].

16.1.2
US Pharmacopoeia

In Chapter <621> [4], the same chromatographic parameters are described as in EP, fortunately also (most) equations are harmonised.

As system suitability test parameters: resolution, plate number, injection precision and tailing factor are described. The plate number is regarded as of less importance, but is acceptable where there are single peaks in a chromatogram. The relative standard deviation should not exceed 2.0% with 5 injections. If it is larger, then 6 injections should be used. Although here the relationship between the reliability of the calculated standard deviation and the number of results can be recognised (see Fig. 8.2b), the acceptance limit is inappropriate, at least for assay applications. Assuming a minimum increase by a factor of 1.5 per precision level, a maximum reproducibility of 4.5% would result, which is for most applications certainly not acceptable. With such a variability, single results may be expected to be found in a range from 91 to 109% (with respect to a true value of 100%, at 95% confidence level).

In contrast to EP, no general acceptance interval for the tailing factor is defined.

With respect to adjustments, differently to EP, supercritical fluid chromatography is not discussed, and gradient elution is not treated separately. Most of the adjustments for LC and GC are identical with the EP, with the following exceptions (cf. Table 16.1):

- 1B2 and 1A2 are not described.

Some conditions are defined additionally:

- Altering the concentration of buffer salts (1C), the pH-range defined in 1B1 must be maintained.
- In the case of ternary mixtures, only one of the minor solvent components is allowed to be adjusted. (1A1).
- The wavelength error of the UV–Vis-detector must be limited to ±3 nm.
- For GC temperature programs, a time adjustment of up to ±20% is permitted.

16.1.3
FDA Reviewer Guidance

This Guidance [5] was intended to support FDA reviewers evaluating the submitted LC-methods, mainly with respect to validation, but also to the design of the analytical procedure. The importance of meaningful system suitability tests for the maintenance of the suitability over the whole lifecycle is highlighted. In Section J, the parameters listed in Table 16.3 are discussed, including recommendations for acceptance criteria.

Table 16.3 System suitability test parameters according to [5].

Parameter	Recommended acceptance criteria
Capacity (retention) factor (k')	> 2
Injection precision (for assay)	RSD ≤ 1% for $n \geq 5$
Relative retention (α)	Not required if Rs is defined
Resolution (Rs)	> 2 for critical peak pair
Tailing factor (T)	≤ 2
Theoretical plate number (N)	Dependent on elution time, usually > 2000
Standard solution at quantitation limit[a]	Somewhat unclear for a check of the quantitation limit. A signal-to-noise ratio of 10 is mentioned, but evaluated as not very practical (not further justified why)

a) Discussed in the text, Section B.

The extent of the system suitability tests depends on the respective application of the analytical procedure. For dissolution, at least k', T and the injection precision are recommended; for assay and impurity determinations, resolution is also recommended.

With respect to system precision, higher variability limits are acknowledged for small impurity concentrations, but no specific recommendations are given. See Section 16.1.1.3 for the (non-)added value of such a parameter.

16.2
Acceptance Limits for the Specification and Precision

Specification limits must include the variability of the samples (or of manufacturing) as well as that of the analytical procedure, of course assuming that both remain within a normal and acceptable range (Eq. 16.6). The following discussion is limited to the estimation of the analytically required part of the specification range.

Specification limits, in general:

$$SL_{u/l} = 100\% \pm BL_{u/l} \pm AV \tag{16.6}$$

where $SL_{u/l}$ = upper/lower acceptance limit of the specification, related to a normalized target value of 100%. $BL_{u/l}$ = upper/lower basic limits, resulting from an (acceptable) sample or manufacturing variability, given in % with respect to the target value and AV = analytical variability of the (whole) analytical procedure.

16.2.1
Assay

16.2.1.1 Based on the Method Capability Index
According to Daas and Miller [11], the analytical variability is assumed to be three times the target standard deviation (TSD), equivalent to the method capability index (see Section 8.1). The target standard deviation is calculated as a pooled repeatability from compendial collaborative trials (Eq. 8.9). In the case of drug substances, the lower and upper basic limits are the sum of the impurities and zero, respectively. For example, assuming a TSD of 0.6% and a sum of impurities of 0.5%, the lower and upper specification limits correspond to 97.7% and 101.8%, respectively. In this approach, as discussed in Section 8.3.4, the variability contribution of the reproducibility is not taken into consideration. With respect to drug substances, this contribution originates mainly from the reference standard, i.e. the same size as for the sample can be expected. Consequently, the standard deviation is increased by a factor of $\sqrt{2}$, which can be assumed to be included in the TSD originating from collaborative trials.

16.2.1.2 **Based on the 95% Prediction Interval (DPhG-Approach)**

According to a consensus paper of the Working Group Quality Control/Pharmaceutical Analysis of the German Pharmaceutical Society (DPhG) [12], an approach taking the overall variability as well as the number of determinations into account is proposed. Combining the equation given in the paper with Eq. (8.12), all variability contributions of the routine analysis can appropriately be included (Eq. 16.7). If the individual contributions are not (sufficiently reliably) known, they have to be used in combination. For example, the first two terms under the square root correspond to the repeatability variance, divided by the number of determinations n. If the inter-series variance is small, the root term corresponds to the overall variance, divided by n (see Eq. 8.11). In a statistical context, Eq. (16.7) describes a prediction interval for future sample assays, which corresponds exactly to the objective of batch release.

Taking the example from Fig. 8.8 (b) with a manufacturing variability of 2.0%, $m = 1$ and $n = 3$, an analytical variability of 3.0% results and consequently specification limits of 95.0 to 105.0% (with $df = 20$).

Analytical variability according to DPhG:

$$AV = t(95\%, df)\sqrt{\frac{s_i^2}{m\,n} + \frac{s_p^2}{n} + s_g^2} \qquad (16.7)$$

where $t(P, df)$ = Student-t-Factor for a statistical confidence P of 95% and degrees of freedom df, related to the precision studies from which the variances were obtained.

s_i^2, s_p^2, s_g^2 = system variance, variance of the sample preparation, inter-series variance, used either directly as squared relative standard deviation, or normalized; m = number of repeated injections in routine analysis; n = number of sample preparations in routine analysis, if the mean is defined as a reportable result (see Section 16.3.3). If each single determination represents the reportable result (i.e. is compared with the specification limits), n must be set to 1.

Provided the mean is the reportable result, the variability of the analytical procedure can be optimised according to Eq. (16.7) by increasing the number of determinations. However, this must be driven by objective requirements, i.e. safety and efficacy considerations. Therefore, it is proposed in the DPhG paper [12] to apply, in typical cases, the traditional specification limits for drug products (95.0 to 105.0%). Also, regarding the statistical uncertainties of the experimentally determined variances, it certainly makes no sense to discuss decimal places! However, Eq. (16.7) is very valuable for estimating at least whether or not the standard limits are compatible with the variability of the analytical procedure. Furthermore, this clear and straightforward approach can be utilized to calculate specification limits in cases of a larger manufacturing or analytical variability, which must of course be justified. This may occur for complex drug products (see Section 8.3.4), complex sample preparations such as derivatization, or very low concentrations (see Section 8.3.6).

16.2.2
Impurity Determination

16.2.2.1 Acceptance Limits of the Specification

In the case of impurities, of course, just an upper specification limit exists, which must be below the toxicologically qualified concentration [13]. According to the ICH guideline ([13], Decision Tree #1), the impurity content should be determined with batches from the development, pilot scale and scale-up procedures. The upper limit of the confidence interval of the thus-obtained mean corresponds to the acceptance limit of the specification.

This approach is questionable, because a "true" value is assumed for batches that do not come from the same manufacturing process. In addition, due to the tightening of the confidence intervals with increasing number of batches (see Fig. 8.2a), the applicant with the smallest number of batches (and thus the least reliability) will be "rewarded".

The DPhG paper [12] proposes, in analogy with the concept of Daas and Miller [11] to estimate the specification limit from the mean of the batch analyses and three times the standard deviation using at least five representative batches from clinical phases II and III.

16.2.2.2 Quantitation Limit and Variability

In order to perform a reliable quantification, the distance from the specification limit to the quantitation limit must be compatible with the analytical variability. A general suitability can be assumed if the quantitation limit is at most 50% of the specification limit, such as in the case of a reporting (disregard) limit of 0.05% and a specification limit of 0.10% for unknown impurities according to the ICH guideline [14]. If a tighter distance is required, Eq. (16.7) can be used to estimate the variability. Owing to the domination of the first term under the square root at low concentrations (see Section 8.3.6), a separation of the variance contributions is not needed (Eq. 16.8).

Suitability of the impurity acceptance limit:

$$SL - QL > AV = \frac{t(95\%, df)\, s}{\sqrt{n_{AP}}} \qquad (16.8)$$

where s, t = standard deviation and Student's-t-factor from precision studies, n_{AP} = number of determinations in routine analysis, QL = quantitation limit of the impurity.

16.2.3
Key Points

- Acceptance limits of the specification must include at least the (normal) variability of the analysis sample (e.g. manufacturing variability) and the analytical variability (provided that the safety and efficacy requirements are fulfilled).

- Usually, this is guaranteed by tradition based limits. However, the compatibility with the analytical variability should be verified.
- In cases of a (justifiable) larger analytical variability, such as at very small concentrations or with very complex sample preparation, the required specification range can be calculated using the DPhG approach.

16.3
Interpretation and Treatment of Analytical Data

The General Information Chapter of USP <1010> [15] describes acceptable statistical approaches to the analysis and evaluation of (analytical) data. Included are the calculation of statistical parameters, identification and treatment of outliers, and comparison of analytical procedures. The USP chapter gives a selection for orientation not prescribed tests.

The structure of this section corresponds to that of the USP chapter, and comments of the author are indicated as such. Equations are only given if they have not already been dealt with in the previous sections, otherwise they are referenced.

16.3.1
Prerequisites

A sensible evaluation of analytical results requires basic prerequisites to be fulfilled, such as a thorough and complete documentation, representative sampling procedures, use of suitable and traceable reference substances, an appropriate qualification of analytical instruments, and properly validated analytical procedures.

16.3.2
Measurement Principles and Variation

Dispersion and distribution of analytical data are discussed, as well as the corresponding parameters such as mean, standard deviation, relative standard deviation, and confidence intervals of the mean (see Section 8.1). Variability can be reduced by averaging. Whether single or repeated determinations are appropriate depends on the intended application and on the given variability.

Comment: The testing for normal distribution requested in <1010> is, because of the small number of usually available data in pharmaceutical analysis, of very limited value. In the case of physico-chemical analyses, "normally" a Gaussian distribution can be assumed, provided systematic effects can be excluded. The latter are preferably checked by simple graphical representation and visual inspection for systematic patterns, such as trends.

The utilization of control samples is described to monitor method variability or as part of the system suitability test. In Appendix A of USP <1010> an example of a control chart is given. An example of a precision study is discussed in Appendix B. Five series of three determinations each are calculated by means of an analysis of variances (see Section 8.3.4). An analysis of the variance contributions is recommended and an equation is given for the standard deviation of the mean or reportable result. This can be used to optimize the overall variability of the analytical procedure. The equation corresponds to Eq. (8.12), but without separation of the repeatability variance into injection and sample preparation variance.

Comment: In Table 1 of USP <1010> Appendix B, standard deviations were calculated from the three repetitions of each series. However, such standard deviations are of negligible value due to the large uncertainty involved (cf. Fig. 8.2b).

16.3.3
Outlying Results

In this section, data are discussed which are very different from those expected, i.e. outlying results. They may be either extreme results, still belonging to the same distribution, or originate from product (sample) failures or analytical errors. In each instance, a systematic investigation for the root cause must be performed. There is a clear distinction between identification ("outlier labelling") and removal ("outlier rejection") of such data. The latter is only allowed if an analytical cause for the deviating result can be assigned (or at least is likely), such as operator mistake, instrument failure, wrong reagents, etc. This is consistent with the requirements of the FDA-Guidance on out-of specification (OOS) test results [16]. The clear definition of the final analytical result ("reportable result") is particularly valuable with respect to the often controversial discussion on averaging: "When assessing conformance to a particular acceptance criterion, it is important to define whether the reportable result (the result that is compared to the limits) is an average value, an individual measurement, or something else. If, for example, the acceptance criterion was derived for an average, then it would not be statistically appropriate to require individual measurements to also satisfy the criterion." [15] Consequently, there can only be an OOS with respect to the reportable result.

Comment: If the mean is defined as the reportable result, the data variability should be checked previously [16], e.g. by acceptance limits for the range or for the standard deviation in order to avoid averaging of non-representative single determinations.

Shortcomings and prerequisites of the statistical outlier tests are discussed in detail, as well as the limited use of their outcome. In Appendix C the generalized extreme Studentized deviate test (Grubbs test), and the tests due to Dixon and to Hampel are described. As the first two are relatively known, only the latter is further outlined here.

16.3.3.1 Outlier Test According to Hampel
This is a so-called "non-parametric" test, because the assumption of a normal distribution is not required, as it is for the other tests.

The results are ordered according to their magnitude and the median is determined, i.e. the value in the middle of the ordered sequence. The absolute difference between the median and each value is divided by the median of these absolute differences (MAD), multiplied by 1.483. If such a normalised parameter exceeds the limit of 3.5, an outlier is indicated (Eq. 16.9).

Hampel's rule:

$$\frac{|x_i - x_M|}{MAD} = \frac{|x_i - x_M|}{\text{Median} |x_i - x_M| 1.483} > 3.5 \qquad (16.9)$$

where x_i = individual result, x_M = median of the individual results, ordered according to their magnitude, MAD = median of the absolute differences between the median and the individual result, ordered according to their magnitude.

16.3.4
Comparison of Analytical Results

Here, statistical tests are described to demonstrate equivalence between two analytical procedures.

16.3.4.1 Precision
The precision of the alternative method must not be less than that of the reference method. This can be demonstrated statistically by comparison of the upper limit of the confidence interval of the variances ratios between the alternative and the reference methods to a previously defined acceptance limit (Eq. 16.10). Because of the large variability in the standard deviations, it is important to use a sufficiently large number of data (see example in Table 16.4., calculation with 3 vs. 9 values).

Upper confidence limit of the variance ratio:

$$CL_{R, \text{upper}} = \frac{s_{\text{alt}}^2}{s_{\text{ref}}^2} F(0.05; df_{\text{alt}}; df_{\text{ref}}) \qquad (16.10)$$

where $s_{\text{alt/ref}}^2$ = variances of alternative/reference method, $F(0.05; df_{\text{alt}}; df_{\text{ref}})$ = Fisher's F-value for an error probability of 5% and the degrees of freedom of the alternative and the reference methods.

Such an equivalence test is preferable to the traditional two-sample F-test, because the latter only tests for a statistical significance. However, even if such a difference is not detected, i.e. if the ratio of the variances of the two methods is less than the critical F-value, this does not mean that the precision is equivalent.

Comment: Furthermore, a significant difference would also result when the alternative method is much more precise than the reference method, certainly not a (good) reason to reject the alternative method.

Table 16.4 Example of a comparison between a LC-assay and a nitrogen determination according to Dumas. For the latter, the amount of nitrogen found is related to the theoretical value (see also Comparison.xls on the Bonus-CD).

No.	LC-assay (%)			N-determination (% assay)		
1	92.93			94.18		
2	91.32			93.22		
3	92.66			92.84		
4	93.50			92.77		
5	92.41			92.58		
6	92.58			92.77		
7	91.27			93.03		
8	93.00			92.96		
9	91.56			92.96		
Usage of data no.	1–3	1–6	1–9	1–3	1–6	1–9
Mean	92.30	92.57	92.36	93.41	93.06	93.03
Variance	0.743	0.518	0.636	0.477	0.345	0.218
RSD	0.93%	0.78%	0.86%	0.74%	0.63%	0.50%
Number of data	3			6		9
Precision						
F-test	1.56			1.50		2.92
• Critical value (95%)	19.00			5.05		3.44
• Significance	no			no		no
Equivalence test $CI_{R, upper}$	12.19			3.37		1.18
Accuracy						
Difference	1.11			0.49		0.68
Equivalence test (95% *CI* of the difference)	−0.25 to 2.47			−0.19 to 1.18		0.14 to 1.21
t-test	1.74			1.30		2.19
• Critical value (95%)	2.78			2.23		2.12
• Significance	no			no		yes

16.3.4.2 Accuracy

Accuracy is demonstrated by an acceptable difference between the means. The proposed equivalence test checks if the confidence interval of this difference (Eq. 16.11) is included in the acceptance interval ($\pm\delta$), defined by the user (Eq. 16.12). In contrast, in the traditional *t*-test the question posed is whether the confidence interval of the difference includes zero. This may lead to situations where either a significant difference does not have any practical importance (Fig. 16.4, Scenario 1; Table 16.4 example with 9 values), or, due to large variability, in practice an unacceptable difference is not detected (Fig. 16.4, Scenario 3). The advantage of the equivalence tests (but on the other hand also their challenge) is that the user can (must) define the measure for the practical relevance. Result variability is included in this test, but at the "user's risk" (Fig. 16.4, Scenario 3), in addition to a small number of data (Table 16.4, example with 3 values). *Vice versa*, with an increasing number of data, the chance of a positive outcome of the equivalence test is increased, whereas that of a significance test is decreased (Table 16.4, example with 9 values). A suitable number of data can also be estimated, as described in Appendix E USP <1010>, together with a calculation of acceptance intervals. (For further applications and discussion of equivalence test in pharmaceutical analysis, see [17–20].)

Confidence interval of the difference between two means:

$$CI_D = \pm t(P, n_1 + n_2 - 2)\, s_p \sqrt{\frac{1}{n_1} + \frac{1}{n_2}}$$

$$s_p = \sqrt{\frac{(n_1 - 1)\, s_1^2 + (n_2 - 1)\, s_2^2}{n_1 + n_2 - 2}}$$

(16.11)

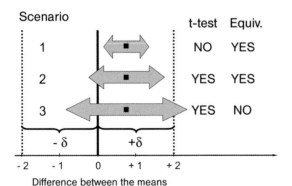

Scenario		t-test	Equiv.
1		NO	YES
2		YES	YES
3		YES	NO

Figure 16.4 Schematic representation of significance (mean or two-sample *t*-test) and equivalence test. The difference between the means is shown by a square symbol, the confidence intervals as double arrows. The acceptable difference δ for the equivalence test is defined by ±2. The "success" of the statistical tests with respect to the comparison is indicated by "YES/NO" for the various scenarios. With respect to the *t*-test and the equivalence test, "NO" means a statistically significant difference and non-equivalence, respectively.

$$-\delta \le (\overline{x}_1 - \overline{x}_2) - CI_D \wedge (\overline{x}_1 - \overline{x}_2) + CI_D \le +\delta \tag{16.12}$$

where $n_{1,2}$; $\overline{x}_{1,2}$; $s_{1,2}^2$ = number of data, mean, and variance of the series 1, 2; $t(P, df)$ = Student's-t-factor for a statistical level of confidence P and degrees of freedom df. For equivalence tests, the question is one-sided, therefore an error probability of 0.10 has to be chosen for a confidence level of 95%; $\pm\delta$ = upper and lower limit of the equivalence interval.

16.3.5
Key Points

- An analysis of variances (ANOVA) is recommended to calculate the precision levels.
- It is made clear that acceptance limits of the specification and the final result of the analytical procedure ("reportable result") are linked and only the latter is relevant for OOS.
- Statistical outlier tests are only appropriate for information purposes, not to invalidate data, because they cannot reveal the cause of the deviation.
- Comparing methods (or data), equivalence tests are better suited than the traditional significance tests such as F- and t-tests. The latter only test for a statistical significance, but not for the practical relevance of such a difference. In the case of equivalence tests, the user defines the practical acceptable difference.

References

1 ICH. Q2(R1): Validation of Analytical Procedures: Text and Methodology Methodology, **2005**.
2 FDA. Guidance for Industry: Analytical procedures and methods validation, **2000**.
3 European Pharmacopeia, 2.2.46 Chromatographic Separation Techniques, 28–32.
4 United States Pharmacopeia 31, National Formulary 26, Section <621> "Chromatography", United States Pharmacopeial Convention, Rockville, **2008**.
5 FDA. Validation of chromatographic methods (Reviewer Guidance), **1994**.
6 G. C. Hokanson, *Pharm. Technol.* **1994**, *18*, 118–130.
7 J. Ermer, Analytical Validation within the Pharmaceutical Environment, in Method Validation in *Pharmaceutical Analysis, A Guide to Best Practice* (J. Ermer and

J. H. M. Miller, Eds.), Wiley-VCH, Weinheim, **2005**, Chap. 1.
8 Revision of Chapter 2.2.46 Chromatographic Separation Techniques, *European Pharmacopeia* 6.4.
9 United States Pharmacopeia 31, National Formulary 26, Supplement 1, Section <1058> "Analytical Instrument Qualification", United States Pharmacopeial Convention, Rockville, **2008**.
10 ASTM. Standard practice for testing fixed-wavelength photometric detectors used in liquid chromatography. Designation: E685-93 (Reapproved 2000), **2000**.
11 A. G. J. Daas, J. H. M. Miller, *Pharmeuropa* **1999**, *11*, 571–577.
12 J. Ermer, H. Wätzig: Setting specification acceptance criteria under consideration of analytical variability. *Reg. Affairs J. (Pharma)* July **2003**, 1–3; http://www.pharmchem.tu-bs.de/

forschung/waetzig/englisch/dphg_
pospapier_eng03.pdf

13 ICH. Q6A, Specifications: Test Pro-
cedures and Acceptance Criteria for
New Drug Substances and New Drug
Products, Chemical Substances, **1999**.

14 ICH. Q3A(R2), Impurities in New Drug
Substances, **2006**.

15 USP. United States Pharmacopeia 31,
National Formulary 26, Section <1010>
"Analytical Data – Interpretation and
treatment", United States Pharmacopeial
Convention, Rockville, **2008**.

16 FDA. Guidance for Industry: Investi-
gating out of specification (OOS) test
results for pharmaceutical production.
October, **2006**.

17 R. Kringle, R. Khan-Malek, F. Snikeris,
P. Munden, C. Agut, and M. Bauer,
Drug Information J. **2001**, *35*, 1271–1288.

18 ISPE. Good Practice Guide: Technology
Transfer, **2003**.

19 J. Ermer, Performance parameters,
calculations and tests, in Method
Validation in Pharmaceutical Analysis,
A Guide to Best Practice (J. Ermer
and J. H. M. Miller, Eds.), Wiley-VCH,
Weinheim, **2005**, Chap. 2.

20 M. Broughton and J. Ermer, Transfer
of Analytical Procedures, in Method
Validation in Pharmaceutical Analysis,
A Guide to Best Practice (J. Ermer
and J. H. M. Miller, Eds.), Wiley-VCH,
Weinheim, **2005**, Chap. 7.

17
Evaluation and Valuation of Chromatographic Data

Stavros Kromidas

17.1
Introduction

As a service provider I am quite aware of the following: in a physical or virtual get-together the last impression should be a positive one – this ensures the addressee's contentment. Thus one never quickly scribbles a new differential equation on the board at 4.15 pm on the last day of a seminar – even just a basic one. No, the whole is rather wrapped up with a "little game" or group work that the participants can complete successfully. The same holds true for a book. Problems, difficulties and so on belong in previous chapters. One sensibly ends with a fairly positive outlook according to the slogan "there are problems everywhere but it still works out somehow" – this ensures the writer receives agreeable nodding from the reader. Or at least what has been said previously is simply summarized.

I am not going to apply this "safe", motivating and ultimately sensible practice here. I will rather try in the following to explain the situation as I experience it in the laboratory, simply and impartially – even if there is a danger that something unpleasant will come up. This is neither stubbornness nor a marketing trick, nothing is further from my mind than just to criticize or point the finger. It is rather about a – certainly subjective – presentation of day to day practice in analytical laboratories as I have experienced it in order to draw some conclusions.

17.2
The Situation – or Why Does So Little Change?

If in private life we talk of "evaluation" of a result (yesterday I had my bathroom retiled) it is a critical view of the value, i.e. the quality of the result in order to be able to judge it. Further, I consciously take into account, or register at least intuitively, several aspects of the result (price, color, clean joints, etc.). This observation leads to a final judgment which could possibly result in the need for action (e.g. the tiles around the bathtub need to be re-laid). Now, we all know that in professional life this natural attitude – in nature instinctively present and hence literally natural – is seldom possible or tolerated. Thus, analyzing the evaluation is often reduced to a "yes/no" decision on a numeric value. It is thus not a matter of "value" but merely a numeric value. Does the number "fit" or not?

Further, we all know that it is not a problem to make it fit: this numeric value for the dead time can be added into the computer in order to obtain the k-value required in the test method; measurements are repeated so that the value is hopefully "in spec" again or it can be manually integrated in such a way that the peak area of the impurity lies below the required 0.05% of the main component. This can sometimes be frustrating or leaves – in particular for the direct users of apparatus – an uneasy gut feeling, which is at some point accepted as the inevitable. One is also unsure if and what is "bad", even the specifications are not always comprehensible, e.g. only marginal changes to the method parameters are allowed whereas one is absolutely free to change integration parameters, how critical is this small peak really, etc.?

Why is the situation like this? Which means, why do we concern ourselves so strongly with the evaluation of the numeric value in the sense of agreement with the requirements and so little with the valuation of the result in the sense of an analytical assessment? We can make a strongly simplified note of the following:

In a globalized, highly competition-orientated professional world, numbers act as simple, comprehensible and accepted values by all parties, yes, even as evidence for a clearly defined "quality". Competition asks for comparability. Comparability asks for objective criteria. For this purpose numerical values are just perfect. The greater the competition the more weight is given to the true to scale values. A truly analytical evaluation is thus inevitably counteracted. Where the numbers come from, whether they are representative, have real relevance etc. is often secondary and is rarely questioned. This is a great shortcoming. In the analytical world (society) concerned with comparability, faith in numbers can quickly prevail. Above is thus written "quality" and not quality because mainly formal aspects are consulted for the definition. Reproducibility, the reproducible finding of numerical values is considered as evidence for constant quality, however, it is, at best, the evidence for constant characteristics – but only those being checked.

Comment: For many years, thanks to GLP and other QM-systems, one has been coerced to replicability and stubborn compliance in the laboratory which was put on a level with quality. One has been concerned with quality control instead of quality. The FDA has now long recognized that this can self-evidently only be a meander in the long run, and in recent years a careful but clearly recognizable change of course has been initiated. Instead of *"Any change is bad"* –, *"Compliance, not science"*-, and *"Blind compliance"* attitudes, now the era of *"Science and risk-based compliance"* and *"Continuous improvement"* is being ushered in. Keywords thereto are: PAT, no GMP-check lists but quality systems, lesser quality controls more quality management.

Now let's get back to what really is and not what could be or maybe what might be some day.

In today's environment – particularly in highly regulated areas – it is out of the question to talk about valuation. This is not really intended and also not demanded. Thus – although we all know better how it works – we do not necessarily try to do something better and more reasonable. In life, especially in professional life, we do little about it because we basically consider it as good. We are only prepared to change an entrenched habit if this either increases our self-confidence considerably or our brain's reward center is activated (experience approval, joy, praise etc.). Otherwise we would rather let it be; after all we do not want to be a "revolutionary". Since, nowadays, "to make mistakes" is equal to weakness, the fear of mistakes is very high, up to the higher ranks of politics, economy and society. It is well known that we do not live in a time where heroes are popular. In a period with permanent time pressure, great emphasis on formalities, provability and alibi mentality etc. we are happy that the whole "works" reasonably. And everything stays the same.

17.3
How Can Something Change and When is it Really Necessary?

"Ability" and "knowledge" should stand at the beginning of a decision. Only when I am able and I know, do I recognize the need for action and can assess the chances and ways of realization. Regarding "learning" and "ability" only the following will be noted: a climate should preferably be created in which these are possible. We know from neurophysiology: humans do not learn and work best under pressure but when they are comfortable. In this (happy) situation the neurotransmitter dopamine is activated; the alertness increases; the brain is ready for learning. If this prerequisite is created, knowledge can be acquired and stored.

The intention of the present book is to somewhat broaden the knowledge of the interpretation of chromatography and the evaluation of the results obtained in the sense of analytical assessment. After knowledge of the validity of a numerical value and the valuation is acquired it will be decided – presuming that the goals are unambiguously clear to all parties – whether there is a need for action and whether there is some leeway.

Four examples of this:

- The *F*-test is widely used. For a real comparison of the scattering of two series of measurements it is generally unsuitable, since in the *F*-test variances (i.e. standard deviations) are used and not the relative standard deviations – where the scattering is in relation to the mean value. If it is required *expressis verbis* by the customer or authorities it should of course be given. This is not being cynical, but it is a reality when stating that one is analysis, the other "business".

- The correlation coefficient is used similarly and is generally accepted: Is it necessary to greatly "strive" to obtain in the case of pure samples two or three "9s" after the decimal point? What does this mean? It is hardly influenced by possible outliers and matrix effects. The relative standard error reacts, however, very sensitively to these types of changes. If I do not want (need?) to obtain in the routine process – independent of the standard of knowledge of the user and other variables – problems and reclamations, I am happy that the correlation coefficient is demanded as a criterion for linearity. With respect to this point I am going to have a quiet life. If, however, it is about knowledge of the quality of a method in need of a calibration (influence of the matrix on the sensitivity, scattering of individual values) then the relative standard error is without doubt the better criterion.

- A HPLC method with which four impurities could be separated was installed and registered in several countries in 1984. Since that time it has become possible to separate nine impurities with the use of more selective columns. The valid requirements of 1984 are followed; thus there is no immediate need for action. Actually nothing is happening. Which word is appropriate here: evaluation, assessment, quality(!) control?

- If I carry out the validation under optimum conditions and not under real conditions and not considering all the influencing factors, I obtain of course small CV values. What is my intention or, asked in a different way, on what basis is the "quality" of my work measured? Does my boss expect three validations in six weeks as the deadlines have already been fixed by the management? Is it possible that I have *de facto* no possibility of carrying out the testing (more) thoroughly? Do I have (or want) to, for whatever reason, always come up with small CVs? Then the "validation" is to be carried out as described above. The pre-programmed problems during method transfer, the inevitable "OOS"-situations in the later routine operation, and the associated fairly high cost from the total operational point of view etc. are accepted, and this after all puts pressure on the quality control – and not on us ...

These examples can be continued arbitrarily.

An approved strategy for increasing the effectiveness and efficiency in a laboratory is based on three steps. First, there is the totally objective, rational, emotionally unproven knowledge. This leads to the acquirements of facts. Then goals should be defined, queried or communicated. Subsequently, the analytical practice should, if necessary, be optimized considering the possibilities. In this third part of the chain I personally see the greatest potential.

Here are some examples:

- During the evaluation of a method one should remember that its "value" is increased by reductions in cost and time. Often a simple *de jure* allowed procedure is enough to enhance a method. For example, even if it is not increased during separation, the flow rate should at least be increased from 1 to 2 mL min^{-1} during flushing of the HPLC column. Or, instead of a 4-mm column a 3-mm column could be used. Both scenarios would result in a reduction of about 50% in the length of time taken and the amount of element used.

- One should use any given leeway, and confront the current general trend "out of fear to be more Prussian than the Prussians". The authorities allow it, they even partly fund it. Crusted structures are often one's own. Thus, the allowed modification of USP- and EP-methods should actually be used (see Chapters 11 to 13), in the case of unknown influencing factors on my result I should (must) multiply these by the coverage factor k and this number should be specified as to be expected/allowed in the routine (see Chapter 4).

- In classical statistics the factor "time" is irrelevant. Consequently, in the daily routine it is typical to: check if the CV is below a threshold and the average value lies within the given limits. Thus, it is a "yes/no"-decision. If the average value is additionally followed as a dependence on time with the aid of a control card, it can be recognized whether a periodic dependence of the average value exists or not. Should this be the case it is due either to aging of the analytical method (chemicals, column) or to periodic changes of the concentration as a consequence of the production process (contamination of the reactants, decomposition, etc.). This can be differentiated with the aid of control samples. Should it be due to production, this possibly results in a greater optimization potential, which would have been recognized with a simple analytical tool.

- One should operate in "worst-case" mentality and practice questioning to arrive at realistic requirements. If, for example, the CV of a method is 0.8% and the specification requirements for the assay were fixed at 98–102%, then it is absolutely clear that single values will be "OOS": the scattering of the method is too high in comparison with the specification requirement. In this instance the requirement of 95–105% would be appropriate – which by the way has been more and more propagated by the FDA in recent years. Or in a situation like this, outliers are not taken seriously; the specification requirements should only be valid for the average value. Further, it can clearly be substantiated how much worse (documented quality loss and concrete monetary loss) the special product will be with a requirement of the scattering of e.g. CV = 1.5% instead of the presently demanded 1%. If this record is missing such requirements should be handled a little more generously. It should be considered with respect to the following: thoughtless requirements make the analysis expensive without increasing the quality of the products and thus its market value. I am, however, aware of the immense difficulty in revising requirements once they have been formulated with correct conceptual arguments.

17.4
Who Can Change Something?

If, during the assessment of a method or, in general, during the assessment of the current analytical practice, a need for action is recognized, a real change would only be possible by the upper management. I actually consider the likelihood of this to be minimal. Why? The management is far removed from the practice; for them the laboratory is a cost intensive necessity which needs to deliver numbers in order to be able to sell the product/service. The laboratory should deliver such numbers/results which are needed according to the slogan: "make sure that it works". For the management these are just numbers, similar to other numbers that they have to do battle with: sales figures, cash-flow, stock prices, etc.

I quote a head of department: "I do recognize that the evaluation of the peak height is more accurate, but we are not going to change something in our practice. Just teach my people how they can analyze better(!) via the peak area so that the whole stays as it always has been. If it does not work they should just repeat the analysis. Using evaluation of the height on all sites would cost me too much time – which I haven't got and, after all, I have more important things to do." Not a satisfactory but quite an understandable attitude.

Many of the users of the equipment generally know quite well what can go wrong – only they are hardly ever in a decision-making position.

The long-serving laboratory personnel and the laboratory management are often responsible for changes: these colleagues have the professional knowledge and, often, with a little courage, creativity and good arguments, the possibility to bring about at least small changes. This includes reconsidering specifications or criteria depending on whether priority is given to the formal requirements of the current method or to analytical "truth", see the examples above. Is it worthwhile with today's workload and time pressures to initiate something if the whole works more or less? It is obvious that the answer has to be an individual one. From my experience I would suggest the following: The first important step is to sort out the facts: number of samples, existing resources, why does someone need it by when, which result, with what precision, etc. From this it is unambiguously apparent whether there is a need for action or not and if there is then to what degree it is realizable. One should only take action if a personal advantage comes with it: strengthening one's own position or self-confidence, recognition, personal satisfaction, enthusiasm, boost of motivation for the whole group, etc. In this way, the chances for success are not too bad.

Index

Quantification in LC and GC: A Practical Guide to Good Chromatographic Data
Edited by Hans-Joachim Kuss and Stavros Kromidas
Copyright © 2009 WILEY-VCH Verlag GmbH & Co. KGaA, Weinheim
ISBN: 978-3-527-32301-2

Contents of the CD

For Chapters 3 to 8 files, which include a spreadsheet (xls) or a chromatogram (cdf), these are contained on the enclosed CD. Some Excel files contain the results of the tables in the book in more detailed form. The change data format (cdf) files are in the AIA format which should be compatible with every data system. Using them you can test your own data system.

The cdf-files are for certain readable with Chromeleon, Empower, EZChrom and LCSolution. To <Open> <data> one possibly has to switch to the ending.cdf. You may need to use an import function. For Empower, the file must be pre-integrated, otherwise an error message occurs. The files on this CD are integrated with LCSolution. If you want to have a look at the AIA files, you can use the AIA-Fileviewer, which can be downloaded free of charge from the home page www.scisw.com.

The file Abbreviations.xls contains the used abbreviations in Chapters 1, 2, 3, 5 and 7 of the book text and the (in few cases different) names used in the Excel files. Since $\bar{x}$ is not allowed as a name in Excel, Mx was used, i.e. average value of x.

The files for Chapter 3 are partitioned in the subchapters of the book. To have a survey of the files, you should open "3. Excel-Files on CD.ppt" and "3. CDF-Files on CD.ppt". For each file a description is given.

For Chapter 4 "4. Instructions.doc" should be read.

Chapter 5 contains the files "5. Excel-Files on CD.ppt" and "5. CDF-Files on CD.ppt".

The description of the cdf-files of Chapter 6 can be read in "6. Deconvolution Files on CD.doc".

Chapter 7 only contains one file "Weighted Regression.xls". The influence of the weighting can be tested. Using WE=0 yields the normal (ordinary) regression.

Quantification in LC and GC: A Practical Guide to Good Chromatographic Data
Edited by Hans-Joachim Kuss and Stavros Kromidas
Copyright © 2009 WILEY-VCH Verlag GmbH & Co. KGaA, Weinheim
ISBN: 978-3-527-32301-2

The description of the two xls-files for Chapter 8 is the content of "8. ReadMe-CDe.doc".

We hope you will find time to try the files which are an important addendum to the book.

The Editors